HISTORY OF THE GREEK ORTHODOX CHURCH IN AMERICA

I believed with full assurance that America was the land I dreamed of, where God intended Greek Orthodoxy to grow.

ARCHBISHOP IAKOVOS

These historical essays are dedicated to the immigrants of the Christian Orthodox faith migrating to the New World, to the millions of parishioners now worshipping in the communities of the Archdiocese and to the future generations which will be carrying on our Greek Orthodox tenets and traditions in the Americas.

HISTORY OF THE GREEK ORTHODOX CHURCH IN AMERICA

Compiled and edited by
REV. MILTIADES B. EFTHIMIOU, PH.D.
and
GEORGE A. CHRISTOPOULOS, M.A.

Published by
Greek Orthodox Archdiocese of North and South America
New York
1984

Prologue

noted historian once wrote, "There are in history no beginnings and no endings. History books begin and end, but the events they describe do not."* At first glance, this might seem a warm salutation to history! But upon scrutiny, this definition is a very stern warning to those of us who live in history and also are called upon to comment upon it.

The history of the Greek Orthodox Church in America is a history of those first immigrants arriving on these shores from Europe, at the turn of the century, up until the well-organized communities of the present time with intermittent stops in between which are chapters in and of themselves. The essays in this volume are stories of the miracle of Greek Orthodoxy and its establishment in this country. Each chapter is an essay which has a beginning, middle and an ending; yet, the *events* which each chapter describes do not have an ending. Let us take but one example: Going back to the humble beginnings of the founding of our communities in this country, one will find the Greek immigrant who had faith in himself and the value of his tradition because he had a sense of dignity and a sense of duty. His faith provided him with a strength for great efforts; his sense of dignity prevented him on many occasions from doing anything that would be embarassing to himself, to his Church, or to his Hellenic origin. So he proceeded to organize his "Koinotis", and he coupled this effort with his loyalties to his city, his state, the nation, by becoming "Greek Orthodox American" in practice as well as in spirit.

This was the scene that was set prior to 1922 when the Greek Orthodox Archdiocese of North and South America was can-

---

*R.G. Collingwood, *The Idea of History*, Oxford, Clarendon Press (New York, 1946), pg. 49 - 52.

I

onically created and approved by the Ecumenical Patriarchate, and when Alexander was chosen as the first Archbishop. The articles written by eminent scholars in this volume dramatically show the division of the history of the Greek Orthodox Church in this country into years of preparation for the coming of the Iakovian years, which are preceded by the Athenagoras/Michael eras. For better understanding of this institutional history, we are forced to split up the stream of events into periods in order to master the material and to reach a fuller understanding of the Greek Orthodox Church's development in this country.

Perhaps the most far-reaching question that all of these essays ask is, what were the most important plateaus in the quest for stability of the Greek Orthodox Church, from the humble beginings, to that great era which I term the Iakovian Era? And what about the highs and lows of this period of our history, the positive aspects and the negative ones? The essays in the first part of the book highlight the "growing pains" between 1922-1932. At that time, one saw the ecclesiastical life of the Greek-American constituents interrupted by the Royalist and Venizelist divisions which objective critics see as a period of deep rooting in spite of the efforts of those whose sole purpose was to create havoc and division. And what about the highlights? Was it during the time of Alexander, the first canonical Archbishop of America, formerly Bishop of Rodostolon? Was it during the time of Metaxakis, who placed the Church of America under the jurisdiction of the Ecumenical Patriarchate, the Mother Church of the world-wide Orthodox? Was it during the time of Damaskinos, Metropolitan of Corinth and Exarch of the Ecumenical Patriarchate whose great vision suggested that Metropolitan Athenagoras become the successor to Alexander? And what about Athenagoras' tenure 1931-1949, who instilled in the Greek Orthodox immigrant a concept of responsibility, and whose admonition that the Greeks of America should unite around their church? Athenagoras made the Greek face the reality that he was going to remain in America; that he was going to participate in the creation of a new Church; and that in doing this, he was going to, by necessity, come into contact with ecclesiastical political authorities in this new country. In essence (and as so well captured by historian Theodore Saloutos) the Greek Church in America would pass from childhood and adolescence to manhood during the magnificent period of Archbishop Athenagoras. And what of the great Archbishop Michael from Corinth, 1949-1958, whose reign marked the changes of leadership of our communities from the older generation to the new generation whose training ground was called GOYA (Greek Orthodox Youth of America)? It was this organization founded by Archbishop Michael that trained thousands of youthful leaders, religious instructors, community activists, and good Greek Orthodox families. GOYA gave the hope and conviction that Orthodoxy was (in the final analysis) the most beautiful religious experience that a person could have!

And where does one begin and end in describing the Iakovian Era, which is the thrust of the final part of this book? The institutions and departments of the Greek Orthodox Archdiocese under the leadership of Archbishop Iakovos, some new and some a continuation of older establishments, testify to the continuity and perseverance of the *Omogeneia*. In essence, our present institutions continue the basic work begun in 1922, and even further back than that with the coming of the first immigrants. All this is a preservation of the "sacred trust" bequeathed to Greek Orthodox Americans who, in turn, were not just founders of communities, but parishes, that became (through the events described in this book) a *Church!* If there was one man who stood out in making Orthodoxy a major faith in this country and in the world it was Iakovos. Our spiritual father is the subject of the middle portion of this volume and the honoree of a special tribute paid to him by President Jimmy Carter. Iakovos' tenure since 1959 has made this last period the bridge between past and future. The Iakovian era is the very embodiment and fruition of the first sixty years so abundantly and so mercifully blessed by Almighty God.

With these preliminary remarks, I have tried to summarize this volume's character and the purpose for bringing together the various scholars who contributed to this collection of essays. There were many people working behind the scenes who made possible the task of writing, collecting and compiling the papers for the past three years. Fortunately, I had the cooperation of many people who deserve equal credit for this book. This was my first assignment upon being appointed director of the Department of Church and Society for the Greek Archdiocese of North and South America.

I wish to express my gratitude to all the contributors who worked closely with me, and my co-worker, Mr. George Chris-

III

topoulos, without whose help and editorial expertise, this book would have never seen the light of day. I am grateful to him for the many weeks he put into this volume making it also, for him, a "labor of love". I am also grateful to staff persons, clergy, laity, as well as many other people who make up the family of the Greek Orthodox Archdiocese of North and South America. In particular, I would like to single out the late Felia Samios, Fr. Alexander Karloutsos, Mr. Chris Demetriades, Fr. Demetrios Constantelos, Ms. Niki Calle, Ms. Terry Kokas, Presvytera Nikki Stephanopoulos, Mr. Christo Daphnides, Mr. George Douris, Mr. John Basil and Mrs. Helen Tryforos. I would also like to give special mention to Mrs. Athena Condos for her loyalty and devotion to me and my department. It is necessary to add a word of regret for the inevitable omission of many worthy topics and areas of activity in our Church, which will undoubtedly be covered by other historians. Such lacunae include the histories of our 500 communities, biographies of hundreds of clerics and lay leaders and such departmental activities of the Archdiocese as Religious Education, Laity, Standing Conference of Canonical Orthodox Bishops in the Americas, Church & Society, Archons, Archives, Spiritual Courts and St. Photios Shrine. Information on them appear in other publications of the Archdiocese and will be disseminated through other books, magazines and other media of the Church and will be the subject of other writers and publishers as well.

Fr. Miltiades B. Efthimiou
New York, June 1984

Editor's Note: Several Greek names have been allowed to stand in the transliteration preferred by the author of each article.

CONTENTS

V

THE CONTRIBUTORS *(In article sequence)*

Rev. Dr. Miltiades B. Efthimiou - is Director of the Department of Church and Society at the Greek Orthodox Archdiocese of North and South America, and Executive Officer of the Order of St. Andrew.

Rev. Dr. Demetrios J. Constantelos - is Professor of History and Religious Studies, Stockton State College, Pomona, NJ and an author of several books.

Rt. Rev. Silas Koskinos - is Metropolitan of New Jersey, Greek Orthodox Archdiocese of North and South America.

Dr. Louis J. Patsavos - is Associate Professor of Canon Law at Holy Cross Greek Orthodox Theological School, Hellenic College, Brookline, MA.

Dr. George Bebis - teaches at Holy Cross School of Theology and specializes in Patristic Theology.

Rev. Dr. George Papaioannou - is Pastor of St. George Greek Orthodox Church, Bethesda, MD.

Rt. Rev. Philotheos, Bishop of Meloa - is an assistant to Archbishop Iakovos, Primate of the Greek Orthodox Church in the Americas.

Mr. Peter T. Kourides - is the legal counsel to the Greek Orthodox Archdiocese of North and South America.

Ms. Julie Charles - is a professional writer living in New York City.

President Carter - is the 39th President of the United States of America.

Mr. Basil Foussianes - has been a long-standing member of the Archdiocesan Council of the Greek Orthodox Archdiocese of North and South America.

Mr. George A. Christopoulos - is formerly a New York Stock Exchange administrator and currently a public relations executive and a multi-faceted professional writer.

Rev. Dr. Stanley Harakas - is Professor of Christian Ethics at Holy Cross Orthodox Theological School, Hellenic College, Brookline, MA.

Mr. Emmanuel Hatziemmanuel - is Director of the Department of Greek Education, Greek Orthodox Archdiocese of North and South America.

Mrs. Stella Coumantaros - is Executive Director of the National Office of the Ladies Philoptochos Society.

Mr. Steve Papadatos - is a leading architect of Greek Orthodox churches.

Dr. Frank Desby - is a leading composer of Orthodox church and liturgical music.

Rev. Angelo Gavalas - is the Director of the Youth Department, Greek Orthodox Archdiocese of North and South America.

Fr. George Poulos - is Pastor of the Church of the Archangels, Stamford, CT.

Fr. Constantine Sitaras - is the former Director of St. Basil Academy, Garrison, NY.

Mr. Allen Poulos - is the former Associate Director of St. Basil Academy, Garrison, NY.

Mr. Spiro Pandekakes - is Executive Administrator of the St. Michael's Home for the Aged, Yonkers, NY.

Rev. Dr. Robert Stephanopoulos - is Dean of the Archdiocesan Cathedral of the Holy Trinity of New York City.

Mr. John Douglas - is on the Archdiocesan Council of the Greek Orthodox Archdiocese of North and South America.

Dr. Nicholas Kladopoulos - is Director of the Department of Registry, Greek Orthodox Archdiocese of North and South America.

Mr. George Charles - is on the Archdiocesan Council of the Greek Orthodox Archdiocese of North and South America.

Introduction

By
Demetrios J. Constantelos

substantial colony of some five hundred Greek Orthodox emigrated to the New World in 1768. The Greek immigrants arrived under the aegis of a Scottish physician named Andrew Turnbull and his Greek wife, Maria. The Greek colony was named New Smyrna and was located in the northeast part of the present state of Florida.

Dr. Turnbull had made plans for both a Greek Orthodox Church and a regular priest to serve the spiritual needs of his settlers. On March 31, 1767, he had petitioned the Board of Trade of England for an annual allowance of £100 for the first Greek Orthodox priest. But, as far as modern scholarship can tell, Dr. Turnbull's plans did not materialize. There is no evidence either that a Greek Orthodox priest accompanied the 1768 Greek immigrants or that a Greek Orthodox Church was established in eighteenth-century America.

In addition to the New Smyrna colonists, many Greek Orthodox merchants, traders, and refugees from Turkish persecution settled in the United States during its revolutionary and early national period. They were dispersed everywhere, from Boston to New York, as far south as New Orleans and as far west as San Francisco. In cities such as San Francisco, they attended services in Russian Orthodox churches.

The earliest Greek Orthodox church in the United States was established in 1862 in the seaport city of Galveston, Texas, and it was named after Saints Constantine and Helen. Even though the church was founded by Greeks, it served the spiritual needs of other Orthodox Christians, such as Russians, Serbians, and Syrians. It passed into the hands of the Serbians, who split with the Greeks. The Greeks then established their own church several decades later; but knowledge of the early years of the Galveston Greek Orthodox community is very limited. Neither the number of Greek Orthodox parishioners there nor the name of the first priest is known. The first known

1

Greek Orthodox priest of this community was an Athenian named Theoklitos Triantafylides, who had received his theological training in the Moscow Ecclesiastical Academy and had taught in Russia before joining the North American Russian Orthodox Mission. Versed in both Greek and Slavonic, he was able to minister successfully to all Orthodox Christians.

Knowledge of the second Greek community in the United States is more extensive. It was organized in 1864 in the port city of New Orleans. Like the Galveston community, the second one was also founded by merchants. For three years (1864-1867) services were held irregularly and in different buildings. Then in 1867 the congregation moved to its own church structure, named after the Holy Trinity. It was erected through the generosity of the philanthropist Marinos Benakis, who donated the lot and $500, and of Demetrios N. and John N. Botasis, cotton merchants who together contributed $1,000.

The church was located at 1222 Dorgenois Street and for several years it became the object of generosity not only of Greeks but of Syrians, Russians, and other Slavs. In addition to Greeks, the board of trustees included one Syrian and one Slav. Notwithstanding the predominance of Greeks on the board, the minutes were written in English and for a while it served as a pan-Orthodox Church.

The early Holy Trinity Church was a simple wooden rectangular edifice 60 feet long and 35 feet wide. The major icons of the iconostasis were painted by Constantine Lesbios, who completed his work in February of 1872. The name of the first parish priest is unknown, but it is believed that a certain uncanonical clergyman named Agapios Honcharenko, of the Russian Orthodox mission in America, served the community for three years (1864-1867). In 1867 the congregation moved to its permanent church and appointed its first regular priest, Stephen Andreades, who had been invited from Greece. He had a successful ministry from 1867 to 1875, when the archimandrite Gregory Yiayias arrived to replace him.

The New Orleans congregation also acquired its own parish house; a small library, which included books in Greek, Latin, and Slavonic; and a cemetery.

The number of churches in the second half of the nineteenth century corresponded to the number of Greek Orthodox communities, which were concentrated in cities. Up to 1891 there were approximately twenty-five hundred Greek Orthodox in the United States; from then on, there was a substantial in-

First Greek Orthodox Church in America, New Orleans, La., 1864.

3

crease in immigration. In 1891 and later many Greek Orthodox churches were founded in large cities such as New York,
Chicago, San Francisco, and Boston, and in many smaller
cities and towns such as Washington, D.C.; Newark, New Jersey; Ipswich, Massachusetts; Omaha, Nebraska; Pensacola,
Florida; and Moline, Illinois. In the course of thirty-one years
(1891-1922), 139 new Greek Orthodox congregations were
organized in the United States and two in Canada.

The following ten churches were the earliest in the United
States after the two in Galveston and New Orleans:

Holy Trinity, New York City (established in 1891)
Holy Trinity, Chicago (1892)
Annunciation (Evangelismos) of the Theotokos,
  New York City (1893)
Holy Trinity, Lowell, Massachusetts (1894)
Annunciation of the Theotokos, Philadelphia (1901)
Saint Nicholas, Newark, New Jersey (1901)
Holy Trinity, Birmingham, Alabama (1902)
Holy Trinity, San Francisco (1903)
Annunciation of the Theotokos, Boston (1903)
Saint Nicholas, St. Louis, Missouri (1904)

Why were the earliest churches named in honor of the Holy
Trinity, the Annunciation, Saint Nicholas? Obviously they
manifest belief in the doctrine of the Holy Trinity and the Annunciation of God's good news—the *evangelismos* of the in-
carnation of Christ and His appearance among men. But the
Annunciation is also a popular holiday among the Greeks because it is linked with Greek Independence Day. The use of
Saint Nicholas' name indicates that a humble and philanthropic, not necessarily an intellectual, Church Father is held in
high esteem by the faithful.

Of the 141 churches founded in the United States between
1862 and 1922,

30 were named after the Annunciation of the Theotokos
24 after the Holy Trinity
19 after the Dormition (koimesis) of the Theotokos
15 after Saint George
13 after Saints Constantine and Helen
 9 after Saint Nicholas
 7 after Saint Demetrios
 6 after Saint John the Baptist
 4 after God's Wisdom (Hagia Sophia)
 4 after Saint Spyridon

4

2 after All Saints and
2 after the Holy Apostles

The rest were named in honor of various other individual saints. While the number of Greek Orthodox communities was substantial, the faithful in the New World were like a flock without a shepherd. For three decades the Church was beset by numerous problems and a great deal of instability. Why?

A

By 1913 the Greek Orthodox population of the United States had increased to nearly a quarter of a million; and by 1922 it was estimated at between 300,000 and half a million. For example, the Massachusetts Bureau of Immigration put the number at 350,000 and the census of 1920 reported 175,972 foreign-born Greeks in the United States. However, considering the new arrivals between 1920 and 1922 as well as those Greek Orthodox born in America between 1862 and 1922, it seems probable that by 1922, when the Church was organized into an archdiocese, the Greek Orthodox in the United States numbered indeed approximately half a million.

Up to the year 1908, the Greek Orthodox communities of the American diaspora were under the spiritual aegis of the Ecumenical Patriarchate of Constantinople. But owing to the uncertainties in Turkey and the wave of Turkish nationalism under the Young Turks, as well as to international political events, Patriarch Ioakim III, in agreement with the Holy Synod of the Patriarchate, issued a tome (letter) on March 8, 1908, which placed the Greek Orthodox churches in America under the jurisdiction of the autocephalous Church of Greece. Nevertheless, the churches in America remained fairly independent, and no effective supervision existed for several years.

The Church in America—it can be referred to as one Church because of the unity of faith, worship, and ethics—was in flux for several reasons. Most of the Greek Orthodox were immigrants whose ultimate goal was not to make their home permanently in the New World, but to accumulate enough savings to be able to return to the motherland. Thus they were not interested in supporting churches in America.

For these and for cultural, linguistic, and national reasons, the Greek Orthodox in America during the first quarter of the twentieth century strongly resisted assimilation. But it was more than ethnic pride that made them resist Americanization. They found in the United States not simply a "melting pot," whatever that expression may mean, but a "pressure cooker." In the first quarter of the century, there was in the United States a wave of antagonism to foreigners, which did little to encourage adjustment or assimilation. There was an evident dichotomy between what the Constitution proclaimed and what actually took place. According to many accounts, the country was interested only in labor and sweat of its new-comers, not in their presence in its established society. Many immigrants resented the paternalism of some religious orga-nizations, immigration authorities, and social or sectarian so-cieties of various kinds, which directly or indirectly were for-cing them to adopt American ways. One of the writers of the period summarized the Greek reaction as follows:

A great deal is being said and written regarding the Amer-icanization or assimilation of the immigrants that seems strange. Some of the heated utterances sound like the nationalistic theories of the Pan-Germans or the Pan-Slavists. If the various races are to be forced to forget all their racial peculiarities and characteristics, customs, us-ages, and language, and to adopt American ways instead, the result will be disappointing. Whenever a people is for-ced to accept, willingly or unwillingly, a certain course of action, the result has usually been the opposite of what was desired. . . . Even the word "Americanization" sounds strange to many ears; it sounds like suppression, force.

The American religious mind of the late nineteenth and early twentieth centuries was determined "to maintain the United States as a homogeneous, evangelical enterprise." The new immigrants were viewed as a threat to the established society. Thus there was a great deal of effort to Americanize them through the values and creeds of the various established re-ligious groups. A modern author rightly emphasizes that:

The immigrant to America. . . was to be simply raw ma-terial for Americanization—a faceless mass of mankind unable to make any contribution to America beyond sweat. Americanization was paternalistic and aimed at impart-ing to the newly arrived the tried and true gospel of the American Civil Religion.

But the treatment of the immigrant by the "natives" was anything but Christian. Leroy Hodges, a commissioner of im-migration, in 1912 described the "evangelical" way as follows:

*Metropolitan of Kerkyra (Corfu) Athenagoras upon his arrival to his
new assigment in the Americas.*

Churches supported by American Protestants located in the immigrant colonies refuse to receive the recent immigrants in their buildings as the native Americans are received, and some of them resort to the practice of holding services for them in barns, stores, and other such places, posing the while before the public as ardent "settlement workers." Some ministers have gone so far as to make the statement that the recent immigrants are a "lot of filthy cattle," with which they do not care to litter up their churches. . . . Not only are they not assisting in the Americanization of the new citizens, but they are engendering an opposition against the institutions upon which the future of the United States rests.

Nevertheless, despite the immigrant's reluctance to accept Americanization, with the passing of time, the establishment of a family, the exposure to new ideas, new institutions, and a new environment, and familiarity with the language, much assimilation was achieved. Besides the threat of being proselytized, the problems the Greek Orthodox encountered during the period under discussion were many and diverse. They had to overcome a great deal of prejudice; and they had to work hard and long in order to improve socially and financially. Since most of them came from the rural stratum of Greek society, from remote villages and small towns, many suffered from an inferiority complex. Others, through ignorance of the language and the laws of the country, found themselves in the courts.

A happy exception in the contest among some denominations to convert the Greeks was the Episcopal Church. For centuries there has been an affinity between the Orthodox Church and the Anglican Communion, and the Episcopal Church provided much help to the early Orthodox when they were in need of buildings for Church services. It was the official policy of the Episcopal Church not to conduct missionary activity among the Orthodox.

In retrospect, one may suspect that many American Christians viewed the Greek Orthodox as pagans. Some years ago I was asked to lecture to a congregation in an "enlightened" New England town. Before I was introduced by the pastor, I was approached by a group of ladies and asked various questions. A delightful middle-aged lady asked me: "Do you still believe in Zeus?" If some "natives" could raise such a question in 1967, one can imagine the kind of ignorance about Orthodox Christianity that prevailed in the early twentieth century.

Many Greek Orthodox, because of an inferiority complex, for business purposes, or in order to avoid discrimination, changed their names (though many Greek names had already been shortened or changed arbitrarily by immigration officers). Thus a Papadopoulos became Brown, a Konstantinides was shortened to Constant, Anagnostopoulos emerged as Agnew, Papanikolaou as Papps, and so on. There are many of Greek origin who have adopted names such as Williams, Johnson, Adams, Peterson, Bell, Kress, Nickolson, Johns, Carr, Larry, Lorant, Pepps, Moore, Stance, Pelican, Ross, Meyers, Allen, Anagnost, Anton, Apostle, and Apostol.

Notwithstanding the proselytizing by several Protestant denominations and other religious groups, as a whole the Greek Orthodox remained faithful to their ancestral faith and their Greek heritage. In order to preserve their culture and traditions they established fraternities, social clubs, Greek language schools, newspapers (two dailies), and magazines.

Many Greeks joined a fraternity before they joined a church—some because they preferred the social life to the religious, others because a fraternity was easy to form and to sustain, while the organization of a church required not only a priest and a church building but regular financial support. Several fraternities played an important role in the organization of the Church. The third paragraph of Article One of a certain fraternity's statement of its objectives declares that one of its purposes was "to preserve the Greek Orthodox Church and to develop and propagate educational and moral doctrines among the Greek compatriots residing in the United States and Canada."

One of the Church's major problems was that she was without any bishops to coordinate the various communities and to direct the destiny of the Church. There was much individualism, dissension, and lack of orientation. The 141 communities were like 141 ancient Greek city-states in an American archipelago. Some churches were social clubs rather than religious communities. To be sure, each congregation usually had its own priest, but priests were expected to administer the sacraments and conduct funerals rather than to lead the congregation, which usually was under the control of a lay board of trustees. The priest, often ill-educated, had come to America for the same purpose as his parishioners, and he was frequently at the mercy of the community, especially the board

9

of trustees, which had appointed him. It has been rightly observed that in Greek parishes until 1922 "congregationalism reigned supreme in an episcopal church."

While undoubtedly many were concerned with the Church as the provider of spiritual and religious values, others were only traditionally Orthodox, attending the Liturgy on Christmas or Easter Sunday, or religious services conducted for some of their friends—a baptism, wedding, funeral, or memorial service. Church attendance was very poor. Nevertheless, it was due to the zeal of certain dedicated laymen that most of the churches were established. For example, Holy Trinity Cathedral of New York, established in 1891, was the work of lay members of the Athena Society. They applied to the holy synod of Greece, which appointed Paisios Ferentinos the first priest of the New York cathedral. Services were held for several years in a hall or in a rented church building until 1904, when the congregation was able to build its own structure at 153 East 72nd Street.

The way the New York cathedral was organized was more or less standard, and it served as the prototype for the establishment of later churches. First, a group of dedicated laymen would get together to determine whether there were enough Greek Orthodox to support a church. As soon as they secured the support of some fifty families, they would apply to the state for a charter of incorporation under titles such as Hellenic Eastern Orthodox Church of _____ or Greek Orthodox Church of _____ . The second step was to apply to the Ecumenical Patriarchate or to the Church of Greece, and sometimes to the Patriarchate of Alexandria, for a priest. The choice of the church to which the application for a priest was sent was often determined either by the political affiliation of the applicants or the place of their origin. Greeks from the mainland usually applied to the holy synod of Greece, while Greeks from Asia Minor, Thrace, and other regions usually directed their application to Constantinople.

As soon as a priest was available the committee would seek to rent a hall or a church building, usually from the Episcopal Church. As a rule the congregation would move to its own building within a few years. Though the Church welcomed all Greek Orthodox, not all were supporting members. As already stated, the concern of many was to save enough to return to the motherland. Thus, though there were many Greek Orthodox, there were few churches and contributing members. This explains why the clergy were poorly paid and the churches in great debt and uncertain about their future.

A turning point in the history of the Church occurred in August of 1918, when the Archbishop of Athens, Meletios Metaxakis, arrived in New York to study the problems of the Greek Orthodox in America. He was accompanied by Bishop Alexandros, the titular Bishop of Rodostolou (who later became Archbishop Alexander of America), by Archimandrite Chrysostom Papadopoulos (a renowned ecclesiastical historian who later became Archbishop of Athens), and by a few laymen. Meletios Metaxakis was determined to bring order out of the chaotic conditions that prevailed in the Church. In the past, several requests had been sent to the mother churches for a bishop and for more concern for the church of the diaspora, but with no definite results. For example, the Greek Consul General in New York City, Demetrios N. Botsis, in a report to the Greek government dated July 15, 1904, made a strong case to the Greek Ministry of Foreign Affairs requesting its intervention with the synod of the Church of Greece for the appointment of a bishop. But no bishop was sent.

The lack of initiative on the part of both the Church of Constantinople and the Church of Greece reveals what happens when a church is not a free institution or when it is tied to the state. Both churches were subject to the internal problems of their respective states, and both were prevented from acting by nonreligious considerations. The Church of Greece was divided between loyalists and liberals, especially in the period of 1914 to 1918, and the good bishops had no time to devote to the problems of the immigrants. The Ecumenical Patriarchate was continuously under the threat of the sword of the Turks. When the Young Turks came to power in 1908, they were determined to expel all Christian minorities, including the historic Ecumenical Patriarchate. It was primarily because of the uncertainties caused by neo-Turkish nationalism that the Patriarchate could not concern itself with the problems of the American Church and issued, in March of 1908, the tome that placed the Greek Orthodox in America under the jurisdiction of the autocephalous Church of Greece.

However, one of the reasons why both the Ecumenical Patriarchate and the Church of Greece were reluctant to appoint a bishop for the Greek Orthodox faithful in America may have been canonical. Canon law forbids the appointment of a bishop in a province or district where a canonical bishop already

11

exists, and there were in the New World Orthodox bishops of the Moscow Patriarchate. It was not the first time that Greeks had been under the spiritual guidance of Russian Orthodox. To be sure, some or even many Greeks may not have been happy under the Moscow Patriarchate. But it is rather unjust to blame "Greek nationalism" for the reluctance of the Greek faithful to accept Russian leadership. It may have been inefficiency, lack of unity, and turmoil within the Russian church that persuaded the Greeks to seek their own leadership.

Whatever the reasons, the first Greek Orthodox bishop arrived in the United States in 1918, and it was in that year that the hierarchy of the church began to organize the Greek Orthodox communities in the States.

B

It was amid chaotic conditions that the first bishop was appointed to proclaim concord and love and to consolidate the Greek Orthodox communities into a united Greek Orthodox archdiocese. Even though Archbishop Meletios stayed in the United States only three months, he had a permanent impact. His work was primarily that of fostering contacts among the clergy, the community leaders, and the faithful. He was a magnetic figure, and he drew to himself opposing factions.

The most important accomplishment of Archbishop Meletios was the creation of the Synodic Trusteeship (Synodike Epitropeia) and the appointment of Bishop Alexander as his own and the synod's personal representative. Thus, since 1918, the Church has been under episcopal leadership. But from 1918 to 1922, several cataclysmic events in the motherland greatly affected the history of the Church in the New World. During this time, Greece was racked by a struggle between the supporters of the crown and those of the political leader Venizelos. The episcopal representative in the United States was tied to the destiny of his superior in Greece. So when Meletios Metaxakis was deposed as Archbishop of Athens because of political developments in Greece his representative in America found himself in a very difficult situation. Upon his deposition from office as Archbishop of Athens, Meletios sought refuge among the Greek Orthodox in America. While Meletios was on his way to New York City, Bishop Alexander was asked to return to Greece. But, with the support of Meletios, Alexander refused to obey the new archbishop in Greece,

The new Archbishop Athenagoras with delegates at the 4th Clergy-Laity Congress, New York, 1931.
(Photo in three parts - left, center and right)

13

placed himself under the jurisdiction of the Ecumenical Patriarchate, and decided to stay and carry on his religious duties.

While in America, Meletios acted as the canonical Archbishop of Athens, and Alexander served as his auxiliary. For almost a year they cooperated closely in the organization of the Church, in the pacification and the union of split communities, and in the advancement of the Greek Orthodox in America.

Meletios was a visionary and a dynamic person. In the short time that he stayed in the United States he accomplished a great deal. On August 11, 1921, he invited the lay and clerical leaders of the communities to the first congress of clergy and laity, which was held from September 13 to 15, 1921, in the Holy Trinity Cathedral of New York. One of the major decisions of this congress was to seek the incorporation of all Greek Orthodox communities into the Greek Orthodox Archdiocese of North and South America. A constitution was drawn up and the charter of the archdiocese was issued by the state of New York on September 17, 1921.

In addition, Meletios aspired to have a clergy educated in the United States not only in theology but also in the English language, able to guide old and young alike, to lead and to preserve the church in multisectarian America. Thus he established Saint Athanasios Greek Orthodox Seminary in Astoria, New York. Furthermore, he organized a "philanthropic treasury" for the poor, he edited a weekly journal, the *Ecclesiastical Herald,* and promoted ecumenical relations between the Orthodox and non-Orthodox Churches, the Episcopal Church in particular. The Greek Orthodox continued to be especially friendly with the Episcopal Church. As early as 1919 Archbishop Alexander attended an Episcopal ordination in Lancaster, Pennsylvania, as an observer. An Episcopal priest and influential author reported that "after the separation of a thousand years, practical union has come; therefore we have a special duty of friendliness to the members of our sister churches of the East. We can cooperate with them in reaching their own people and help them to be faithful to their great church."

The same writer records that Archbishop Alexander had expressed "his willingness to enter into an agreement by which our clergy may be licensed to minister to his people where their priests are far away."

Because of his excellent reputation and his activity among non-Orthodox Church organizations and agencies, Archbishop

Meletios was awarded several honors by American institutions, including universities.

One of Meletios' efforts was to mediate on behalf of the Christian minorities in Turkey, which since 1908 had been undergoing relentless persecutions. The genocide inflicted upon the Armenians by the Turks had moved many Americans but little had been done to save the Christian minorities in Anatolia. Charles Evans Hughes was a champion of the Christian minorities there as long as he was outside the government. But when as secretary of state he was pressed for action against Turkey, Hughes forgot his idealistic principles and became a "political realist." In a letter dated December 8, 1921, Meletios pleaded with him to protect the minorities in Turkey.

Despite the fact that so many Greek Orthodox had numerous personal and family financial obligations, they responded generously to several humanitarian causes. Of course, for many, one of the obligations was to provide dowries or financial assistance for their sisters back home. It was not unusual for brothers to remain single or to postpone their marriage in order to accumulate enough money for their sisters. Parents and brothers in the motherland also received liberal contributions from their sons and relatives in America.

The Greek Orthodox responded generously to several drives in behalf of war refugees, victims of earthquakes, and various philanthropic causes. Among the needy, the Greek Orthodox included brethren of the Russian Church who had fled their motherland during and following the 1917 revolution. Bishop Alexander issued a special encyclical stressing the need for assistance to all, including people outside his jurisdiction.

Another turning point in the history of the Church in America was the election of Archbishop Meletios Metaxakis to the throne of the Ecumenical Patriarchate of Constantinople, on November 25, 1921, as Meletios IV. In his farewell encyclical to the Greek Americans, dated December 31, 1921, Meletios appealed for mutual forgiveness and concord among the Greek Orthodox and enjoined them to unite around the person of Bishop Alexander. He arrived in Constantinople on January 24, 1922.

Meletios was an impatient and sometimes careless man. Of course, he had realized that in the Church of America the Ecumenical Patriarchate possessed a substantial power, both moral and economic. Thus one of Meletios' major decisions as patriarch was to annul the 1908 tome, thereby bringing the

15

Church of America under the jurisdiction of the Ecumenical Patriarchate. This was confirmed by the Patriarchate's synod and was conveyed to the Church in America on March 1, 1922. Conditions had not changed either in Greece or in Turkey. The argument that the Church in America was again placed under the Patriarchate because conditions in Greece were ominous is not convincing. The transfer was the arbitrary act of a dynamic person who knew how to turn events to his advantage.

Two months later, on May 11, 1922, Patriarch Meletios announced to Bishop Alexander that the Holy Synod and he had elected Alexander Archbishop and had elevated the Church in America to an archdiocese. The archdiocese was to include three diocesan districts, with their sees in Boston, Chicago, and San Francisco.

Archbishop Alexander had been in the United States for four years and had firsthand knowledge of the problems as well as the potentialities of his flock. But from the very beginning he had encountered several major problems. When he had refused to acknowledge the new Archbishop of Athens and his synod (which had deposed Archbishop Meletios Metaxakis) and to return to Greece as he had been ordered, the synod of Greece had appointed the Metropolitan of Sparta and Monemvasia, Germanos Trojanos, as its exarch in North and South America; he arrived in New York in June 1921, while Archbishop Meletios was still in the country. Thus there were two archbishops: Alexander, who owed his authority to the Ecumenical Patriarchate of Constantinople, and Germanos, who had been appointed as exarch by the Church of Greece.

Priests and communities were now divided not only between Royalist and Venizelist churches but also between churches that belonged to two ecclesiastical jurisdictions. Even though in the beginning the breach seemed to widen, the majority of the congregations finally rallied to Archbishop Alexander, who gained more and more communities after the organization of the Church as an archdiocese. In the meantime the relations between the Ecumenical Patriarchate and the Church of Greece improved, and Metropolitan Germanos Trojanos was recalled to Greece in January of 1923. This helped tremendously to heal the schism between the communities, most of which came under Alexander's jurisdiction, while a few remained independent.

Another difficulty for Archbishop Alexander was to bring

16

Archbishop Athenagoras surrounded by delegates and clergy attending the Fifth Clergy-Laity Congress in Chicago, October 30, 1933.

the scattered communities, which had been accustomed to self-government and complete independence from any authority except that of the membership, under the aegis and authority of a centralized archdiocese. Closely related was the archbishop's right to appoint priests to the congregations. But this right collided with a practice of long standing, according to which the board of trustees of a congregation sought out and appointed its priest. It was very difficult for the laity to understand the theological teaching concerning the nature of the Church, the episcopal office, and its function. After decades without a bishop and in isolation from the mainstream of Orthodoxy, the Greek Orthodox in America had grown into independent congregations. It took more than twenty years of hard work to bring the communities into the fold of the archdiocese.

In June 1923, a few months after the departure of Germanos Trojanos, the Ecumenical Patriarchate elected two bishops for the Church in North America. Philaretos Ioannides was appointed Bishop of Chicago for the Middle West and Ioakim Alexopoulos was consecrated Bishop of Boston for the New England States. They were not meant to be auxiliary bishops under Archbishop Alexander. All three were to constitute a small synod, which was augmented four years later when, in 1927, Kallistos Papageorgakopoulos was consecrated and appointed Bishop of San Francisco for the Western States.

Notwithstanding the new arrangement, with one archbishop and three bishops, the decade of the 1920s continued to be turbulent. The election of the three bishops did not satisfy the congregations that had sided with the Royalist party. In fact, all bishops were considered Venizelists. The division assumed large proportions when a Royalist Metropolitan named Vasilios Kombopoulos arrived in the United States to head the Royalist communities. Metropolitan Vasilios acted unilaterally and established the Autocephalous Greek Orthodox Church of the United States and Canada.

The question has been raised again and again: What kind of man was Alexander? Was he the proper hierarch to lead widely dispersed and highly individualistic congregations in America? These questions cannot be answered in a brief essay, but a few observations may be made. To be sure, he has been criticized severely, but he has been held responsible for things that were beyond his control, and a close study of his encyclicals and private correspondence reveals that Alex-

18

ander was devoted to the Church as well as to the Ecumenical Patriarchate. Even though much of his time was devoted to administrative duties, to exerting every effort to unite congregations and bring dissident communities under the aegis of the archdiocese, Archbishop Alexander was greatly concerned with humanitarian and philanthropic programs. Broadminded and tolerant, he was a good man. It has been rightly observed that "Alexander fought courageously. If he lost heart and erred at the end of this painful decade, it was because every struggle has an end and a victim."

Alexander had some failures. Despite many efforts to maintain Saint Athanasios Seminary, it had to close its doors two years after its opening for lack of financial support. It was not that the Greek communities could not support a seminary; rather it was the dissensions in the American Church that affected the financial condition of the school. He failed also to unite all the communities and to assert the authority of the archdiocese. But who could have succeeded under the circumstances? Following more than thirty years of ecclesiastical anarchy, how was it possible for one or even four bishops to gather the widely scattered flock into one fold? Nevertheless, Alexander laid the foundations, and the unity of the Church he sought was achieved much more easily under his successor.

During Archbishop Alexander's primacy the second and third Clergy-Laity Congresses were held. The second was held in New York in August 1922, and the third in Chicago in October 1927. Those assemblies served not only to enhance the authority of the archdiocese but also as links between congregations and their members.

The first quarter of the twentieth century was full of dramatic happenings for the Greek nation and its Church. In particular, the third decade was a decade of cataclysmic episodes that permanently affected the life of millions of Greek Orthodox. After more than 3,000 years in Asia Minor, the Greeks were permanently uprooted from there. Chauvinistic Turkey became determined to expel from her territories all Christian minorities, whether Armenians, Greeks, or others. Thus hundreds of thousands were expelled from their homes, while many more were put to the sword.

More than once, Archbishop Alexander rallied his flock in America to help refugees, orphans, and the many destitute who fled the Turkish madness. From 1919 to 1930, the last year of his stay in the United States, Alexander wrote numerous

19

encyclicals and promoted many philanthropic causes. These documents reveal that the Archbishop appealed for material contributions to the Christian minorities expelled from Pontos and other regions of Asia Minor. He wrote encyclicals and personal letters to help orphans and orphanages; poor students, needy professors, and destitute priests; the Ecumenical Patriarchate; the Near East Relief Fund; victims of earthquakes in Greece; sanatoria; the American Red Cross; and unemployed and poor miners in the United States. He also had a deep concern for the poor, for schools, and for educational programs.

Of course, there is no way of knowing how effective all those appeals were, and there is no concrete evidence of how much actual material assistance was given to the needy. What is known, however, is that those documents reflect a man full of empathy and active concern.

Archbishop Alexander also established in the archdiocese the Office of Relief of the Greek Archdiocese of North and South America, with a board of twelve and himself as chairman. The purposes of the office of relief, located at 215 West 23rd Street in New York, were to provide assistance to all those entering or leaving the country; to provide financial as well as other support to immigrants who for some reason were held in custody or in prison; to find housing for those newcomers who desired to stay in New York until ready to move on to their destination; to counsel immigrants concerning the laws of the country and to assist those in conflict with the law; to organize classes and meetings for the orientation of the newly arrived; to visit hospitals and give help to the sick. How effective this office was and to what degree it carried out its functions is very difficult to say.

In 1929, there were close to 200 Greek Orthodox congregations in the United States. Of these, 133 were under the jurisdiction of Archbishop Alexander; 50 were under the so-called Autocephalous Greek Orthodox Church of the United States and Canada, headed by the deposed Archbishop Vasilios Kombopoulos; a few were totally independent; and a few more, which followed the Julian or Old Style Calendar, were under either the Patriarchate of Jerusalem or that of Alexandria. Of course, the main problem was to unite the two major bodies. And this became the common concern of the Church of Constantinople and Athens.

Patriarch Photios II of Constantinople, in agreement with Archbishop Chrysostom Papadopoulos of Athens and with the cooperation of the Greek government, decided to recall all bishops from America, including Archbishop Vasilios Kombopoulos, who was to be restored to his canonical status. In order to facilitate the transition and prevent new schisms, the Patriarchate appointed Damaskenos, the capable Metropolitan of Corinth, as exarch of the Ecumenical Patriarchate and as interim head of the canonical Greek archdiocese. He proved a shrewd arbitrator.

Damaskenos arrived in the United States on May 20, 1930, and after he had visited the President of the United States and other civil authorities, he issued an irenic encyclical to all Greek Orthodox congregations, explaining the reason for his arrival. He met with Archbishop Alexander and the other bishops, to whom Damaskenos conveyed the Patriarchate's decision.

Alexander at first appeared conciliatory but soon hardened his position. In fact, on May 26, 1930, he issued his own encyclical, expressing his feelings of disappointment over Damaskenos' tactics. He stressed that the exarch's activities were unwarranted and appealed to his flock for intervention by writing to Patriarch Photios, the Greek government, and to Damaskenos on his behalf.

Nonetheless Damaskenos, who proceeded methodically and rather cautiously in the very beginning, was determined to carry out his mandate. When Alexander refused to resign, Damaskenos cabled the Patriarch and the synod asking for more drastic measures against Alexander. The result was that Alexander was officially deposed by the Patriarch.

Furthermore the Patriarchate accepted Damaskenos' second recommendation, which called for the election of Athenagoras, Spyrou, Metropolitan of Kerkyra (Corfu), as the new archbishop of the American archdiocese. In addition, Damaskenos proposed a new constitution, which favored a more centralized archdiocese, the abolition of the synodic system, and the appointment of auxiliary bishops rather than autonomous heads of diocesan districts.

Damaskenos' final report reveals that the Church in America was evolving along independent lines, that there was a great deal of assimilation, and that there was a great need for educated clergymen who would inspire not only religious enthusiasm but a sense of ethnic pride and of loyalty to the Ecumenical Patriarchate. He had an excellent awareness of the tendencies in Greek Orthodoxy, and his lengthy report reflects the issues, trends, dangers, and orientation of the Greek Ortho-

dox in the New World. While Alexander favored more autonomy and was cognizant of the American realities, Damaskenos emphasized the need for patriarchal control and stronger ties between the Greek Orthodox in America and the motherland.

Deposed, left with few friends, and disillusioned, Alexander
yielded to the decision of the Patriarchate and agreed to leave
the United States. He was appointed Metropolitan of Kerkyra,
whose metropolitan in the meantime had been elected to succeed him in America.

This change took place at a time when conditions were greatly
improving in the Greek communities in America. Indeed, it
illustrates what has been observed by historians: That not only
major revolutions but social changes of a less drastic nature occur when things start to improve, and not under the most de-
pressed conditions. Archbishop Alexander rightly complained
that Damaskenos had arrived when things were becoming better.

Whatever the future evaluation may be of the archbishop,
the fact is that Alexander "fought the good fight" and accomplished a great deal under ominous circumstances. The Greek
Orthodox should cherish his memory in grateful acknowledgement of a very dedicated ministry.

In the early years of its establishment, the Church gave the
immigrants moral support, cultural reinforcement, and a sense
of belonging together, of dignity and self-esteem. Through
its priests and services, the Church stood by the immigrant
in the hours of birth and death, gave comfort in sickness and
encouragement in moments of anxiety and catastrophe.

Elected on August 13 1930, the new archbishop arrived in
New York City on February 24, 1931. As already stated, the
climate in the Church had improved considerably. Dissident
churches still existed, but on the whole the congregations were
eager to forget the conflicts of the past and their political commitments, and to cooperate for the progress of all.

Athenagoras was a diplomat, often a politician, and an excellent public relations man. One of his first undertakings upon
arrival was to visit President Herbert Hoover, other political,
religious, and civic dignitaries, and of course, his own flock.
Early in his ministry he visited some 50 congregations. His
strength derived from the way he approached and captivated
people. He had little sympathy for abstract theology and theoretical arguments. His encyclicals reveal him a pragmatist. He
imposed his authority and won over even his adversaries by a

fatherly manner rather than by any profound thought or arguments.

One of the major changes introduced in the Church was a new constitution. The synodic system that had governed the Church from 1922 to 1930 was abolished, and the bishops were no longer autonomous heads of dioceses but titulars, serving as auxiliaries to the archbishop. During Athenagoras' primacy there were five diocesan districts, each with one auxiliary bishop, in New York, Boston, Chicago, San Francisco, and Charlotte, North Carolina. By 1931, there were 220 priests with an equal number of parishes and a Greek Orthodox population of some 750,000 persons. In 1934, there were 250 priests, more priests than congregations.

During the Great Depression several small Greek Orthodox churches were closed down because many of their parishioners had moved to larger cities seeking employment. Other churches were threatened with foreclosures or bankruptcies because of large mortgages. On several occasions the Archdiocese moved to prevent such failures. For example, the communities of Clinton and Lawrence in Massachusetts were threatened with extinction in 1936; but local clergy-laity congresses, such as the fourth local congress of New England, held in Brockton, MA, as well as appeals from Archbishop Athenagoras, provided the necessary money for the security of these two communities, which have survived to the present day.

There is evidence that the Church as an organized body made some efforts to relieve the poor and help the needy during the Great Depression, but there is no statistical evidence concerning the type of work that was initiated by the archdiocese. The best indication of Athenagoras' concern for the destitute and the needy is a moving encyclical issued on October 20, 1932, in the name of the Central Council of the Ladies Philoptochos Society. This encyclical, signed by the Archbishop as president of the society, ten members of the board, and two secretaries, was an excellent example of the Church's concern for human beings. It summarized the theological teachings about philanthropy and gave direction to the various chapters of the Society and to the congregations on a number of important matters, for example, where and how to seek out and help the unemployed, the sick, the orphans, the widows, the homeless, those in prison, the aged, the young immigrants who had no one to turn to for guidance, those in conflict with the law because of ignorance, and others in need.

Of course, every archbishop has issued many similar appeals, and it is difficult to say whether these appeals had any result. It was up to the individual priest and individual chapters of the Philoptochos Society to respond to the needs of their people, and the Church must have responded according to her means and ability to the needs of her people during the Great Depression.

Nevertheless, the Archdiocese was criticized for not doing enough to help the poor and the unemployed. One Greek critic of the clergy wrote:

What have the clergy done for the various victims of the depression? All other churches and various social organizations have established centers for the care and relief of the poor; only our glorious orthodoxy sleeps under the mandrake and satisfies itself with a few appeals and pompous pretensions. And, what shall we say when as we are told, clergymen are engaging in profitable enterprises, neglecting their high calling to become real estate and stock market manipulators?

In his 18 years of ministry in America, Archbishop Athenagoras left behind several landmarks that continue to determine the destiny of the Church in America. The problems of dissent, congregational independence, and schism continued to exist for several years after his arrival. But as a result of his prodigious efforts, his enormous patience, tolerance, and persuasiveness, and his numerous contacts with and visits to the local churches, most of the dissident congregations joined the ranks of the Archdiocese. Only a few communities, which followed the Julian Calendar (Old Calendarists), remained outside the archdiocese's fold.

For several years during Athenagoras' primacy, economics was one of his major problems, not only because of the Depression and the Second World War, which paralyzed many communities, but also because of the amateurish way in which the Archdiocese had organized its finances. To sustain itself and maintain its programs the archdiocese had to rely on small parish contributions, collections on special occasions, and donations from a few prosperous individuals.

It was during the Eighth Clergy-Laity Congress, held in June 1942 in Philadelphia, that the grounds were laid for a stable though poor economic system. The congress adopted the *monodollarion*. Each Greek Orthodox family became ob-

24

Archbishop Athenagoras with representatives of the Greek War Relief seen with Spyros P. Skouras (left) and Stephen C. Stephano next to him.

The Archbishop with Sophocles Venizelos at the Greek-Turkish Conclave at the Hotel Pierre in New York.

25

ligated to contribute $1.00 a year to the archdiocese through the local parish. By modern standards the contribution was extremely low; but it provided the Church with a stable income that allowed the archdiocese to plan and partially pursue its goals. Through this system the archdiocese was able to provide salaries, very small to be sure, and support its institutions.

Athenagoras, however, is best remembered for his efforts to establish educational institutions and to provide theological training for the clergy. There were several priests with a theological degree from one of the four theological schools of Greece and the Greek Orthodox diaspora (University of Athens, University of Thessaloniki, Theological School of Chalki in Turkey, and the Holy Cross Theological School in Jerusalem). But most of the parish priests in America did not have any formal theological or even college education. Some had mastered the English language, but most had remained parochial and untouched by the new environment. There were many good immigrant priests who taught by their faith and devotion rather than by their education or special theological training.

But the membership of the Church no longer consisted only of Greek-speaking immigrants. A great number of the faithful were native born, and English was their mother tongue. The educated among them were not comfortable in a Greek-speaking church. Conditions in the Church had improved, but she continued to lose ground. Many of those who had moved upwards, both socially and educationally, left the Church. The archdiocese realized the need for educated leadership on the local level.

Holy Cross Greek Orthodox Theological Institute opened its doors in September 1937, with 14 students and three instructors. It was placed under the deanship of the dynamic bishop of Boston, Athenagoras Kavvadas, and was located in Pomfret Center, Connecticut. It was meant to be a preparatory seminary, not a full-fledged theological school, for it possessed neither the capital nor the necessary faculty. Its graduates were to continue their theological education in one of the theological schools of Greece or the Ecumenical Patriarchate. The Second World War, however, frustrated the original plans, and Holy Cross was reorganized as a professional five-year college and seminary. A turning point in the school's history occurred in 1946, when it moved to its present location in Brookline, Massachusetts. Undoubtedly, the establishment of Holy Cross was one of Archbishop Athenagoras' achievements, though much credit must be given to the school's first dean, who single-handedly directed its destiny for more than ten years.

The purpose of the theological institute was to give the fundamentals of theological education and much practical training for the pastoral ministry. Saint Basil's Teachers' College was established in 1944 to train teachers for Greek-language schools and for the secretarial needs of the parishes. Saint Basil's was until very recently located in Garrison, New York. It is now merged with Hellenic College, in Brookline, MA.*

Admittedly, for many years both schools provided a limited education beyond high school and made no other pretensions. But both have contributed valuable services to the Orthodox Church in America. The theological school has made much progress in recent years and has received full accreditation as a graduate school of theology. Today more than three hundred Orthodox clergymen and several professional theologians are graduates of Holy Cross. Many of these have gone beyond their B.A. and B.D. training, receiving M.A.'s, Th.M.'s or other master's degrees from some of the country's finest institutions, such as Harvard, Boston University, Yale, Columbia, Fordham, Rutgers, Princeton Theological Seminary, the University of Michigan, and Duke, while several have received Th.D.'s or Ph.D.'s.

From 1931 to 1948, when Archbishop Athenagoras was elevated to the office of Ecumenical Patriarch, some one hundred new congregations were organized and several new church buildings were erected. In his ministry, the archbishop was assisted by several very dedicated bishops, one of the most industrious and totally committed being Germanos Polyzoides. He wrote many educational books in both English and Greek and hundreds of articles, and served as editor of the Church monthly, as a professor at Holy Cross, as chancellor of the archdiocese. Bishop Athenagoras Kavvadas, despite his eccentricities, was a remarkable individual. He directed the seminary and Saint Basil's and left the mark of his personality on both. A man with a mind of his own, he proved totally loyal to Archbishop Athenagoras. Bishop Kallistos, who survived

---

*Editor's Note: St. Basil Academy continues to function; the author's reference is to a merger of a section of that school into Hellenic College.

the changes of 1930, remained the auxiliary bishop for the Western States, which he served until his death in 1940. Less colorful but just as faithful was Bishop Eirenaios Tsourounakis, who served the Midwestern States between 1943 and 1954.

Among the most important institutions organized by Archbishop Athenagoras was the Ladies' Philoptochos Society, which was set up in 1931, with central offices in the Archdiocese and chapters in every congregation. Unfortunately, statistics are not available, but the Philoptochos Society has served as the right arm of the Archdiocese and of each congregation in their philanthropic and social welfare responsibilities. During the Depression and Second World War years, every Philoptochos chapter contributed valuable services. The amount of work, the style, and the kind of social welfare involvement differed according to the experience and nature of each chapter. Some tended to be parochial, others cosmopolitan, in their outlook and their involvement.

The local congregation was a microcosm of both the strengths and the weaknesses of the nationwide church. Much depended on the vision, the experience, and the faith of the parish priest and the parish board of trustees. The numerous beautiful churches that came into being during the Athenagoras era were the result of local initiative—including financial initiative. Some communities were far more advanced spiritually, intellectually, socially, and financially than others.

The establishment of a theological school to train American-born students for the priesthood did not imply that the Church would soon adopt English as a liturgical tongue. The language problem existed then as it exists today. Slowly, however, but steadily, more and more English was used not only in parish life but also in the Archdiocese and its institutions. The Church had been bilingual from as early as the 1920s. Even then there were advocates of complete Americanization (whatever the term may mean) of the Church, as there are now.

During Athenagoras' primacy, emphasis was placed on the teaching, the use, and the perpetuation of Greek not only in liturgical and sacramental life but also in many other functions of the local parish, the archdiocese, and its institutions. Athenagoras' outlook was ecumenical, but he saw no incompatibility between a Greek-speaking Church and an English-speaking environment. It was a deep appreciation of the Greek language and culture rather than chauvinism or unrealistic

28

romanticism that made him an ardent advocate of Greek in the life of the Church.

In order to preserve the language and the Greek heritage, the Church insisted on the founding of a Greek school in every community. In 1931 Archbishop Athenagoras established the Supreme Board of Education, which was expected not only to coordinate Greek education in the various parishes but also to promote the teaching of Greek in public schools. This board has been the least successful department of the Archdiocese. It has engaged in much rhetoric and little initiative or concrete activity. For several years the board was also in charge of catechetical education, but the two branches of education were later divided into two separate departments. Both received a new impetus and were more effectively organized under the primacy of Athenagoras' successor.

In the course of almost two decades, Athenagoras made many friends among the non-Orthodox. He was largely respected by President Harry S. Truman. When Athenagoras was elected Ecumenical Patriarch in 1948 and had to move to Istanbul, President Truman provided his private plane to transport the new Patriarch to his See, in acknowledgement of the importance of the office that Athenagoras was to occupy. Both the Orthodox Church in America and the Ecumenical Patriarchate became better known and more influential because of the stature and contributions of Athenagoras.

C

Before the departure of the new Patriarch for Istanbul, Athenagoras Kavvadas, the Bishop of Boston, was appointed *locum tenens*. While many had expected Bishop Kavvadas would be elected the new archbishop, Patriarch Athenagoras bestowed upon him the titular dignity of Metropolitan of Philadelphia and soon after appointed him to act as liaison between the Ecumenical Patriarchate and the Church of Greece. It was not the best reward for a bishop who had been loyal and industrious for several decades. Bishop Kavvadas was not elected archbishop of the Church in America because he was suspected of independent tendencies. The Patriarch, well known for his arbitrary and political maneuvers, was afraid that Kavvadas might seek the establishment of an autocephalous church in the Americas.

Bishop Athenagoras Kavvadas was succeeded as *locum tenens* by another widely experienced man, Germanos Polyzoides, titular Bishop of Nyssa, who had served the Church in America since the early 1920s. While many had expected that, since Kavvadas had been passed over, Germanos Polyzoides would surely become the new head of the Church in America, it was announced that Metropolitan Timothy of Rhodes had been elected the third archbishop of the Greek Archdiocese. But Timothy died while in Istanbul preparing for his new position. For the second time, Bishop Germanos of Nyssa was appointed *locum tenens* and for the second time he was by-passed. Why did the Patriarch not appoint one of his former close collaborators to the office of archbishop? It seems that injustice was done to both Athenagoras Kavvadas and Germanos Polyzoides.

On October 11, 1949, Michael Konstantinides, Metropolitan of Corinth, was elected archbishop of the Greek Orthodox Church of the Americas, and he arrived in New York on December 15, 1949. He was an excellent choice. Michael was a theologian with a profound mind, deep faith and commitment, and wide experience. He had served the Church in various capacities and had headed the Church in Corinth for more than ten years.

Nevertheless, the immediate reaction to Michael's election was not favorable, especially among several young archimandrites in America. Some were afraid that it would take a long time for Michael to become acquainted with the problems of the Church here. Others accused him of fundamentalism and austere manners.

Archbishop Michael proved to be a great and inspiring leader. While others laid the foundations and yet others erected the building, Michael provided the goods with which the structure was filled. In his ministry he stressed religious renaissance and emphasized the need for a rediscovery of the spiritual, or religious, dimensions of the Church as an organization and as the organism of Christ, on both the national and local levels. He was a genuinely religious man with deep Biblical interests. From an early age he was devoted to learning and writing. He was the author of several books and the translator into modern Greek of such books as *The Imitation of Christ,* by Thomas a Kempis, and *My Life in Christ,* by John of Krostadt. He knew other Christian traditions as well as Orthodoxy. He had spent several years in Russia and 12 years in England.

Archbishop Athenagoras Cavadas, first Dean of Seminary in Pomfret, Connecticut, leaving for new post as Prelate of England. Seen in picture is Cathedral Dean, Rev. Basil Efthimiou, next to Cavadas; the then Fr. Tsoucalas (extreme right), presently Archbishop Iezikiel; the then Fr. Kokkinakis (who later became Prelate of England and successor to Cavadas); seminarians and Cathedral Board members. In back row, 6th from left is former Fr. Coucouzes, present Archbishop of North and South America. Seen in picture are: Fr. B. Gregory, Fr. P. Payiatis and Fr. A. Sarris.

Archbishop Michael watching Greek Independence Day Parade.

31

A forceful and Christocentric preacher (though many complained of his lengthy encyclicals), Michael left behind the memory of an Archbishop who inspired his flock with religious faith, zeal, and commitment. From the very beginning he proved innovative. Within one year he visited 107 congregations in the United States and Canada, and six months later he visited the Greek Orthodox of Latin America, who had never before been visited by an Orthodox bishop. Within months following his visit to South America, a bishop was ordained for the Orthodox there.

Through Archbishop Michael's efforts the 10th Clergy-Laity Congress, held in November 1950 in St. Louis, adopted the *dekadollarion*, the $10 annual obligation of each Greek Orthodox family to the archdiocese. It was a salutary financial step, which contributed to the reorganization of the Archdiocese. The finances of the Church have been on a firm ground ever since. Under Michael's initiative the department of religious education was totally recognized and placed in the hands of professional religious educators. The first serious Sunday School manuals were published in the 1950s.

Another "first" of Michael's decade was the establishment of a formal national organization whose ultimate goal was to bring the young into the fold of the Church. Michael was particularly proud of the Greek Orthodox Youth of America (GOYA), and rightly so. Many of today's leaders on the archdiocesan as well as the local level were raised in the spirit of Michael's GOYA. Michael's sincerity, his idealism, and his devotion to Christian values are indelibly imprinted in the minds and hearts of the new generation of Greek Orthodox. In him there was no guile.

The first home for the aged of the archdiocese was established on the initiative of the clergymen's association of metropolitan New York; Archbishop Michael contributed not only his moral support but also financial assistance. His humanitarian concerns could fill a book, but the drives he inspired to help the victims of the earthquakes that almost destroyed the Ionian Islands in 1954 are especially memorable.

For more than two decades Holy Cross Greek Orthodox Theological School was no more than a priests' preparatory school. Those were difficult years, but Michael intended the school to become a center of theological activity as well. It was under his initiative that a dean with better academic credentials was appointed, several new and professionally trained faculty members were added, and a theological journal was established. The new dean was the Rev. Nicon Patrinacos, who holds a doctorate from Oxford.

Michael was in favor of the Greek language in the liturgy; nevertheless it was during his primacy and through his own encouragement that the English sermon was introduced in Orthodox worship and Sunday School texts and instruction became bilingual.

Archbishop Michael had been involved in inter-Orthodox and interchurch relations from as early as 1927, when he represented the Ecumenical Patriarchate at the Faith and Order Conference held in Lausanne. He had participated in dialogues between Orthodox and Anglicans and among various Orthodox jurisdictions, and he encouraged both inter-Orthodox and interchurch relations. In the 1954 General Assembly of the World Council of Churches held in Evanston, IL, Archbishop Michael was elected one of the six presidents.

Perhaps one of the most controversial subjects during Michael's primacy was the effort of the Archdiocese to impose upon all parishes the Uniform Parish By-laws. Several congregations were disturbed because they viewed the Archdiocese's plans as in violation of their own by-laws and as a further evidence of bureaucratic centralization. But all the Archdiocese was trying to do was to eliminate chaos and introduce more harmony between the Archdiocese and the parish. What Archbishop Athenagoras had tried to do through his personality and politics, Archbishop Michael tried to do through the adoption of a common constitution and common by-laws.

During the decade of the 1950s, not only the Greek but other ethnic Orthodox churches became better known. Several states granted official recognition to Eastern Orthodoxy as a major faith, and for the first time the United States Congress adopted a bill that recognized the Eastern Orthodox in the Armed Forces as separate from Protestants and Catholics. The Orthodox in the Armed Forces were allowed to include "E.O." (Eastern Orthodox) on their tags. The fact that, for the first time in American history, a Greek Orthodox Archbishop was invited to offer prayers at the inauguration of President Dwight D. Eisenhower was an important event for the Orthodox, who had been overlooked for many decades.

But it was not only through the efforts of the Archdiocese that more and more recognition of the Orthodox was achieved. Local initiative was just as important, if not more so. Several new church buildings, such as Hagia Sophia in Los Angeles and the Annunciation Church of Milwaukee (designed by Frank Lloyd Wright) did a great deal to inform the public of Orthodoxy's presence in the United States. All in all, the decade of the 1950s was both innovative and extremely fruitful for the Church, which became the most important institution of the Greek-American community at large.

The growth, spiritual awakening, and recognition of the Church's presence in America that began in the early 1950s increased in the early years of the 1960s. The leader in the new impetus was none other than the new Archbishop, Iakovos. Following the untimely death of Archbishop Michael in 1959, the Ecumenical Patriarchate elected the young and dynamic Bishop of Melita, Iakovos Coucouzes, to become the new head of the American Church. He was enthroned in New York on April 1, 1959. . . .

Every aspect of the Church's life and mission in the last twenty years has revealed movement forward. Archbishop Iakovos has been a dynamic, energetic, and progressive leader. The Archdiocese's most successful efforts have been in the establishment of new congregations and the erection of innovatively designed church buildings, as well as religious education.

The decline of the American city and the emergence of suburbia, the soaring crime rate, the housing shortage, the spread of slums, racial tensions, and the prosperity of the Greek Orthodox contributed to their exodus to the suburbs. Many new and prosperous congregations came into being in the suburbs of cities such as New York, Philadelphia, Baltimore, Boston, Detroit, and Chicago. The annual rate of community growth has been about five percent; there are no accurate statistics of membership growth.

The erection of new and architecturally innovative church buildings, such as the Annunciation Church in Milwaukee, St. Nicholas Church in Flushing, N.Y., and the Ascension Church in Oakland, California, continued. While some congregations preferred to exist in their own cocoon, a few, especially those whose membership was composed overwhelmingly of third and fourth generation Greek Orthodox and many proselytes, became more cosmopolitan and socially involved. Nevertheless the Church as a whole continues to be an immigrant Church, concerned much more with itself and its own problems than with the outside world.

Banquet honoring Fr. Coucouzes in Boston, Mass. Seen in photo are Tom
Pappas, Archbishop Michael and Fr. George Bacopoulos.

Greek delegation of Chicago presenting a Byzantine cross to the late Mayor
Richard J. Daly. L. to r. Tom Brown, Nicholas J. Melas, the late Bishop Timothy
of Chicago and Archon Pierre A. DeMets.

The Archdiocese expanded both materially and organizationally. Several new departments were organized. The department of economic development and the department of religious education have indeed made many forward steps. The Philoptochos Society too has been reorganized along more professional lines.

Archbishop Iakovos has made tremendous efforts, providing leadership in resolving many internal and external issues, including racial conflict and inter-Orthodox and interchurch relations. He has issued a great number of substantive encyclicals and has delivered inspiring speeches. Nevertheless much remains to be achieved in several areas in the life of the Church, such as liturgical renewal, social issues and family concerns, missionary work or involvement in society's problems, and issues facing clergymen either as celibates or as family men. There are scores of divorced clergymen and not a few, either married or celibate, who have left the Church, which does not provide professional counselling.

One healthy phenemenon in the Church's life of recent years is the emergence of criticism and the urge for internal dialogue, a result not only of the Logos movement but also of interest on the part of intellectuals and ordinary laymen. For example, there was a great deal of reaction against the Archdiocese's pronouncements concerning the language problem, the development of the Church's institutions, such as Hellenic College, and the handling of the emergence of the Orthodox Church in America and of ecumenical relations.

But whatever the problems and the criticism, the institutional Church, whether as individual parish or as an archdiocese, moves on, and many of its faithful face the future with confidence. Indeed, when the conditions and problems of the Church of the first half of the 20th century are compared with those of the third quarter, it must be admitted that much has been accomplished.

Greek-Americans In Crisis

THE PERIOD OF UPHEAVAL IN THE GREEK ORTHODOX
CHURCH OF AMERICA (1918-1923) AND THE SYNODICAL
EXARCH, THE BISHOP OF MONEMVASIA AND
LAKEDAIMONIA, GERMANOS TROIANOS.

By Metropolitan Silas of New Jersey

he political irregularities which existed in Greece during World War I divided the country and resulted in the great catastrophe of Asia Minor. That created further repercussions in the political life of Greece and a crisis in the Greek Church in the Diaspora.

The Church leaders vying for favor with the Greek royalty or caught up in the controversies of the day, pursued actions which did not honor them. On December 12, 1916, in the Ares Field[1], Athens, Eleftherios Venizelos, the Prime Minister of Greece, was anathematized by the President of the Holy Synod, Metropolitan Theoklitos. This unusual ecclesiastical action against a political leader, clearly indicated the magnitude of the civil strife that existed among the Greek people.[2]

Unfortunately, the Greek Government did not inhibit that unpatriotic act. A series of events followed. The Allies in May 1917, compelled King Constantine to relinquish the throne to his second son, Prince Alexander. With the exile of Constantine, Venizelos returned to Athens from Thessaloniki and, in the presence of Prince Alexander, took the oath of office as Prime Minister. Theoklitos, the Metropolitan of Athens, did not attend Venizelos' inauguration, sending the palace chaplain as his representative.[3] It was not possible for Theoklitos, who had presided over Venizelos' anathematization only a few months before, to attend his inauguration as Prime Minister.

At the same time, the Minister of Church Affairs and Public Education introduced a bill to the newly-instituted Parliament (which previously had been disbanded by King Constantine) restructuring the charter of the Holy Synod. This bill demanded that a special session of the Spiritual Court defrock Theoklitos and the Metropolitans who participated in the Venizelos' anathematization. Among those who received a six-month suspension, was the Bishop of Monemvasia and Lakedaimonia, Germanos Troianos.

The position of Metropolitan of Athens vacant, the Holy Synod elected Meletios Metaxakis,[4] then Metropolitan of Kitios (Cyprus), Theoklitos' successor. Assuming his new responsibilities, Metaxakis began the reorganization of the Church with a special emphasis on the Church in America, which, in accordance with the Patriarchal Synodical Decision (Tome) of 1908, was placed under the guardianship of the Church of Greece.

During World War I, Metaxakis had undertaken a special mission which brought him to the United States.[5] In his entourage were such distinguished men as Chrysostomos Papadopoulos (subsequently Archbishop of Athens), Alexander Demoglou, the Bishop of Rodostolon, Archimandrite Alexander Papadopoulos, chief secretary of the Holy Synod of the Church of Greece, and Hamilcar Alevizatos, Professor of the Theological School of the University of Athens.[6]

Metaxakis' trip to America was necessitated by national and ecclesiastical reasons. His remarks before the Holy Synod of the Church of Greece during the meeting of January 13, 1920, included the following:

> "During my visit to America, I was informed of the presence of a Russian Bishop on American soil without the permission of the Ecumenical Patriarchate. The Patriarchal Tome of 1908 directed the immediate assignment of a Greek Bishop in America. However, I learned in America that for a decade, diplomatic pressures prevented the implementation of the Patriarchal Tome. Upon my arrival, I waited for the Russian Bishop to come to me; however, he did not. In order to give him the opportunity, I sent Archimandrites Chrysostom and Alexander to him, He, in turn, reciprocated by sending an Archimandrite to visit me. I then realized that he expected me to visit him, thus recognizing him as the canonical Bishop in America, under whose jurisdiction the Greek Church ought to belong. I held a press conference with the Greek and English language newspapers, in which I

38

Patriarch Meletios IV (Metaxakis), the founder of the Greek Orthodox Archdiocese of North and South America, on a pastoral visit to the United States. He is shown here in Chicago surrounded by lay and clerical leaders as follows: Charles George, Archon Paul Demos, Archdeacon Germanos Polyzoides, Rev. Constas H. Demetry, and Andrew Karzas.

quoted Orthodox teaching relative to lands outside the existing Patriarchal boundaries that canon law places them under the jurisdiction of the Ecumenical Patriarchate. Thus the Church in America is under the canonical authority of the Ecumenical Patriarchate, and only by its authority can certain actions be taken. Our presence in America is by virtue of the permission granted by the Ecumenical Patriarchate in the Tome of 1908, rendering us the only canonical jurisdiction. No other such permission has been granted. We are aware only that the Patriarch of Antioch requested the permission of the Patriarchate to send the Bishop of Seleucia to America for the needs of the Syrian Orthodox. Prior to this, Efthymios, who was ordained by the Russians for the Syrians, but never recognized by the Patriarchate of Antioch, was abandoned by the Russians. This event reinforced our position regarding canonicity in America. The position which Efthymios held was subordinate to the Russian Bishop. Throughout our presence in America, the Russian Bishop attempted indirectly to impose this position of hegemony, yet never openly or officially."[7]

Upon his arrival in America, Meletios Metaxakis sent his historic seven-page encyclical to the faithful, from Washington, D.C., "Greetings in the Lord and again unto you, greetings. This apostolic greeting I send to you from the capital of the United States."

Before his departure for America on August 4, 1918, the Holy Synod of Greece had introduced and passed a decision for the establishment of the Holy Archdiocese of America. For the sake of history, I include the text of the charter which sets forth the following:

"As per canonical order and as has been the practice for centuries, all Orthodox communities outside the canonical geographical boundaries of the Holy Churches of God are under the pastoral governing of the Most Holy Apostolic and Patriarchal Ecumenical Throne. Since it has become more practical for administrative reasons to unite the Greek communities of Europe, America, and elsewhere, to the Most Holy Autocephalous Church of Greece, it has proved necessary to transfer the pastoral responsibility for these communities to the Holy Synod of the Church of Greece."

It was by the Patriarchal Tome of March 18, 1908, that the Apostolic and Patriarchal Throne transferred these commu-

nities to the Church of Greece. The specific wording of the
Tome reads:

*"In our efforts to better serve the pastoral and adminis-
trative needs of all Greek Orthodox residing in North and
South America, and to provide for frequent episcopal
visitations to the faithful, as stated in the Patriarchal Tome,
. . . . . . the Holy Synod of the Church of Greece, established a Diocese by Synodical decree to be known as
the 'Archdiocese of America' which shall be considered
as one of the Dioceses of the Autocephalous Church
of Greece. This Diocese shall be governed according to
the existing canonical and legal rules and regulations. The
Archbishop of America shall be elected and replaced in
like manner as are elected and replaced all others in the
diakonia of the Archdiocese and Dioceses of the Autocephalous Church of Greece. All Greek Orthodox of permanent and/or temporary residence in North and South
America are to recognize him as their spiritual and ecclesiastical leader and it is through him that they are to communicate with the Holy Synod of the Church of Greece."*[8]

The Archdiocese of America, therefore, was first established by the Church of Greece on August 4, 1918. Upon his
election as Ecumenical Patriarch in 1922, Meletios Metaxakis
returned the Orthodox Churches of the Diaspora to the jurisdiction of the Ecumenical Patriarchate and by a Synodical
Decree established the Archdiocese. He then proceeded with
the election of its first Archbishop, Alexander Demoglou, Bishop of Rodostolon. This election was in recognition of his services during the difficult years of 1920-1922. The civil recognition of the Archdiocese of America by the State of New
York (its legal incorporation), occurred on September 19, 1921.[9]

The Holy Synod of the Church of Greece not only established the Archdiocese of America, but also under the presidency of Metropolitan Meletios Metaxakis approved and
signed a decree appointing the Metropolitan as an Exarch of
the parishes in America. We submit the text of the decree
for your information.

*"The Holy Synod of the Church of Greece having ap-
proved that the Metropolitan of Athens and President
of the Holy Synod personally visit the Orthodox communities in America to bless them and strengthen them
in faith, hope and love, hereby appoints him Exarch*

to organize communities and to assume the responsibility of temporarily organizing the Greek Orthodox Churches of America for the sake of caring and responding to their needs. As Exarch of the Holy Synod, His Excellency has absolute authority to act in America as its express representative in all that the Holy Canons permit and demand of the Episcopal office, on behalf of the Body of the Church. The Metropolitan of Athens is given full authority by the Holy Synod to act as he deems necessary for the benefit and progress in Christ, not only for the Churches of America, but for all communities in Europe which he may visit according to need. Hence, this synodical decree is granted."[10]

Thus, Meletios Metaxakis, the Metropolitan of Athens, became the first Exarch of the Archdiocese of America. He was assisted in the difficult task of organizing the Church in America by Alexander Demoglou, the auxiliary bishop of Rodostolon. At the time of his arrival in America, Bishop Alexander was technically a bishop under the administrative jurisdiction of the Metropolitanate of Vellas and Konitsa as administrator. This Metropolitanate belonged to the Ecumenical Patriarchate, which had elected Alexander Bishop of Rodostolon in 1907.[11]

Upon the recommendation of Metaxakis—a recommendation made while he was still in America—Bishop Alexander was appointed Synodical Representative to the newly established Archdiocese of America.[12] This position had previously been offered to and declined by Archimandrite Chrysostomos Papadopoulos, a professor at the University of Athens who had been a member of Metaxakis' delegation to America.

The Church in America thus belonged administratively to the Holy Synod of the Church of Greece and the majority of its membership was comprised of faithful from the Royal Kingdom of Greece. This period (1916–1931) was one of the most politically divisive times in all of modern Greek history. It was inevitable that the political disputes between Royalists and Venizelists would spill over into the life of the Greek Orthodox Church in America. The Greek community in America was bitterly divided between these two parties and parishes generally aligned themselves behind one or the other. It became imperative that the priest serving a community belong to the political party of the parish he was to serve. Homilies were not based upon the respective Gospel readings, but rather upon the political affiliation of the clergyman delivering the sermon. In Greece, the administration of the Church belonged to whatever faction was in political power at the time. However in America, due to the separation of Church and State, the Church was divided and each faction projected a bishop and ecclesial structure. The same situation was to be found among the Greek language newspapers: the Royalists were supported by the *Atlantis*, the Venizelists by the *National Herald*.

The Archdiocese of America was established by Meletios Metaxakis, a capable man with vision and foresight. However, he was deeply involved in Greek politics. Because of his political affiliation with Eleftherios Venizelos, Metaxakis was appointed the Archbishop of Athens after the passing of a law which altered the constitution of the Holy Synod of the Church of Greece.[13] Nevertheless, this fact did not minimize the effectiveness of Metaxakis, whose labors contributed to the historical development and maturity of both church and nation. This affiliation of his mission with the political life of the Greek nation resulted in his dependence on the strength and capability of the political parties that he followed. Meletios remained in office as the Archbishop of the Church of Greece as long as Eleftherios Venizelos was in power.

This situation changed when the political adversaries of Metaxakis came to power. New constitutional changes were introduced, this time to depose Metaxakis. Stripped of all his authority, Metaxakis found refuge in America—a country which he greatly loved and admired. Having a large following in America, Metaxakis began working to further organize the Archdiocese which he had previously established. Under his auspices, the St. Athanasios Theological School was founded in Astoria and the first Clergy-Laity Congress was convened at the Holy Trinity Cathedral, then located on East 72nd Street. The Archdiocese was incorporated on September 19, 1921. Its charter was signed by Metaxakis, as president of the corporation, and by his deacon, Germanos Polizoides, the present Metropolitan of Ierapolis, who served as secretary of the new corporation.

Metaxakis struggled to authenticate the injustice of his removal as Metropolitan of Athens and the canonicity of his position. Since the Athenian episcopal throne was now held by Metropolitan Theoklitos, it was no longer possible for Alexander, who had been appointed by the deposed Metaxakis, to remain as Synodical Representative. On February 16, 1921,

Theoklitos appointed the Reverend Nicholas Lazaris,[14] his nephew by marriage, to the position of Synodical Representative. Alexander was recalled by the Holy Synod, but refused to return to Greece.

During the March 4, 1921 meeting of the Holy Synod, the Greek Ministry of Foreign Affairs forwarded a telegram from the Greek Embassy in Washington and the Greek Consulate in New York to the bishops gathered in session. This telegram, Protocol Number SP257, notified the Holy Synod that Alexander of Rodostolon had informed both the Embassy and the Consulate that he had broken communion with the Church of Greece and now recognized the Ecumenical Patriarchate as his canonical superior. Alexander further stated that he had the support of the then Locum Tenens of the Ecumenical Patriarchate, Dorotheos, in taking these actions. A second telegram, from the Reverend Nicholas Lazaris, who was then pastor of the Annunciation Church of New York, was also read. Listed as Protocol Number SP258, it informed the Synod that many clergy of the most influential communities remained loyal to the Church of Greece. Still a third telegram, signed by "the General Assembly of Clergy" and listed as Protocol Number SP259, was presented at this meeting. This telegram informed the Synod that Alexander's proclamation of independence from the Church of Greece was a representative decision acceptable to them. Furthermore, they requested that the Synod not send another bishop to America as this would cause bitter dissension and still further division.

In response to this confusing array of telegrams, the Holy Synod sent two telegrams via the Ministry of Foreign Affairs: one, destined for the Greek Embassy in Washington, requested that the faithful be notified by the press that henceforth, the clergy are to refer to the Synod for decisions concerning any and all ecclesiastical matters and that the Greek people in America remain faithful to their Church and mother country; the other was sent to the Consul in New York urging Bishop Alexander and the clergy following him to repent so as not to feel the wrath of the holy canons for their apostasy.[15]

At this same meeting, the Holy Synod appointed Anthony, the Bishop of Patras, to replace Alexander. However, Anthony declined for reasons of health. The position was then assigned to Bishop Germanos Troianos of Monemvasia and Lakedaimonia, who was appointed Exarch by the Synod, with the understanding that he was not to ordain either deacons or priests. Germanos also refused.[16]

With the refusal of Germanos Troianos, the Holy Synod on March 31, 1921, appointed Meletios, the Bishop of Messenia, to conduct an investigation of Alexander. It was further decided that two bishops—Spyridon of Arta and Synesios of Thebes— would be sent to America for the express purpose of rectifying its ecclesiastical chaos.

This situation changed yet again after the publication of an article by Troianos in the Athenian newspaper, *Politeia*. Writing, in response to an earlier report by Alexander printed in another paper, *Embros*, Troianos indicated the strength of his convictions regarding the current state of ecclesiastical matters in America. This article reflects his strong character, theological ability and willingness to fight for what he felt to be true. Troianos wrote:

"The bishop in question, by using a variety of shallow arguments and rash statements analogous to those which he used to create the chaotic ecclesiastical and political situation among the Greek Orthodox communities in America, now endeavors to justify his totally uncanonical and unpatriotic actions by eluding his responsibilities and refusing to answer to that same Holy Synod which he had previously represented under the false assumption that he is now under the jurisdiction of the Ecumenical Patriarchate. It is unheard of and totally inappropriate for any self-respecting citizen, let alone a bishop, to reject, blaspheme and revolt against the superior whom he once recognized and from whom he once sought instructions. If the voice of his episcopal conscience were not so overshadowed by base thoughts and calculations of personal interest, and if his heart was warmed and moved even in the least by devotion to his Church and love for his country, Bishop Alexander of Rodostolon, even if he had been unjustly wronged, should not have acted treasonably, becoming in this time of national crisis the cause of scandal and division amongst our honorable countrymen abroad.[17]

This article was a manifesto against Alexander. In it Troianos criticized Alexander's actions as impudent, uncanonical and unpatriotic. He considered him to be disrespectful to his canonical ecclesiastical authority and unsuitable for the episcopacy. Troianos continued:

45

"Even if he is without ulterior motives, surely his self-proclaimed theory, by which every subordinate can, of his own volition, choose to recognize or reject his canonical superior, is unheard of and childish. If this were true, any subordinate could, at any moment, create anarchy. Which canons, which laws, which regulations, which prior example, which moral conscience gives the authority to Alexander or anyone else to respect and recognize a certain superior one day and on the very next day to reject and revolt against that same superior, merely for the sake of safeguarding his position."[18]

It is in fact true that after the removal of Metaxakis as Metropolitan of Athens, Alexander had continued to accept and execute the directives of the Holy Synod until the beginning of January, 1921. This was the conclusion of Bishop Meletios of Messenia in a report to the Synod dated May 5, 1921.[19]

Alexander's change in stance was quite possibly due to the presence of Meletios Metaxakis in America after his deposition. In these politically charged times Troianos, writing about Alexander of Rodostolon, was indirectly referring to Metaxakis and the canonicity of his election as Metropolitan of Athens.

The publication of this article and the subsequent acceptance by Troianos[20] of the position of Exarch to America prompted the resignations of Bishops Spyridon of Arta and Synesios of Thebes as representatives of the Synod to the Greek community in America on May 25, 1921. Troianos was introduced to the Greek community in America by an encyclical of the Holy Synod of Greece which was addressed to all of the Greek Orthodox faithful in North and South America:

"Certain individuals—both clergy and laity—who have become instruments of the most evil enemy of our faith, have endeavored to spread evil in the vineyard of the Lord and are responsible for battle and division among the Christian Churches of America, poisoning the conscience of the faithful, who, until that time had been peaceful and united."

To combat this anomalous situation, the Holy Synod had decided the following:

"To send to you as our Synodical Exarch, His Eminence Bishop Germanos of Monemvasia and Lakedaimonia, a man strong in the faith, with a reverent spirit and devotion to the God-led Church of Christ. He has great zeal

46

*for the truths of the Fathers of our Church and is capable
of teaching you the truth. A man of self denial, he will
guide the apostate clergy to the true and illuminating path
of our Lord.*"[21]

The first Synodical Exarch for the Archdiocese of America
was Meletios Metaxakis; the second was Germanos of Monemvasia and Lakedaimonia.[22] It must be noted that Alexander
of Rodostolon was not appointed as Exarch of the Holy Synod,
but merely Synodical Representative. He had the privileges
of an auxiliary bishop with added administrative responsibilities. Any decisions regarding matters of substance had to be
made by the Holy Synod in Greece, whose president was the
Exarch of America. With the removal of Metaxakis as president of the Holy Synod, Alexander automatically ceased to
exercise his responsibilities. If Meletios Metaxakis had proceeded with the election of an Archbishop for America and
if the Holy Synod had elected Alexander of Rodostolon to this
post, matters would have been different. As the bishop of an
Eparchy, he would have administered the Church in America in accordance with the canons of the Orthodox Church and
the regulations regarding religious corporations in the state
of New York, where the Archdiocese had been incorporated.

As noted earlier, the Holy Synod established the Archdiocese of America on August 4, 1918. After the refusal of Chrysostomos Papadopoulos, Metaxakis recommended the appointment of Alexander of Rodostolon as "Synodical Representative" to America. Alexander's appointment became official on October 17, 1918, and he was then directed in his
actions by Metaxakis.

Metaxakis had been sent to America by the Holy Synod because of his strength of will and organizational abilities. On
his first trip to America, Metaxakis had visited the majority
of the Greek communities, met with the leaders of other denominations and had established particularly warm relations
with the Episcopalians. But Germanos Troainos was also endowed with many talents and virtues. A man of strong faith,
devoted to the Orthodox tradition, he was a talented, exceed-
ingly expressive speaker, capable of charity in his opinions,
all of which made him easily respected and loved by those
who came in contact with him.

The "revolution of Alexander of Rodostolon," as it was term-
ed by the Church of Greece, began February 26, 1921. We must
bear in mind that at this time, the Old Calendar was still in use

47

by the Holy Synod of Greece and the Greek State, while in America, the Church used the Gregorian or New Calendar. Consequently, in accordance with the America date, the Synod sent a telegram recalling Alexander on February 19, 1921. However, it was probably delivered by the telegraph office on February 20th. Six days later, on February 26, 1921, Alexander circulated his famous encyclical, Protocol Number 2615, in response. In it he addressed "the clergy and reverent faithful of the Greek communities in the United States," informing all Greek Orthodox communicants of the Church in America that:

> "We, as your canonical ecclesiastical authority, duly appointed by the Synod and established here, do hereby sever all ties and communion with the ecclesiastical authorities in Greece. . . . . and declare that until this situation is rectified in accordance with the canons, we recognize the Ecumenical Patriarchate as our only canonical authority and henceforth shall refer all matters regarding our Church to the Patriarchate.[23]

Upon sending this encyclical, Alexander sought the advice of legal counsel. On March 4, six days after the circulation of Alexander's encyclical, he sent his secretary, Mr. D. Valakos, to the law offices of a certain Mr. Labrieskie. Mr. Valakos gave his attorney a translated copy of Alexander's encyclical and asked whether the Bishop should relinquish the files, records and property of the Archdiocese to the new Synodical Representative. Initially, Labrieske recognized that the new Synodical Representative had the right to demand all property and files. But upon further study of the situation and the encyclical itself, Labrieskie concluded that the American court system was not in a position to determine the canonicity of the current Holy Synod of Greece and would, most likely, reject the petition of the newly appointed Synodical Representative. According to Labrieskie, everything hinged upon the stand of the clergy and laity and he recommended the convening of a local American Synod to decide the situation. In his opinion, the Greek Orthodox Church in America had to slowly discard any foreign interference and, in time, gain its independence.[24] His document, with the aforementioned remarks, is dated March 5, 1921.

In studying this encyclical and the recommendations made by Labrieskie, the following is noted: (1) Alexander lacked administrative ability. He should have sought legal counsel before the decision of the Holy Synod to recall him and before sending his encyclical. (2) Labrieskie's recommendation that he seek support of the clergy and laity is ecclesiastically un-canonical. If for personal reasons a cleric disagrees with his superiors, it is his responsibility to resign rather than revolt and cause a schism. (3) Alexander was unaware of the position of the Patriarchate in this matter and had not consulted with the Patriarchate prior to the circulation of his encyclical. In fact, the Patriarchate sided with the Church of Greece. On August 20, 1921, Nicholas Caesarea, the Locum Tenens of the Ecu-menical Throne, wrote the following to the representative of the Greek government in Constantinople:

"Responding to Protocol Number 5765, the letter of your esteemed Honor written on the 9th of this month concern-ing the Church in America, we have the honor to reply and ask that you receive this letter on behalf of our church. It is the opinion of the Synod that the Church in Amer-ica continues to be governed by the Patriarchal Synodical Tome of 1908 which places the Greek Orthodox Churches under the jurisdiction of the Holy Synod of the Church of Greece. To this day, the Ecumenical Patriarchate has never interfered in the affairs of the Church in America, nor have we given any directive to either cleric or lay-man to act or intervene in the administration and organi-zation of the Greek Orthodox churches and communi-ties there: We remain with warm wishes and deep esteem, the Locum Tenens of the Ecumenical Throne, prayer-fully in Christ, Nicholas of Caesarea."[25]

In this conflict, therefore, Germanos Troianos had an added weapon which made matters even more difficult for his ad-versary. Germanos of Monemvasia arrived in New York with a fighting spirit ready for the ensuing struggle and immediately began consolidating his position by organizing and strength-ening the communities under him, resolving any difficulties or problems that he found. Wishing to give moral substance to his efforts, he began his offense with an encyclical address-ed to the "Reverend Clergy of the Greek Orthodox Churches of America and Canada" in which he stated:

"We wish to inform you and we require of you, that, upon receipt of this encyclical, you make the above known to your faithful immediately. . . Further, you will hence-forth recognize me as your only canonical ecclesiastical authority. . . . I require of you that you remain completely

*willing to execute your responsibilities. . . . I further di-
rect that you studiously refrain, in word or in deed, from
interfering in political problems of a general or local nature. . . Finally, I direct that you remit to me within fifteen
days (1) the canonical documents of your ordination and
(2) the canonical document assigning you to the parish
you are now serving. The Synodical Exarch, Bishop Germanos of Monemvasia and Lakedaimonia."*[26]

This encyclical depicts the man: his willingness to challenge,
his ability to compete and his faith in overcoming the opposition. It also shows that Germanos Troianos was a leader of
strong character with an ability to command. Indeed, the encyclical clearly demonstrates that Troianos knew what he
wanted and how to obtain it. In a situation as anomalous and
tumultous as this, he knew that only dynamic direction and
strong will could bring about the desired results.

In reorganizing the offices and finances of the Exarchate,
Troianos attempted to organize a clergy association to care
for his immediate needs. He sought financial aid from the communities under his jurisdiction to help defray the cost of his
offices, which were centrally located in New York City at 12
West 76th Street. A hard and exacting worker himself, Troianos was equally demanding of his clergy. When a few of them,
perhaps due to the anomalies of the existing situation, returned
to Greece without his permission, he wrote to the Holy Synod
requesting their punishment:

*"In view of the unbecoming behavior of certain clergy
under our exarchal jurisdiction, behavior which deserves
strict ecclesiastical discipline, and since they are further
guilty of abandonment, having left their posts without
my knowledge, opinion or permission. . . . . . I humbly
request that by encyclical decree, the Holy Synod forbid these clergymen from celebrating any sacraments
and that. . . . . the eminent Metropolitans forbid these
clergymen to participate in any sacraments."*[27]

Heated conflict continued between the communities and
many sought to resolve their differences in civil courts. On
August 22, 1921, Troianos wrote to the Holy Synod of the
Church of Greece that:

*"Due to the violent reaction against our canonical author-
ity by some ecclesiastical councils and clergy close to
Meletios Metaxakis, the former Metropolitan of Kitios,*

and Alexander of Rodostolon, civil proceedings before the American courts have developed between ecclesiastical councils and clergy appointed either by the Holy Synod or myself. These trials have arisen probably at the initiative of the aforementioned clergy. Metaxakis threatened such court action in his report written to the Greek government via the Royal Embassy in Washington. Alexander himself has been present at many of these hearings and as a witness had lied under oath. Such court proceedings are now underway in Chicago, Atlanta, Manchester, and Chicopee Falls. Fortunately, one such case in Toronto, Canada, ended yesterday in our favor."[28]

Here, then, is the result of the bitter political rivalry: court proceedings which merely served to minimize religious feelings, augment hatred, intensify division, and lead to an alienation which resulted in ruin and total separation. These court proceedings were so harmful that they sealed the division and magnified ecclesiastical disarray. These court problems demanded the attention of the Synodical Exarch, taking him from his pastoral, administrative and liturgical responsibilities and forcing him to remain closed in his office, concentrating his efforts on gathering information and material for the defense of his case.

However, Troianos did find time to visit large Greek centers, such as Chicago and Washington. During his visit to Chicago, he celebrated the Liturgy at the Holy Trinity Church where, in his own words, he addressed "an overflowing crowd of parishioners." Writing to the Synod of the Church of Greece, he continued:

"The response to my speech was encouraging. Many faithful visited with me and had even prepared a tremendous welcome upon my arrival at the train station. Many confided in me that, though they were Venizelists, they had been convinced by the truths they had heard from me and would remain faithful to our ecclesiastical authority."[29]

At a dinner held in his honor at the LaSalle Hotel, some five hundred people were present. Writing in this same report to the Holy Synod of Greece, Troianos added:

"An additional six hundred persons were unable to enter due to lack of space. I am convinced that my visits to Chicago and Washington will radically change the situation in the Greek centers of America. Unfortunately, these

51

pastoral visits are most difficult for the following critical reasons: (1) due to the enormous work of the Exarchate, my presence in New York is imperative; (2) the distances are too great and the trips are very taxing; and (3) the cost of such trips is enormously expensive. However, I need to emphasize that I shall endeavor to overcome any difficulties and fulfill my mission as best I can."[30]

The above excerpts emphasize the magnitude of his efforts and the ways in which he planned to address the many problems confronting him. However, at the same time, Alexander of Rodostolon was also at work. He turned over command of the conflict to Meletios Metaxakis, a dynamic individual with excellent organizational skills.

In an anonymous report entitled "Memorandum Regarding the Ecclesiastical Situation of the Greeks in America," the presence of Germanos Troianos in America as the Exarch was declared uncanonical for the following six reasons:

(1) *"The Bishop of Rodostolon had already been canonically appointed as the Synodical Representative.*

(2) *In his encyclical announcing his presence in America, Troianos writes that he is here to administer the Church of America with the approval of the Greek government. Consequently, his mission in America is simultaneously ecclesiastical and political. The Greek Church of America, however, established on the basis of American law and structured as a religious corporation, cannot have relationships with domestic and foreign political authorities;*

(3) *Troianos presents himself as a representative of Theoklitos and his Synod, who forcibly took over Synodical authority in Greece, contrary to the holy canons;*

(4) *According to the Patriarchal Tome of 1908, the appointee of the Synod cannot assume pastoral responsibilities in America without first receiving the blessings of the Ecumenical Patriarchate.*

(5) *Germanos, as the Bishop of Sparta, co-celebrated the Liturgy with the defrocked Theoklitos. Thus as the representative of Theoklitos in America, Germanos himself was defrocked.*

(6) *Although he had personally acknowledged Meletios as his canonical Metropolitan, by his actions Germanos now declares that Meletios was never the Metropolitan of Athens."*[31]

This eleven-page memorandum was most likely written by Metaxakis, who was at this time struggling to prove the canonicity of both his election and the appointment of Alexander of Rodostolon as Synodical Representative.

The reasons put forth in this memorandum for regarding Troianos' appointment as uncanonical cannot, upon critical examination, be accepted. First, Germanos Troianos did not arrive here as Synodical Representative, but rather as Synodical Exarch. On August 1918, the Holy Synod of Greece had decreed that Metaxakis had

> *"the exarchal responsibility of organizing the communities in America and the over-all care for and the temporary administration of the Greek Orthodox Churches in the New World. Metaxakis has the privilege to act as Exarch in America with express authority granted by the Holy Synod to direct the faithful by virtue of his episcopal authority in all matters permitted by the holy canons."*[32]

When he became Metropolitan of Athens, Metaxakis held the Archdiocese of America for himself. Since the Church of Greece had established the Archdiocese of America on August 4, 1918, Metaxakis should have, upon his return to Greece as Metropolitan of Athens, proceeded with the election of a bishop to this newly vacated position. Had this been done, it is possible that the ensuing ecclesiastical anarchy would not have occurred. Whoever had been elected Archbishop of America would have been the primate and not merely the representative of the Metropolitan of Athens as was the Bishop of Rodostolon. However Metaxakis, the capable and dynamic president of the Synod, continued to retain full power as Exarch of America and administered the Archdiocese through his Synodical Representative, Alexander.

When Meletios was removed from his throne as Metropolitan of Athens by decree and deprived of all administrative duties, he was also deprived of the privileges given him by the Synod as Exarch of the Archdiocese of America. Germanos of Monemvasia and Lakedaimonia was not assigned to the position held by Alexander of Rodostolon, but was appointed Exarch in the vacant position previously held by Meletios. As an auxiliary bishop, Alexander had assisted Metaxakis in the administration of the Archdiocese of America and had been appointed the Synodical Representative. However, the appointment of an auxiliary bishop to an administrative position has nothing to do with the concept of canonicity.

Second, the criticism that Troianos "comes to shepherd the Church of America by approval of the Greek government" and hence his presence here is not "strictly ecclesiastical but simultaneously political" is not accurate. In the Decree of the Holy Synod of the Church of Greece, voted upon the Synod of August 4, 1918, and chaired by Meletios Metaxakis, it is written that

> *"The Holy Synod. . . . . .formulates and establishes by Synodical decision an episcopal jurisdiction to be known as the Archdiocese of America. This Archdiocese will be considered one of the episcopates of the Autocephalous Church of Greece. In regard to its relationship with the Holy Synod and its faithful, it shall be governed by the existing canonical and legal rules and regulations of the Holy Synod. The Archbishop of America shall be elected and replaced in the same manner as all other archbishops and bishops of the autocephalous Church of Greece."*[33]

The architect of this decree was Metaxakis himself, who, in 1918, agreed to abide by

> *"the legal provisions concerning the election of a bishop in the Church of Greece: (1) the ratification of his election by the government; (2) an oath, before the King by the newly elected bishop; and (3) the publication of the formal approval of the election and installation of the new bishop in the government newspaper."*[34]

What does the content of this oath given by the bishop before the King include? Does he not swear to respect the constitution, laws and traditions of the Greek nation? Is it not true that this act makes him an employee of the State in the deepest sense of the word? Didn't Meletios himself ascend the throne by civil decree after first swearing an oath before the King as the highest authority of the country? It would have been proper not to mention, either in the Act of Incorporation of the Archdiocese or by Germanos Troianos, that the State was responsible for acting upon the election and installation of the Bishop and, especially, a Bishop who is to serve in a State with religious liberty and where separation of Church and State exists.

The author of the anonymous memorandum is mistaken when he claims that the Church in America is administered on the basis of American law and that the Archdiocese, as a religious corporation, is administered by the laws which govern secular matters. The author is ignorant of American law. The act of religious incorporation gives the Church, as a legal entity, all of the privileges extended to it by American law. However, the act of incorporation does not establish the parish, the diocese or the Archdiocese as a Church. The Church was founded by our Savior, Jesus Christ. The parish, diocese and Archdiocese are established by the bishop, who by virtue of apostolic succession, has the authority to do so. Third, it is a fact that Germanos Troianos, the Bishop of Monemvasia and Lakedaimonia, received his authority as Exarch of America from the Holy Synod of the Church of Greece, whose president at that time was Metropolitan Theoklitos. It would have been preferable if the "anathema" had not happened, but the perplexities of history cannot be overcome by human strength or pressure. Fourth, the author of the memorandum was correct on this point. According to the Tome of 1908, Troianos should have visited the Phanar. But would Metaxakis have gone to the Phanar while Exarch? Had Alexander of Rodostolon, who was from nearby Chalcedon, visited the Phanar after his appointment? It was wrong for Troianos not to have gone to the Phanar. He did not respect the existing laws which he so often quoted in order to censure Alexander of Rodostolon. Fifth, although a weak argument, it should also be stated. Although Theoklitos was supposedly permanently defrocked by a synod convened by political authorities during a revolutionary period, the actual history of the case is different. After the restoration of political order, it became clear Theoklitos had not in fact been defrocked, but simply denied the throne of Athens, which was then given to Chrysostom Papadopoulos. Consequenlty, the fifth point lacks both a logical and canonical basis. Sixth, finally we must accept that Germanos Troianos displayed duplicity in his political stance. Since Troianos had recognized Meletios as Archbishop of Athens, he should not have denied the validity of his election, even though Meletios had been relieved of his duties for whatever reasons. It would have been more honorable for Germanos Troianos to have referred to Meletios as "the former Archbishop of Athens" during the year that Meletios Metaxakis spent in America (1921-1922). The memorandum shows the depth of the chasm that had split the Church in America. It depicts the leaders of the two factions as force-

55

ful men for whom reconciliation was impossible.

During the period of service in America, Germanos Troianos had an additional advantage: the Greek diplomatic services, which represented the Greek government, worked with him as Synodal Exarch for the success of his mission. This assistance was not to be taken lightly. The people needed the Greek consulates for all matters concerning Greece. These consulates recognized Troianos as the Exarch of the Holy Synod and indicated their confidence in the Exarchate by ratifying all of its decisions.

During this period, the majority of the clergy, who came from Greece or the ancient Patriarchates, supported Alexander of Rodostolon. Because of this, Troianos sent the following telegram to the Holy Synod via Greek consular channels:

"Contact the Foreign Ministries and Consulates of Greece and Asia Minor, asking them not to issue visas to America for Orthodox clergy without any prior consent. Further, request all Patriarchates take necessary steps against the entry of questionable clergy."[35]

The unending pressure and anxiety, the constant struggle with few results, injured his health, forcing Troianos to ask to be recalled by the Holy Synod. The surviving documents also indicate that he was faced with financial difficulties. We know of these difficulties from the encyclicals of the clergy association and the encyclical of Germanos to the communities.

The Greek government had allotted a stipend for the support and expenses of the Exarchate. Perhaps due to the war that was then being waged in Asia Minor, Greece found it difficult to continue this support. The minister of Foreign Affairs of the Greek government, in a letter dated May 4, 1922, and addressed to the Ministry of Ecclesiastical and Public Education, expressed the need to send a replacement for the Bishop of Monemvasia and Lakedaimonia. In this letter, Protocol Number 12809, he writes:

"Since Troianos was adamant in regard to his recall, it is requested that you (i.e. the Ministry) ask the Holy Synod for the immediate appointment of a replacement. Without a bishop in the Archdiocese of America, our national and ecclesiastical affairs are in grave danger. We bring to your attention that the recall of Troianos is necessary for financial reasons. It is impossible for our Ministry to continue with the daily support of this special mission."[36]

The above letter indicates that the Exarchate received a daily allotment which the Greek government now found difficult to continue. The Ministry of Foreign Affairs became aware of the establishment of an ecclesiastical fund, which was believed capable of sustaining Troianos' efforts. For this reason, the Government sought a bishop who would not demand "this daily allotment." The Greek Embassy in Washington informed the Ministry of Foreign Affairs by telegram "of the decision of the Exarch, Germanos, to return to Greece." Of course, the Greek government viewed the paying of this daily allotment to Germanos with irritation. Concurrently, Alexander of Rodostolon pursued his efforts without any financial help from Greece and was also able to pursue various other works, including the establishment of the Theological School of St. Athanasios.

In any event, news of the decision to recall Germanos Troianos spread quickly throughout the Greek community in America. He had even decided the date of his return—"after Holy Easter." His friends and followers began sending telegrams, pleading that he not leave for fear their cause would be jeopordized. The *Atlantis* newspaper sent the following telegram to the Ministry of Foreign Affairs on June 23, 1922:

> *"The recall of the Exarch, Troianos, has left a negative impression. We consider the continuation of his mission necessary to uphold the prestige of the Government and the Church. The followers of Metaxakis will be encouraged in their machinations. Please notify the Synod. Atlantis newspaper.*[37]

The Holy Synod of the Church of Greece, under pressure from the Ministry of Foreign Affairs, proceeded in the replacement of Troianos. Synesios of Thebes was chosen by the Synod as his replacement. This was made known to Troianos by the Greek Ambassador in Washington.

Several days later, Troianos wrote (it is not known to whom this letter was written—possibly to the Ministry of Foreign Affairs):

> *"A few days ago, I was informed of the Minister's telegram regarding the appointment of the Bishop of Thebes as my successor. This information was not divulged to me by the Holy Synod. I awaited with much agony the appointment of a replacement and the news brought me much relief. I was, however, surprised by the response*

here. All the faithful are vociferously demanding that I not depart. They foresee insurmountable difficulties if I do." [38]

He concludes with the following words:

"In view of this response by the people, I have promised that, if directed, I would submit and remain for the sake of the work to be done. These thoughts I bring to your attention. Germanos of Sparta."[39]

Aware that a new Exarch was to replace him, Troianos now requested to remain in Amerca. The following cryptic telegram was then sent by the Ambassador in Washington, on July 31, 1922:

"In continuation of telegram 88, I am honored to bring to your attention that the Exarch, Germanos, has agreed to remain here to continue his work. Please inform me of your final decision so this issue can be resolved (signed) Bouros."[40]

It appears that the work of the Synodical Exarch did not bear fruit. The Government continued to pay the daily allotment. This worried the Minister of Ecclesiastical Affairs, Mr. Siotis, who realized that the pressure of Germanos Troianos as Exarch in America was no longer politically expedient.

The Minister did not change his mind in regards to the work, mission, and worthiness of the cause. Most likely this was due to the fall and retreat of the Greek Army in Asia Minor. Another reason was the pressure capably exerted by the Venizelists in America against Germanos Troianos. These factors prompted Siotis to write to the Holy Synod on October 11, 1922.

Mr. Siotis was correct in recommending the dissolution of the Exarchate so that the Hellenes in America could be united. Writing to the Holy Synod, the Minister suggested the following:

"I am enclosing the telegram of the Greeks in America concerning the recall of the Metropolitan of Monemvasia and Lakedaimonia. We ask that you proceed with his recall since the presence of the Exarch in America impedes relations between the two Churches. We believe that this will greatly help in uniting the Greeks in America, which will also be beneficial to our Greek nation."[41]

How situations change by the events of history! The following day, the Holy Synod decided the following:

"Taking into account the Minister's document, numbered 39436 and dated October 11, 1922, regarding the dissolution of the American Exarchate, the Holy Synod unanimously decides to recall the Synodical Exarch in America and directs that henceforth the clergy there will refer all matters to the Holy Synod. (signed) Metropolitan of Athens, Theoklitos, President, Hydra and Spetse Prokopios, Kalavrita Timothy, Arta Spyridon, Paronaxia Ierotheos, Secretary Archimandrite Germanos Roumbanis, October 12, 1922."[42]

After a fourth telegram from the Holy Synod, the Bishop of Monemvasia and Lakedaimonia sent an encyclical to the communities and priests that the Exarchate has been dissolved and he was no longer serving as Exarch. He then requested a three-month leave of absence from the Holy Synod for health reasons.

The Holy Synod endeavored to convince Troianos[43] to return to his eparchy without success.[44] In the sequence of events that ensued, the Minister of Ecclesiastical Affairs and Public Education forwarded a previously submitted petition of Metropolitan Germanos to the Holy Synod. The Minister further recommended that Germanos' request not only be denied again, but that he be reminded that the Holy Synod would proceed according to the Holy Canons and the Laws of the State regarding "his blatant refusal to fulfill his mission."[45]

On March 10, 1923, Chrysostomos Papadopoulos was elected Archbishop of Athens and all Greece. The new Archbishop sent many telegrams trying to return Germanos to his eparchy. Troianos, however, informed Ambassador Tsamadon that he has ceased serving as Exarch and had asked for a three month leave of absence, a period needed for the restoration of his health. His request was denied by the Holy Synod, which demanded his return to Sparta on four occasions. He in turn, continued petitioning for the leave of absence. It is true that, in his petition on January 25, 1923, submitted by his brother Gerasimos Troianos, editor of the newspaper *Ethnos* and his power-of-attorney in Greece, he again asked the Holy Synod to grant him three months leave of absence. He cited the necessity to travel to Europe to care for a chronic eye disease, which, having deteriorated over the years, required immediate attention and treatment.[46]

The Holy Synod refused his petition again and assigned

the Metropolitan of Corinth Damaskinos to proceed with an inquiry. This decision was reached by the Holy Synod of the Church of Greece, presided over by Chrysostomos Papadopoulos, and probably prompted by the fact that Troianos had been criticized for appearing in churches and speaking against Venizelos, Metaxakis, and Alexander. During a Sunday Liturgy, he had also spoken of his political beliefs when visiting the community of SS. Constantine and Helen in Milwaukee, Wisconsin.

Archimandrite Chrysanthos Kaplanis, a clergyman of the Exarchate of the Church of America, wrote a letter to the Holy Synod on January 4, 1923, which included the following:

"The Synodical Exarch, believing that his recall was a result of pressure upon the Holy Synod by the Revolutionary Committee, does not recognize the decision of the Holy Synod. Having declared the Greek Church in America autocephalous and independent, he has assumed its leadership and no longer recognizes the Reverent Holy Synod."[47]

Although Archimandrite Kaplanis may have over-stated, it does seem that because of the existence of the revolutionary government of Gonatas, Germanos avoided a return to Greece. He had aligned himself with King Constantine and others of the Royalist faction and feared repercussions. In America as Father Kaplanis wrote, in spite of the specific directive not to proceed with ordinations of clergy without the permission of the Synod, the Synodical Exarch "unnecessarily ordained numerous clergy, who were totally uneducated and of low social standards."[48] It should be stated here, in all fairness, that Troianos did indeed have an eye disease which resulted in eventual blindness.

The Synodical Exarch, the Bishop of Monemvasia and Lakedaimonia, Germanos Troianos, was a man of many talents. A prolific writer and graceful speaker, with an excellent theological background, he was able to captivate the masses. He led them demagocically with his ideas, which he tried to impose, regardless if they were incorrect or not in the best interests of the Church or the nation. His strong character often led him to extremes. Troianos' description of Alexander of Rodostolon as "shallow, uncanonical, unpatriotic, and a revolutionary Hierarch with a limited conscience,"[49] exceeds mere pettiness and reveals his innermost reflections and darkest moods. Alexander and Metaxakis should have respected the Holy Synod, the highest level of ecclesiastical authority, although they did not, they did proceed and act rationally and in a more Christian manner than did Germanos Troianos.

The course of history is directed by God, not humanity. God witnesses the anomalies and difficulties in history and overcomes them by giving humanity the proper solution.

The national catastrophe brought about the uprooting of the Greeks in Asia Minor, the land of Ionia, where Christianity flourished for ages, since the sermons of Paul and of the beloved disciple, John the Theologian. This Hellenistic community was used by Divine Providence as the center for the development and promulgation of Christianity and the Ecumenical Synods.

What a grand coincidence that Metaxakis, the founder of the Archdiocese of America, the visionary man of deeds, was to lead Orthodoxy as Ecumenical Patriarch. He is the same person, who with the Tome of 1922, returned the Church of America to the Mother Church, the Ecumenical Patriarchate, where it naturally belongs. This Patriarchate, in the process of history has led her faithful in the path of salvation, free from ethnic pressures, looking not to national interests but rather embracing and projecting the ecumenicity of Orthodoxy.

In contrast, the Church of America, which sixty years ago received her first Archbishop, Alexander Demoglou, was led by political concerns to potentially destructive division and decay. But, He, who perceives the needs of time and history, the all-knowing God, sent to the divided and shattered Church of America, a man, who as a tool of Divine Providence, as another Moses, led this fragmented Archdiocese to unity, progress, and advancement. Indeed, for eighteen years, Athenagoras Spyrou deeply planted the roots of the Orthodox Church in America. Therefore today, due to our divinely arranged administrative and spiritual bond with the Ecumenical Patriarchate, we have that healthy religious and socially involved organization known as the Greek Orthodox Archdiocese of North and South America.

FOOTNOTES

[1]Spyros Markezenis, *The Political History of Modern Greece*, volume 4, p. 182.

[2]Markezenis, p. 183.

[3]Markezenis, p. 220.

[4]*The Official Journal of the Greek Government*, copy 137, 1917.

[5]*Ecclesiastical Truth*, volume 28, 1908, p. 14.

[6]Metropolitan Vasilios Atessis of Lemnos, *A History of the Bishops of the Church of Greece: 1833 to the Present*, p. 214.

[7]Archimandrite Theocletos Strangas, *History of the Church of Greece from True Sources: 1817-1967*, volume 2, p. 847.

[8]Strangas, pp. 845-846.

[9]Rev. Demetrios Constantelos, *Understanding the Greek Orthodox Church: Its Faith, History and Practice* (New York, 1982), p. 138.

[10]Strangas, p. 846.

[11]Atessis, p. 214.

[12]Strangas, p. 847.

[13]*The Official Journal of the Greek Government.*

[14]Strangas, p. 984.

[15]Strangas, p. 985.

[16]Strangas, p. 985.

[17]Strangas, p. 985.

[18]Strangas, p. 985.

[19]Strangas, p. 990.

[20]Strangas, p. 987.

[21]Strangas, p. 987.

[22]Archimandrite Meletios Galanopoulos, *Biographical Sketch of His Eminence Metropolitan Germanos Troianos of Sparta* (Athens, 1933).

[23]*Archives of the Greek Orthodox Archdiocese*, the Metaxakis file.

[24]Metaxakis file.

[25]Strangas, p. 960.

[26]Strangas, pp. 989-990.

[27]*Archives of the Church of Greece* (Holy Synod), file 10, protocol number 1814. October 27, 1921.

[28]Protocol number 357, October 22, 1921.

[29]Protocol number 1390, August 12, 1921.

[30]Protocol number 1390.

[31]Metaxakis file.

[32]Strangas, p. 846.

[33]File 10, August 7, 1921.

[34]Strangas, p. 817.

[35]File 10, 1922.

[36]Protocol number 12809.

[37]Greek newspaper "Atlantis" (New York, June 23, 1922).

[38]File 10, 1922.

[39]File 10, 1922.

[40]File 10, July 27, 1922.

[41]File 10.

[42]File 10, July 31, 1922.

[43]File 10, October 11, 1922, protocol number 29426.

[44]File 10, October 12, 1922.

[45]File 11, January 21, 1923.

[46]File 11, 1923.

[47]File 11, January 11, 1923.

[48]File 11, January 25, 1923.

[49]Strangas, p. 985.

Ἵνα αὐτῇ ἀξίᾳ δοξ... τῷ ... εὐΐ οικουσ..
Κωνσταντινουπόλεως Ν. Ῥώμης
καὶ Οἰκουμενικὸς Πατριάρχης,.

Ἀρ. Πρ. 2388

Ἐξ ἑνὸς καὶ τοῦ αὐτοῦ Ἁγίου Πνεύματος καταυγασθέν-
τες οἱ τῆς εὐσεβείας διάφοροι πατέρες καὶ διορρήμονες δι-
δάσκαλοι, ὥστε μὲν τῇ ἑαυτῶν καὶ τῇ διδασκαλίᾳ ὥσπερ
ἀστέρες ἀρχέφωτοι τὸ νοητὸν ἐκἀπεργυναν τῆς εὐσεβείας
στερέωμα, θεσμοῖς δικαιοτάτοις ἀνωσόφως καὶ τὴν συνέχου-
σαν τάξιν ἐν τοῖς ἐκκλησιαστικοῖς ἐσελυπώσαντο, πάντα πρὸς
τὸ ἄριστον καὶ συμφέρον ἐπὶ τῷ τελευταίῳ θεμελίῳ τῶν Ἀποστόλων, ὄντος ἀκρογωνιαίου αὐτοῦ Ἰησοῦ Χριστοῦ τοῦ Κυρίου
ἡμῶν διαλαξάμενοι. Διὰ τοῦτο δὲ καὶ ὅσα περὶ τὴν τάξιν
καὶ τὴν διοίκησιν ἡμῶν ἐν τοῖς ἐκκλησιαστικοῖς οἱ μακάρι-
οι πατέρες ἐθεσπίσαντο καὶ ἐνομοθέτησαν, καὶ ταῦτα οὐχ
ἧσσον ἐνστερνιζόμεθα καὶ εὐλαβούμεθα καὶ εἰς τὸν αἰῶνα
ἀκράδαντα καὶ ἀσάλευτα παραμένειν βουλόμεθα πανταχοῦ
τοῖς πατρικοῖς ὅροις ἑπόμενοι καὶ τούτοις ὡς οἴακι, καὶ
γνώμονι ἀσφαλεῖ χρώμενοι καὶ καθοδηγούμενοι. Ἐν οἷς
ἐστι καὶ τὰ εἰς τὴν πνευματικὴν προστασίαν τῶν κατὰ τό-
πους πατριαρχικῶν ἀφορῶντα. Κἂν τούτοις γὰρ οὐκ εἰμὴ καὶ

ὡς ἔτυχεν, ἀλλὰ κατὰ τὴν ὑπὸ τῶν πατέρων ὡρισμέ-
νην τάξιν καὶ ἁρμονίαν προβαίνομεν καὶ πρὸς αὐτὴν
τὰ εἰσερχόμενα συμβιβάζομεν, τὸν κρίκον τῆς ἁρμονίας καὶ τῆς εὐτάξεως εὐκαίρως μὲν καὶ εὐλόγως, εἴ που
δέοι, ἐπεκτείνοντες ἢ πρὸς τὰς ἀνάγκας ῥυθμίζοντες,
ἀλλ᾽ οὐ μέντοι ῥηγνύντες ἢ συγχέοντες. Οὐ γὰρ προσδι-
κας ἀντιβαινούσας νεωτεριστικῶς τὰ παραδιδόμε-
να καὶ σεμνὰ θέσμια καταχέομεν, ἀλλὰ διατάξεσι
συμφώνοις ἐν ἁρμονίας τὰς ἐπιγιγνομένας ἀνάγκας
πρὸς τὴν κανονικὴν τάξιν συμβιβάζομεν καὶ διαρα
αἴνομεν τὸν δεσμὸν τῆς ἁρμονίας καὶ τῆς εὐτάξεως
ἐπὶ μᾶλλον κραταίνοντες καὶ προβιβάζοντες.
Ὃ δὴ καὶ νῦν ποιοῦμεν διὰ τοῦ παρόντος ἡμετέρου
Πατριαρχικοῦ καὶ Συνοδικοῦ Τόμου περὶ τῶν ἔξω
τῶν καθωρισμένων ὁρίων τῶν ἐπὶ μέρους αὐτοκεφάλων ἐκκλησιαστικῶν περιφερειῶν διεσπαρμένων
ἐν Εὐρώπῃ τε καὶ Ἀμερικῇ καὶ ταῖς λοιπαῖς χώραις
ὀρθοδόξων Ἑλληνικῶν ἐκκλησιῶν, ὧν ἄχρι τοῦδε ἀσα-
θὴς καὶ ἀκαθόριστος ἐτύγχανεν ἡ τάξις μιᾶς κανονικῆς ἐπωνυμικῆς ἀρχῆς. Εἰ καὶ καλῶς γὰρ κατὰ τὰ
ἄλλα αἱ εἰρημέναι ἐκκλησίαι τὰ ἑαυτῶν ὡδὶ ἀπ᾽
ἀρχῆς ῥυθμίσαι καὶ διατάξασθαι ἐφιλοτιμήθησαν
τὴν εὐσέβειάν τε καὶ τὴν κατὰ τὴν λατρείαν εὐταξίαν
καὶ τὴν ἐπωνυμικὴν δὲ εὐτάξιαν σχηματικῶς ἐν τῇ ξένῃ
καὶ ἀπαραμειωτέως περιέσωσαν καὶ διεφύλαξαν,
ὅμως διὰ τὸν τρόπον τῆς ἑαυτῶν γενέσεως καὶ
συστάσεως, ὡς οὐκ ἀπ᾽ ἀρχῆς τινος δηγνοῦν ἐκκλησιαστικῆς ἢ ὠφελικῆς οὐδὲ κατὰ σύστημά τι

αὐτῶν. ε´.) Πᾶσαι αἱ ἐν τῇ διασπορᾷ ἐνορί-
αι εἰς ἔνδειξιν τῆς πρὸς τὴν καθ᾽ ἡμᾶς Μεγάλην
τοῦ Χριστοῦ Ἐκκλησίαν συναφείας καὶ ἐνότη-
τος καὶ τῆς υἱικῆς αὐτῶν εὐλαβοῦς ὀρέξεως
καταβάλλουν κατ᾽ ἔτος ὑπὲρ τῶν ἀναγκῶν
αὐτῆς ποσόν τι, ὁριζόμενον ἐφ᾽ ἑκάστης κατὰ τὴν ἰδίαν αὐτῆς προαίρεσιν.

Ταῦτα οὕτω συνοδικῶς ἐν Ἁγίῳ Πνεύματι ἀποφη-
νάμενοι καὶ ὁρίσαντες περὶ τῶν εἰρημένων ἐν
τῇ διασπορᾷ ὀρθοδόξων Ἑλληνικῶν ἐνοριῶν
εἰς μόνιμον ἀσφάλειαν αὐτῶν καὶ ἀπαρέγκλη-
τον τήρησιν κυρούμεθα αὐτὰ διὰ τοῦ παρόντος ἡμετέρου Πατριαρχικοῦ καὶ Συνοδικοῦ Τόμου, οὗ ἂν τὸ ἔγγραφον καθ᾽ οἷ ἐώθη καὶ ἐν τῷ
ἱερῷ κώδικι τῆς καθ᾽ ἡμᾶς Μεγάλης τοῦ Χριστοῦ
Ἐκκλησίας. Ὁ δὲ Θεὸς πάσης χάριτος ὁ καλέσας
ἡμᾶς εἰς τὴν αἰώνιον αὐτοῦ δόξαν ἐν Χριστῷ
Ἰησοῦ, ἀδιάπτωτον μὲν καὶ ἀδιάσπαστον φυλάξαι
πάντοτε ἐν τῇ ἁγίᾳ αὐτοῦ Ἐκκλησίᾳ τὸν δεσμὸν
τῆς ἀγάπης καὶ τῆς ἑνότητος, αὐτὰς δὲ καὶ τὰς εἰρημένας ἐν τῇ διασπορᾷ ὀρθοδόξους ἐκκλησίας κατευθύναι, στηρίξαι, σθενῶσαι, θεμελιῶσαι
ἐν τῇ πίστει καὶ τοῖς ἁγίοις αὐτοῦ παραγγέλ-
μασιν. Αὐτῷ ἡ δόξα καὶ τὸ κράτος εἰς τοὺς αἰ-
ῶνας τῶν αἰώνων. Ἀμήν.
Ἐν ἔτει σωτηρίῳ αϡλη´ Μαρτίου ἐπινεμήσεως ϛ´
† Ὁ Κωνσταντινουπόλεως ...

† ὁ Κυζίκου Ἀθανάσιος

† ὁ Νικομηδείας Φιλόθεος

† ὁ Ταϊγανρόγ Ἄνθιμος

† ὁ Ἰκονίου Ἀθανάσιος

† ὁ Χίου Κωνσταντῖνος

† ὁ Μαρωνείας Νικόλαος

† ὁ Δραμινίτζης Γρηγόριος

† ὁ Γρεβενῶν Ἀγαθάγγελος

† ὁ Δρυϊνουπόλεως Λουκᾶς

† ὁ Λέρου ὁ καλούμενος Γερμανός

† ὁ Σερβίων ἢ Κοζάνης Κωνστάντιος

Tomos of March 18, 1908, issued by the Ecumenical Patriarchate, which officially transferred the Greek American Church to the jurisdiction of the Holy Synod of Greece.

(Shown here are the beginning 2 pages and the last 2 pages)

History of the Charters

The Structure of the Archdiocese According to the Charters of 1922, 1927, 1931 and 1977

By
Lewis J. Patsavos

he four charters of 1922, 1927, 1931 and 1977 by which the Archdiocese has been administered during its 60 year span are the focus of this chapter. They reflect the growth and development of the Greek Orthodox Archdiocese in America from its inception to its current adulthood.

The first two charters display the idealism of a Church struggling to establish itself in a new land by upholding its traditional synodal form of administration. The reality of turmoil and dissension, however, necessitates a new form of administration to meet the challenge of preserving unity. The Charter of 1931 displays the pragmatism of a Church committed to preserving its identity, even though for a time adopting a monarchial administrative structure foreign to its essence. Nevertheless, true to the Orthodox canonical tradition of adaptability, the leadership of the Church recognized the need to return to its original synodal administrative structure. It recognized the fact that the reason for which the previous structural change was initiated no longer justified its continued existence. It also recognized the truth that the Church as the Body of Christ can flourish only when all its members are allowed to exercise their various functions.[1] Where there is participation and ownership in an administrative model, there is also a vibrant organization. The Charter of 1977 displays the optimism of a Church daring to meet the challenge of the future, because it does so with faith in Christ.

The methodology employed in our reflection is to allow the documents to speak themselves. Beginning with the Charter of 1922, selected articles or pertinent sections from each are presented in paraphrase. Where appropriate, commentary is given. Owing to the unavailability of an official translation of the first three charters in English, it was determined preferable to proceed in this manner. Following the first charter, only those articles or pertinent sections from each which differ from the immediately preceding charter are presented. The Charter of 1977, having circulated from the beginning in both Greek and English, alone affords ample opportunity for general comment within the restraints of this brief analysis.

A. THE CHARTER OF 1922

Purpose

Following the incorporation of the Greek Archdiocese of North and South America in 1921, its first charter was granted the following year. The 27 articles of the charter begin by defining the purpose of the Archdiocese and continue by articulating the way in which this purpose is to be achieved. The purpose, as stated in Article 2, is "to nurture the religious and moral life of American citizens of the Orthodox faith, who are either themselves Greek or of Greek ancestry." In the execution of the means towards achieving this purpose, the practice of the Great Church of Christ in Constantinople is to be normative.

Administration

Article 3 of the charter established the relationship of the Archdiocese to the Ecumenical Patriarchate. It is to be a supervisory relationship based upon the canonical and historic right of the latter. The following article delineates the geographical boundaries within which the administration of the Archdiocese is to take place. They comprise four diocesan districts: New York (the chief See of the Archdiocese and therefore headed by the Archbishop), Boston, Chicago, and San Francisco. The latter three Dioceses are each headed by a Bishop whose title bears the name of the city in which he resides. Mexico, Central and South America are placed under the administrative jurisdiction of the Archdiocese in New York. The communities in Canada are divided among the remaining three Dioceses to which they are most proximate.

Concerning the Archbishop

Once enthroned, the Archbishop is permanently installed in his See and therefore cannot be transferred. In the event the Archiespicopal See becomes vacant, one of the three remaining Bishops may be elected Archbishop. Until the election, however, which must take place within three months, the bishop having seniority of ordination acts as *"locum tenens"*.

Upon the mandate of the Synod of the Archdiocese, an extraordinary meeting of the Ecclesiastical Assembly is convened by the "locum tenens", who is its presiding officer. The purpose of the meeting is to select three candidates from the list approved by the Holy Synod of the Ecumenical Patriarchate. It is the prerogative of the Synod of the Archdiocese to elect the Archbishop. Once informed by the Synod of the election, the Patriarchate must then grant its approval. This electoral procedure was to be followed for the succession of the then Archbishop, who had himself already been duly elected by the Holy Synod of the Ecumenical Patriarchate. The consecration of the Archbishop is reserved to the Ecumenical Patriarch.

Concerning the Bishops

The Bishops, like the Archbishop, may not be transferred. In the event an episcopal See becomes vacant, one of the remaining Bishops acts as "locum tenens", with geographical proximity to the vacant see being the chief consideration. The Archbishop acts as "locum tenens" for the Sees of Boston and Chicago, the Bishop of Chicago for the See of San Francisco.

The procedure established for the initial election of Bishops to the three newly created Dioceses is as follows: Subsequent to ratification of the present charter, an extraordinary meeting of the Ecclesiastical Assembly of the Chicago Diocese first is convened by the Archbishop, who is its presiding officer. The purpose of the meeting is to select three candidates for this Diocese from the list of eligible clergy of the Archdiocese. Qualifications of candidates for the episcopal office include: a diploma from a duly recognized Orthodox School of Theology; an irreproachable life; ecclesiastical experience; and the approval of the Holy Synod of the Ecumenical Patriarchate. It is the prerogative of the Synod to elect the Bishop.

The above electoral procedure was to be repeated for each newly created diocese. Once Bishops were thus elected for all three episcopal Sees, future elections were to take place in the manner prescribed for the Archbishop. Consequently, the future election of Bishops would be the prerogative of the Synod of the Archdiocese. In the case of the Bishop, only the permission by the Ecumenical Patriarch is necessary for his elevation.

Allowance is also made for the situation arising in the event a second episcopal see becomes vacant before it is possible to fill an earlier vacant see. In such a case, it is evident that the two remaining Bishops do not suffice to constitute a synod. Therefore, the election for the first vacant see takes place by the Holy Synod of the Ecumenical Patriarchate and the two Bishops of the Church in America. Candidates are the same selected earlier at the extraordinary meeting of the Ecclesiastical Assembly. The election for the second see is then held by the now fully constituted Synod of the Archdiocese according to the normal procedure.

Each Bishop is entitled to a one month leave from his responsibility annually. Request for a longer leave, of not more than four months duration, must be approved by the Synod of the Archdiocese. Each Bishop exercises in his episcopal see the full extent of authority foreseen for the episcopal office by the holy canons and the agelong practice of the Church. Rights and obligations of Bishops include: ordination of priests, consecration of churches, assignment of priests and deacons, supervision of parish administration, issuance of marriage permits and writs of divorce, and distribution of the holy myrrh received by the Archbishop from the Ecumenical Patriarch. During the Divine Liturgy and other sacred services, the Bishops are to be commemorated by the priests and deacons under their jurisdiction; the Bishops, in turn, commemorate the Archbishop; and the Archbishop commemorates the Ecumenical Patriarch.

Administrative Bodies of the Archdiocese
I. Synod of the Archdiocese

The Archbishop and three Bishops comprise the Synod of the Greek Archdiocese of North and South America. As decreed in the holy canons, it must convene twice a year, before Easter and in the fall, wherever determined by the Archbishop. The Synod of the Archdiocese has all the authority and responsibility inherent in the "provincial synod", as defined by the holy canons. It is accountable to the Holy Synod of the Ecumenical Patriarchate for the inviolate preservation of the doctrines and canons of the Eastern Orthodox Church. In the event of a tie vote in the Synod, the tie is broken by the vote of the Archbishop, who is its presiding officer. In the absence of the Archbishop, the Synod is presided over by the Bishop having seniority of ordination.

70

✣ Μελέτιος Ἐλέῳ Θεοῦ Ἀρχιεπίσκοπος Κωνσταντινουπόλεως Νέας Ῥώμης καὶ Οἰκουμενικὸς Πατριάρχης ⁓

Ἀριθ. Πρωτ.
252.

Μετά

Τῆς περὶ ἡμᾶς Ἁγίας καὶ Ἱερᾶς Συνόδου

σκεφθέντες ἐπικυροῦμεν τὰ ἀκόλουθα :

Καταστατικόν
Τῆς Ἑλληνικῆς Ἀρχιεπισκοπῆς Ἀμερικῆς
Βορείου καὶ Νοτίου.

Ἄρθρον Α΄.— Ἱδρύεται θρησκευτικὸν Σωματεῖον ὑπὸ τὴν ἐπωνυμίαν : «Ἑλληνικὴ Ἀρχιεπισκοπὴ Ἀμερικῆς Βορείου καὶ
Νοτίου» χάριν τῶν αὐτόθι Χριστιανῶν, τῶν ἀνηκόντων εἰς τὴν
Ἁγίαν Ὀρθόδοξον Ἀνατολικὴν Ἐκκλησίαν καὶ ἐχόντων ὡς γλῶσσαν λειτουργικὴν ἀποκλειστικῶς ἢ προτιμωμένην τὴν Ἑλληνικὴν
ἐν ᾗ ἐγράφησαν τὰ Ἅγια Εὐαγγέλια καὶ τὰ λοιπὰ βιβλία τῆς
Καινῆς Διαθήκης.

Σκοπός.
Ἄρθρον Β΄.— Σκοπὸς τῆς Ἐκκλησίας ταύτης εἶναι νὰ οἰκοδομῇ τὸ

θρησευτικόν καί ήθικόν βίον τῶν Ἑλλήνων καί τῶν ἐξ Ἑλληνικῆς κατα-
γωγῆς Ὀρθοδόξων Ἀμερικανῶν σολιτῶν ἐπί τῇ βάσει τῶν Ἁγίων Γραφῶν, τῶν
ὅρων καί τῶν κανόνων τῶν Ἁγίων Ἀποστόλων καί τῶν Ἑπτά Οἰκουμενικῶν Συνόδων τῆς ἀρχαίας ἀδιαιρέτου Ἐκκλησίας, ὡς οὗτοι ἑρμηνεύονται ἐν τῇ πράξει τῆς ἐν Κωνσταντινουπόλει Μεγάλης τοῦ Χριστοῦ Ἐκκλησίας.

Διοικητική Ὑπαγωγή.

Ἄρθρον Γ΄.— Ἡ Ἑλληνική Ἀρχιεπισκοπή Ἀμερικῆς Βορείου καί Νοτίου
διατελεῖ κανονικῷ καί ἱστορικῷ δικαιώματι ὑπό τήν Ἀνωτάτην Πνευματικήν καί Ἐκκλησιαστικήν Ἐξουσίαν τοῦ Οἰκουμενικοῦ Πατριαρχείου Κωνσταντινουπόλεως.

Διοικητική Διαίρεσις.

Ἄρθρον Δ΄.— Ἡ ὅλη Ἀρχιεπισκοπή διαιρεῖται εἰς τέσσαρας Ἐπισκοπικάς
περιφερείας.

1ον Τήν τῆς Νέας Ὑόρκης. Αὕτη περιλαμβάνει τάς Πολιτείας:

Νέας Ὑόρκης μετάς ἐν τῇ πόλει τῆς Νέας Ὑόρκης καί Βρούκυν κοινότητας καί τάς κοινότητας Schenectady, Syracuse, Rochester, Buffalo,
Endicott.

Κοννεκτικούτης, Stamford, New Haven, Ansonia, Waterbury, New
Britain, Danielson, Norwich, Thompsonville.

Νέας Ἰερσέης, Newark, Orange, New Brunswick, Trenton, Paterson.

Πεννσυλβανίας, Philadelphia, Reading, Bethlehem, Altoona, Wilkes-
rre, Pittsburg, East Pittsburg, Vandergrift, Erie, New Castle, Monessen,
Voodlawn, Ambridge.

Ἄρθρον ΚΕ΄. Πᾶσα διάταξις οἱουδήτινος Ἐκκλησιαστικοῦ Ὀργανισμοῦ ἀντιβαίνουσα εἴτε πρὸς τοὺς Νόμους τῶν Ἡνωμένων Πολιτειῶν εἴτε πρὸς τοὺς Ἱεροὺς Κανόνας τῆς Ὀρθοδόξου Ἐκκλησίας, εἶναι αὐτοδικαίως ἄκυρος.

Ἄρθρον ΚΣΤ΄. Μέχρι τῆς ἐκλογῆς τῶν Ἐπισκόπων καὶ τοῦ καταρτισμοῦ τῆς Ἱ. Συνόδου, διὰ τὴν Γενικὴν Συνέλευσιν, τὸ Πνευματικὸν Δικαστήριον, τὸ Διευθῦνον Συμβούλιον καὶ τὴν Ἱερατικὴν Σχολὴν ἔχουσιν ἰσχὺν αἱ διατάξεις τοῦ προηγουμένου προσωρινοῦ Καταστατικοῦ.

Ἄρθρον ΚΖ΄. Τὸ παρὸν Καταστατικὸν ἐψηφίσθη ἐν Γενικῇ Συνελεύσει, ἐκυροποιήθη κατὰ τὸ Δ΄ ἄρθρον τοῦ προσωρινοῦ Καταστατικοῦ τῆς Ἑλληνικῆς Ἀρχιεπισκοπῆς Ἀμερικῆς Βορείου καὶ Νοτίου, ὑπόκειται δὲ εἰς ἀναθεώρησιν ἂν κριθῇ τοῦτο εὔλογον μετὰ διετῆ ἐφαρμογὴν ἀπὸ τῆς νομίμου κυρώσεως αὐτοῦ. — Νέα Ὑόρκη, 11 Αὐγούστου 1922.

Ἐν Μυτιλήνῃ ἀναθεωρηθὲν τῇ ....

ὁ Πατριάρχης Κωνσταντινουπόλεως ....

ὁ Δαμασκηνὸς ....

† ὁ Κωνσταντινουπόλεως ....

First charter of the Greek Orthodox Archdiocese of North and South America, drafted in New York, and approved by the Ecumenical Patriarchate, August 11, 1922. (Above documents are excerpts from the original which is 16 pages long.)

II. Local Ecclesiastical Assembly

Each Diocese has its own Local Ecclesiastical Assembly, which is comprised of all the clergy of the Diocese and of one lay representative from each incorporated parish. Lay representatives are elected by their respective parish councils. Each Local Ecclesiastical Assembly is convened by the Diocesan Bishop, who is its presiding officer, or his representative. It meets regularly each year in May, and in extraordinary session whenever the Bishop deems necessary. A quorum consists of the Bishop (or his representative) and twelve members, of whom at least six must be clergymen.

III. General Ecclesiastical Assembly

The General Ecclesiastical Assembly of the entire Archdiocese is comprised of the Archbishop, the Bishops and 24 selected members, 12 from the clergy and 12 from the laity. Up to six clergy and lay members are elected by each of the four Ecclesiastical Assemblies. The General Ecclesiastical Assemby meets regularly every two years in September, upon the invitation of the Archbishop, and in extraordinary session following the decision of the Synod of the Archdiocese. The General Ecclesiastical Assembly is presided over by the Archbishop or, in his absence, by the Bishop having seniority of ordination. A quorum consists of the Archbishop or Bishop presiding in his place, at least one other Bishop and 12 members, of whom six are clergymen.

Clergy and lay members of both the Local Assemblies and the General Assembly may be represented by corresponding members of the clergy and the laity who have been duly authorized. Besides their own vote, representatives of absent members also cast the votes of those whom they represent. The exception to this procedure is the casting of votes for the selection of Bishops, for which only those present are eligible to vote. Once representation has been duly authorized and recognized by the Assembly, it may not be revoked and is valid for as long as the Assembly is in session. In the event the competent Assembly does not establish rules of composition or procedure, this is to be done by the local Bishop or his representative.

Local Ecclesiastical Assemblies are empowered to supervise the management of all ecclesiastical affairs and to enact legislation together with the local Bishop for the effective administration of all ecclesiastical institutions. This legislation must be in harmony with the holy canons and the laws of the

United States and requires the approval of the Synod of the Archdiocese to be valid. By-laws of individual parish charters, if contrary to the holy canons or laws of the United States, are invalid. The General Ecclesiastical Assembly reaches decisions and approves measures which foster common action throughout the Archdiocese towards achieving its stated religious, moral and social goals.

IV. The Administrative Council

Within the Archdiocese and each of the three Dioceses there is an Administrative Council. The Administrative Council of the Archdiocese is comprised of the Archbishop, who is its presiding officer, four clergymen assigned to the Archdiocese, and four distinguished laymen. The former are appointed by the Archbishop; the latter are proposed by the Archbishop and approved by the Local Ecclesiastical Assembly. The same holds true, with one exception, for the Administrative Council of each of the Dioceses. In the event of a shortage of priests in the vicinity of the Diocese, three clergy and three lay members suffice.

The appointment of both the clergy and lay members of the Administrative Council extends for a term of two years. The vice president of the Council is the highest ranking clergyman, who at the same time retains his voting rights. Both the treasurer and the secretary are elected by the Administrative Council from its membership.

The jurisdiction of the Administrative Council includes all matters for which the wider body is competent. An exception are those matters, which according to the canons are the exclusive prerogative of the Bishop, or when he is acting in concert with the spiritual courts. The Council also has the right to enact legislation within the sphere of its jurisdiction which does not conflict with the present charter.

Spiritual Courts

Each Diocese has its own spiritual court comprised of at least two priests and the local Bishop as president, or his representative. This court hears all canonical offences of the clergy in the first instance, with exception of those for which the penalty is deposition. Cases found by preliminary hearings to bring with them the possibility of deposition are forwarded to the Synod of the Archdiocese.

Decisions of the spiritual courts of the Dioceses calling for a suspension of more than two months may be appealed before the Synod of the Archdiocese within 31 days from their date of issue. Decisions of the Synod of the Archdiocese calling for deposition of suspension of more than one year may be appealed before the Holy Synod of the Ecumenical Patriarchate within 91 days.

Theological Seminary

The Greek Archdiocese of North and South America maintains a seminary for the education of its clergy. This seminary is headed by a Board of Trustees, of which the Archbishop is president. It is comprised of four clergymen and three laymen, who are assigned by the Archbishop or by the Synod, once it is formed, from among the most distinguished members of the clergy and the laity of the Archdiocese in New York. The administrative structure of the seminary and its program of studies are organized by the Board of Trustees and approved by the Archbishop. Once formed, the Synod of the Archdiocese will grant this approval. The seminary has its own fund, the resources of which are used exclusively for the purpose of educating both clergy and teachers.

Concluding Articles

Ordained clergy of the Greek Archdiocese of North and South America are assigned to their ministerial posts without interference by the civil authorities. The basis for all administrative procedure are the holy canons, as well as the regulations adopted according to them and the laws of each state in which the Church's jurisdiction extends. Any regulation of an ecclesiastical organization, which is contrary either to the laws of the United States or to the holy canons of the Orthodox Church, is invalid.

Until the election of Bishops and the formation of the Synod, decrees regulating the General Assembly, the Spiritual Court, the Administrative Council, and the Theological Seminary are those of the previous temporary charter.[2] The present charter was voted in General Assembly, convened according to Article 4 of the temporary charter of the Greek Archdiocese of North and South America. It is subject to review, if judged reasonable, within two years from the time of its formal validation.

B. THE CHARTER OF 1927

Introductory Remarks

The Charter of 1922 closed with a statement expressing intention of revision within two years of its validation. In fact, the revision did not take place until five years later. The revised charter was voted upon by the Third General Assembly of the Archdiocese of North and South America, which met at the Cathedral of St. Basil in Chicago, October 12-14, 1927.[3]

There is little difference in language between the revised charter and its predecessor of 1922. Most of the original wording is retained intact as is the order of topics addressed in each article. The revised charter is a refinement of its predecessor, but also includes in several articles provisions not foreseen previously. These provisions assure a more effective operation of the administrative structure of the Archdiocese. It is not the intention of this charter to introduce a radically new administrative structure. Rather, it tries to apply lessons learned from the experience in anticipation of future developments in the life of the Church in America. In the following discussion of the Charter of 1927, only those articles which vary from their earlier counterparts will be commented upon.

Purpose

As set forth in the Charter of 1922, the purpose of the Archdiocese is "to nurture the religious and moral life of American citizens of the Orthodox faith, who are either themselves Greek or of Greek ancestry."[4] This charter, no doubt reflecting the need felt by its drafters to perpetuate their identity in the western hemisphere, goes further. Not only is the purpose of the Church in America "to preserve and to propagate the Orthodox Christian faith," but also "to teach the original language of the Gospel."[5] Indeed, this expanded purpose was an indication of the growing awareness that the Greek Orthodox Church in America was here to stay and, therefore, had to plan for the future.

Administration

Article 4 introduces the same four diocesan districts as in the previous charter with one notable exception. It refrains from naming the cities and towns in which parishes already exist but mentions only the states comprising each diocesan district. Furthermore it refers to "communities to be established in the future," obviously in anticipation of parishes to be organized in the years ahead. Where the same state includes some parishes belonging to one diocesan district and some parishes belonging to another, the geographical boundaries of the diocesan districts are delineated with exactness and precision.[6]

Concerning the Bishops

Article 6 makes reference to episcopal leave of absence, which it limits to three months. In the event a Bishop must leave the country, canonical permission must be requested from the Ecumenical Patriarch and Holy Synod of the Patriarchate. In the case of the Archbishop, canonical permission is requested directly; whereas in the case of the Bishops, it is requested through the Archbishop.

Synod of the Archdiocese

In that article, allowance is made for the convocation of the Synod of the Archdiocese in extraordinary session at a time and place to be determined by the Archbishop. Similar action can also be taken by a majority of the membership of the Synod whenever deemed necessary.[7]

Spiritual Court

The article dealing with this ecclesiastical institution operating in each Diocese stresses that only the Bishop has a decisive vote. All other members have a consultative vote. Furthermore, when the spiritual court is presided over by a priest representing the Bishop, then the entire membership has a decisive vote.[8]

Article 12 under "Spiritual Court" recalls that the Ecclesiastical Authorities are responsible for upholding the integrity of marriage with reference to both its ecclesiastical and spiritual aspects. The omission of this article from the Charter of 1922 and subsequent results apparently necessitated its inclusion in the new charter. One can only surmise that secular legislation governing issues of marriage and divorce may have contributed to the need for this addition.

Local Ecclesiastical Assemblies

Article 13 defines with more precision the lay composition of the Local Ecclesiastical Assembly. Whereas earlier it was not stated who might be elected to the Ecclesiastical Assembly, this apparent need is now appropriately addressed. A wide variety of possible candidates is suggested. These include members of the parish council or of the community itself, as well as Orthodox Christians of another city or of the city in which the Ecclesiastical Assembly takes place. What is of utmost importance is that candidates be in good standing with the Greek Orthodox Church.

What constitutes good standing with the Church is not further articulated. Consequently, it must be assumed that this refers to a person's membership in the Church in both the broad and narrow sense. In the broad sense, one is a member of the Church through baptism and subsequent communion in the faith; in the narrow sense, by meeting obligations determined by each local parish. Stressing the obvious, i.e., good standing with the Church as a prerequisite for election to the Ecclesiastical Assembly, leads one to believe that this was the direct result of the turbulence which had already begun to surface in many of our communities at this time.[9] Another variation concerns the frequency of the meetings of the Local Ecclesiastical Assembly. Unlike the earlier regulation calling for annual meetings in May, the new regulation calls for biennial meetings in September besides the extraordinary sessions foreseen by both charters.

General Ecclesiastical Assembly

As with the Local Ecclesiastical Assembly, so too with the General Ecclesiastical Assembly. Suggestions are now made as to who might be elected to this latter body. Possible candidates suggested include either persons from among the membership of the four Local Ecclesiastical Assemblies or elsewhere. Again the need for candidates to be in good standing with the Church is stressed. Furthermore, rather than every two years, meetings are to be convened every three years in September. Representation of clergy and lay members of both the Local Assemblies and the General Assembly (due to distance) is allowed. For the third time it is stressed that a layperson serving in this capacity must be in good standing with the Church. When there is voting, the representative of absent members also disposes of their votes, but never casting more than three votes including the representative's own.

Competencies of Ecclesiastical Assemblies

Article 15 remains the same in both charters. However, Article 16 of the Charter of 1927 makes mention also of educational and philanthropic institutions of the Diocese. This is most appropriate in view of the expanded purpose of the Church in America stated in Article 2: "to preserve and to propagate the Orthodox Christian faith" and "to teach the original language of the Gospel." Legislation enacted for the effective administration of all institutions must be in harmony not only with the holy canons and laws of the United States, but also with the decisions of the General Ecclesiastical Assemblies of the entire Archdiocese.

The Local Ecclesiastical Assembly has the right to elect representatives and two alternates for each representative to the General Ecclesiastical Assembly. It also has the right to elect lay members and their alternates to the Mixed Council of the Diocese. Section 1 of Article 16 concludes with the following requirement: In order for the charter of any parish to be valid, it must have the prior approval of the local Bishop. This addition is especially significant when seen in the light of events alluded to earlier. It was necessary to consolidate parishes under the authority of the Bishop. This was undoubtedly a measure taken to strengthen the Bishop's authority which might otherwise be challenged by those seeking parochial autonomy.

Article 17 is new. It describes the level of authority exercised by each Ecclesiastical Assembly. Decisions of the General Ecclesiastical Assemblies superseded those of the local Ecclesiastical Assemblies, which in turn superseded those of the General Assemblies of the communities. Differences between the General Ecclesiastical Assembly and the Synod of the Archdiocese in matters of jurisdiction are settled by the Ecumenical Patriarchate.

Election of Archbishop and Bishops

It must be remembered that the creation of Dioceses and subsequent election of Bishops for the Archdiocese of North and South America were initiated by the Charter of 1922. The procedure established for the future election of both the Archbishop and Bishops is upheld in Article 18 of the new charter. One additional qualification of candidates for either of these offices is understandably absent from the earlier charter. It is the need for a fruitful ministry of at least five years for Bishops, seven years for the Archbishop. Particularly with regard to the office of Archbishop, it is stressed for the first time that candidates might be selected from among the other Metropolitans of the Ecumenical Throne currently in office.

Theological Seminary

Article 23 of the Charter of 1922 (introducing the establishment of a Theological Seminary) is replaced in the current charter by Article 24. In keeping with the stated purpose in Article 2, the present article foresees the establishment of churches, schools, philanthropic institutions, and missions, as well as the circulation of publications and other legitimate means to foster its goal. It concludes by highlighting the Theological Seminary which functions to meet the needs of the Archdiocese.

Concluding Articles

The next-to-the-last article reminds both clergy and laity of what otherwise would appear self-evident. The need to return to this matter, however, in isolation strongly suggests the existence of abusive behavior. The language is strong and straightforward, leaving no doubt as to its intention. "Neither cleric or layperson may hold office or even be a member of the Greek Orthodox Church of America if that person does not belong to the (universal) Greek Orthodox Church of Christ, and none may remain in office or even be a member of the Church in America if that person ceases to be in good standing."[10]

In conclusion, Article 28 certifies the approval and ratification of the charter by the Holy Synod of the Ecumenical Patriarchate. Furthermore, it officially announces its enactment, which, however, is to be eventually replaced by a Collection of Regulations of the Greek Orthodox Church in America to be compiled by a special committee. According to the decision of the General Ecclesiastical Assembly, this committee will be elected by the Synod of the Archdiocese. The Collection of Regulations will be enacted without being voted upon by another General Ecclesiastical Assembly once approved and ratified by the Ecumenical Patriarchate.

C. THE CHARTER OF 1931

Introductory Remarks

The Charter of 1927 was shortlived, having been replaced in 1931 by another charter, under which the Archdiocese was administered up to 1977. Much has been written and said about this document: that it is "monarchical" compared to the democratic charters of 1922 and 1917;[11] that "it was imposed by the Ecumenical Patriarchate and the Greek Government";[12] that it "(was) installed in specific and bald denial of American constitutional notions. . . contrary to American Orthodox Christians' democratic expectations".[13]

One cannot fully appreciate the Charter of 1931 without a knowledge of events surrounding the life of the Church at that time. An investigation into the Church's state of affairs during the early years of its existence on the American continent makes sad reading.[14] Waves of immigrants from Greece seeking a new life in America brought with them the political rivalries of their homeland. As a result, communities were divided and the legitimacy of the existing ecclesiastical authority was contested. In addition, clergy of questionable credentials, who had been ordained by bishops representing opposing factions, contributed to the already chaotic situation permeating the Church in America.

This was the situation encountered by the then Archbishop Athenagoras, who on February 24, 1931 arrived in this country as the new spiritual leader of the Greek Archdiocese. His perseverance and vision, as well as his administrative ability, contributed significantly to the eventual stabilization of the Church here. The seeds of dissension had been scattered long before his arrival. What now appeared necessary was the consolidation of authority into one source in order to preserve unity. The cause of unity must indeed have been the main concern of all those burdened with the responsibility of drafting the new charter. It would be difficult, otherwise, to explain the drastic departure of its model of administration compared with that of the preceding charters.

Administration

In the first place, no mention is made of Dioceses. The only administrative unit mentioned is the Archdiocese. The Archdiocese is headed by the Archbishop, for whom an Auxiliary Bishop is foreseen to assist in administrative tasks. Both the Archbishop and Auxiliary Bishop proposed by him are to be elected by the Holy Synod of the Ecumenical Patriarchate.

Little is said about the qualifications of either the Archbishop or Auxiliary Bishop. The main requirements are that they possess a diploma from a recognized Orthodox School of Theology, that they had served at least five years in the previous degrees of the priesthood, and that they be at least 30 years of age. Interestingly, the last two requirements are absent from the earlier charter.

Article 8 introduces the establishment of an archdiocesan office, about which nothing more is said other than that its purpose is to be outlined by a special regulation. Another special regulation is foreseen for the creation of Ecclesiastical Assemblies to assist in the realization of the purposes for which the Archdiocese exists. To be included in this regulation among other details for the proper functioning of these Ecclesias-

83

tical Assemblies are their composition, time and place of convocation, and competency.

There exists also another body to reinforce the Archdiocese in the realization of its goals, and especially in the management of ecclesiastical property. It is the Mixed Council,
which is to function according to the directives of another special regulation. The Mixed Council is charged with the establishment of a General Ecclesiastical Fund and a pension fund
for the clergy. Pending circulation of the proposed special
regulation, the Mixed Council of the then current Archdiocesan Administration continues to function according to Articles 21 and 22 of the Charter of 1927. In the event a member of this temporary Council is absent, the Archbishop appoints a replacement.

Parish Council

In order to achieve on the local level the goals outlined for
the Mixed Council, the Parish Council is introduced for each
community. Once again, a special regulation is anticipated
to define its competency, responsibilities and general operation. Given the tumultuous situation which gave rise to a new
charter, the concept of a local community affairs council must
be acknowledged as an ingenious way to promote stability.

In the event that there are more than one church and community within the same parish, a special regulation can be
made to create parochial districts. The relations of these districts to the Archdiocese are to be determined by the same
regulation which created them.

Ecclesiastical Agencies

In accordance with the stated purpose of the Archdiocese,[15]
the establishment and/or promotion of several institutions is
officially sanctioned through their inclusion in the charter.
These include Missions to meet the needs of Christians who
have not been organized as religious corporations,[16] a Board
of Higher Education under the chairmanship of the Archbishop to oversee the promotion and organization of the schools
of the Archdiocese,[17] and a Department of Religious Education for the religious upbringing of the youth of the Church.[18]
Mention is also made of spiritual courts[19] and of the competency of the Ecclesiastical Authorities in dealing with the ecclesiastical and spiritual aspects of marriage and divorce.[20]

Concluding Articles

For all the above, as well as for all the preceding administrative institutions and procedures, special regulations are foreseen. Henceforth, these regulations will play an indispensable part together with the charter in the orderly administration of the Archdiocese. They are to be drafted by committees appointed and chaired by the Archbishop and will be binding following their ratification by the Ecumenical Patriarchate. They will define the composition of the various Ecclesiastical Councils, their responsibilities and competency, and the mode of operation of the organizations to which they relate. The right to append any necessary completion or detail to the present charter is recognized to the Mixed Council. Any provision of the special regulations or decision of the Mixed Council at variance with the charter is invalid.

The need to affirm the right of the canonical and lawful Ecclesiastical Authority to assign the clergy of the Archdiocese to their ministerial posts[21] is a clear indication of the uncanonical activity of unauthorized hierarchs. There is no doubt however, as to which Ecclesiastical Authority is considered legitimate in view of the statement which follows. The practice of the Church as interpreted by the Ecumenical Patriarchate is considered normative. There follows the same prohibitive statement met several times in the Charter of 1927 concerning those persons, both clergy and lay, who cease to be in good standing with the Church.[22] Provisions of any ecclesiastical organization at variance either with the laws of the state or the holy canons are by right invalid.

Finally, the concluding article confirms the charter's composition according to the provisions of Article 28 of the previous charter. It confirms its ratification and validation by the Holy Synod of the Ecumenical Patriarchate and thereby proclaims its enactment. It then allows for possible amendments in non-essential provisions to be initiated by a special committee appointed by the Archbishop. Such amendments however, must first be ratified by the Patriarchate to be valid.[23]

D. THE CHARTER OF 1977

Introductory Remarks

The Charter of 1977 by which the Archdiocese is presently administered is the result of efforts begun several years earlier to decentralize the Church's cumbersome administrative system. Over the years since the Charter of 1931, the Church in America had expanded in a way the early immigrants to this country perhaps never dreamed possible. Together with this expansion grew also the ever weightier responsibilities of the chief hierarch, the Archbishop of North and South America, the geographical extent of whose episcopal authority is without precedent in the annals of church history.

The practical, although less than canonical, solution found to the problem of administering such a vast ecclesiastical province was the proliferation of Auxiliary Bishops. Under the Charter of 1931, which foresaw only one Auxiliary Bishop,[24] the Auxiliary Bishop functioned more or less as the administrative executor of the Archbishop. And so it was with the ten Auxiliary Bishops strategically situated throughout the United States, Canada and South America. Up to 1977, they executed the administrative decisions of the Archbishop within their archdiocesan districts.

It must be acknowledged that the Charter of 1977 is a noble attempt to adjust to the ever growing needs of the Church of the late 20th century. These needs demanded a participatory form of administration congruent with the conciliar nature of the Church. And it is to the credit of our church leadership that the transition to a synodal form of administration has finally been initiated. There have been critics of the present charter charging that it does not go far enough. Yet one thing is certain: It is living proof of the coming of age of the Church in America. This recognition however is given without putting at risk the Church's hard earned unity and cohesion for the sake of experimentation. It will be up to future generations to determine whether the vision of those who drafted the Charter of 1977 was justified.

A comparison of the present charter with the Charter of 1931 reveals an ever-increasing awareness of the Church's mission in the western hemisphere. Whereas the first two charters speak of outreach to Orthodox who are Greek born or of Greek parentage alone, the Charters of 1931 and 1977 leave open the possibility of inclusion of a much broader membership with the approval of the Ecumenical Patriarchate. The former charter speaks of Orthodox communities in America of a different nationality, the latter of "other Orthodox groups, parishes and dioceses that have voluntarily submitted to (the) jurisdiction (of the Archdiocese of North and South America)".[25] There is the additional statement in the latter document which speaks of service to "all of the Orthodox living in the western hemisphere".[26] Although this statement is to be understood figuratively, it nevertheless presents us with the moral imperative to pursue fervently pan-Orthodox unity in America.

Also characteristic of the present charter is the pastoral tone of its stated purpose, as well as its reference for the first time to ecumenical activities. Gone are the days in which the chief concern of the Church was the zealous preservation of the faithful alone, and the nurture only of Greek-speaking Orthodox. The Church of the 70s and 80s has a much broader scope of its role in the pluralistic society of which it is a part. No longer will it suffice for the Church to remain an entity unto itself, removed from contacts with other Christian and even non-Christian bodies. The Charter of 1977 commits the Church in America to dialogue and involvement in the ecumenical movement, however not upon its own initiative, but upon the directives always of the Ecumenical Patriarchate.

Administration

Assisting in the administration of the expanded role of the Archdiocese as delineated above are the ecclesiastical assemblies and councils, in which both clergy and laity participate, encountered in all three previous charters. The most significant innovative feature of the present charter, is the Synod of Bishops, which was restored after 46 years. The Synod of Bishops functions as a modified provincial synod. The most important modification of the provincial synod's traditional prerogatives concerns the election of Bishops.

Election of Archbishop and Bishops

With regard to the election of the Archbishop, this is clearly the exclusive prerogative of the Holy Synod of the Ecumenical Patriarchate. In this process, the Synod of Bishops together with the Archdiocesan Council has only an advisory voice. In the election of other Bishops, which is also the prerogative of the Ecumenical Patriarchate, the participation of the Synod of Bishops is expanded to include its nomination of three candidates, from whom one is elected Bishop by the Holy Synod of the Patriarchate.

The election of Bishops, and especially of the Archbishop, has long been the means by which the Patriarchate has extended its supervision over the Church in America. The issue is a sensitive one and ought not to be seen in isolation from the whole question of Orthodox unity in the western hemisphere. Consequently, it ought not to be considered utopian that this issue will one day be re-examined within the wider context of pan-Orthodox unity. Therefore, it would be premature to assess the procedure for electing Bishops and the Archbishop solely on its own merits. Perhaps the added weight and prestige afforded by the involvement of the Ecumenical Patriarchate in the process will prove ultimately beneficial. In any event, it does not appear likely that the Patriarchate is about to abdicate its role in the election of Bishops. On the other hand, it must not be forgotten that this role was secured to it in a much more direct way by the Charter of 1931. That administrative structure, it must be remembered, foresaw no involvement, other than that of the Archbishop, for the election of Bishops.[28] Seen in this light, the current charter shows progression, as well as promise for the future.

Decentralization

The main characteristic of the Charter of 1977, as indicated, is its contribution to the decentralization of the administrative structure of the Archdiocese. This was accomplished primarily by restoring the collective authority of the Synod of Bishops and the individual authority of each Bishop in his own Diocese. Once again however, the extent of authority to be exercised is modified by a provision of the charter. This trend to decentralize the concentration of authority within the See of the Archdiocese is set in motion by ceding to each diocesan Bishop certain rights and responsibilities previously accorded only to the Archbishop. Article 8 of the charter lists these rights and responsibilities.

The overriding consideration of the above arrangement is to promote initiative at the diocesan level while at the same time preserving the bond of unity and cohesion which has sustained the life of the Archdiocese. This is pursued by reserving to the Archbishop the right to supervise and coordinate the activities of the Bishops in accordance with the provisions of Article 7. The key to the success of this undertaking must be sought in the balance of authority exercised by the Archbishop and the Bishops. There must always be a "modus operandi" that facilitates the Bishop to initiate needed change and that takes into account the responsibility of the Primate for the general well-being of the Church at large. In this transitional period of decentralization, the present charter seeks such a "modus operandi" in the manner just prescribed. As suggested by Article 24 however, revision of the present order may be initiated upon request of the Clergy-Laity Congress at any time, as the need arises.

Liturgical expression to the present order is given through the order of commemoration. The order of commemoration introduced by Article 9 is noteworthy. It prescribes a dual commemoration of both the Archbishop and diocesan Bishops by the clergy. Despite the reaction such a practice was bound to cause, it does in fact reflect the two levels of authority currently operative within the Archdiocese.

Role of Laity

As in the previous three charters, the laity is well represented in the administrative bodies of the Archdiocese. These include the Archdiocesan Clergy-Laity Congress and Archdiocesan Council on the archdiocesan level, and the Diocesan Clergy-Laity Assembly and Diocesan Council, which are their local counterparts, on the diocesan level.

The extent of lay participation in administrative matters is impressive. Article 4 extends to the lay membership of the Archdiocesan Council participation in the designation of both the number and boundaries of diocesan sees. This is a right which, according to the canonical tradition of our Church, belongs essentially to the hierarchy. Nevertheless, history provides us with many examples of collaboration between Church and State in the designation of diocesan sees.[28] The Charter of 1977 thus affirms the need for greater participation of the laity in the organizational life of the Church as it relates to the world at large. It in this way gives recognition to the fact that the laity shares together with the clergy in the kingly (administrative) office of Christ. One must applaud the leadership of our Church for preserving the consciousness of this prerogative inviolate up to the present day.

The other area in which increased participation in administrative matters is accorded the laity by the Charter of 1977 is the election of Bishops. Article 13 grants the Archdiocesan Council together with the Synod of Bishops an advisory voice in the election of the Archbishop. Article 14 grants the right of consultation with the Synod of Bishops for the nomination of three candidates, of whom one will be elected Bishop. Involvement of the laity in the election of its spiritual leaders is an indisputable fact. It began with the election of Matthias to replace the traitor Judas[29] and thereby set the pace for what followed during the next three centuries. Abuses, however, and the ever-increasing influence of secular rulers in the affairs of the Church resulted in the abolition of the laity's God-given right to elect its spiritual leaders.[30] Once again sufficient historical precedents exist as a reminder of what was once an inherent right of the laity.

It must always be remembered that the governing of the Church is exercised primarily by the Bishops in Christ's name, not in their own name or in the name of the laity. Assisting the Bishops in their task are the lower ranks of the clergy and the laity, always within the limits set by the holy canons. The Bishops should not act without the laity, neither should the laity act without the Bishops and other clergy. As evidenced both in scripture and in the writings of the Fathers, the laity participated in the governing of the Church from the beginning, under the leadership of the Apostles and their successors.

There are nevertheless, areas in the organizational life of our Church, in which the laity participates today, which need probing to determine whether in fact they correspond to the spirit of the Church's original practice. One does not doubt the sincere intentions of the hierarchy to recognize to the laity their due by involving them in the decision-making process of church administration. It is of paramount importance, however, that there be a clear understanding of roles and pre-rogatives and limits. Furthermore, and herein lies the key to harmonious relationships within the Church, only pious, ac-tively practicing laypersons should be accorded the privilege of assisting in Church administration. Nominal Christians should at all costs be excluded. Vigilance of course, is neces-sary. This is particularly necessary because the lay element everywhere is becoming more and more secularized, at least from the point of view of Christian knowledge and educa-tion. Consequently, it is in constant danger of going astray from the canonical path, while at the same time gaining for itself more and more ecclesiastical rights.

On the other hand, the laity has acquired more self-aware-ness that the Church belongs to it as well as the clergy. To-gether they are the Church in the world. Neither "klerikokra-tia" (predominance of the clergy) nor "laikokratia" (predom-inance of the laity) should dominate the Church. Since au-thority in the Church is characterized by service, all in the Church are servants of God. The various problems which be-set the Church locally and universally today can only be re-solved with the participation of the laity. What is needed is a definition of lay participation in the Church through guidelines which are in harmony with the work of the clergy on the model of the early Church. This can only be achieved by returning to the image of the Church as the Body of Christ, of which mention was made in the introduction. In Corinthians I, St. Paul drew a picture of the unit which should exist within the Church if it is to fulfill its proper function. He said a body is healthy and efficient only when each part is functioning as it should.

Only with the awareness and application of what has been said above by the clergy and laity alike will it be possible to achieve a renewal and strengthening of our Church. This after all, is the goal of all four charters of our Archdiocese. We must avoid the extremes of both clergy and lay superiority and retain the mutual dependence of the ministerial priesthood of the clergy and royal priesthood of all the faithful. Then will our Church, both locally and universally, indeed be what it is theologically—the salvation of the world in Christ Jesus.

FOOTNOTES

[1] Cor. I, 12:12-31.

[2] The temporary charter mentioned in Article 26 of the Charter of 1922 was the charter by which the Archdiocese was administered while under the jurisdiction of the Church of Greece.

[3] Charter of 1927, Art. 28.

[4] Charter of 1922, Art. 2.

[5] Charter of 1927. Art. 2.

[6] Charter of 1927, Art. 4.

[7] Charter of 1927, Art. 7.

[8] Charter of 1927, Art. 10.

[9] For an historical account of the life of our communities during the formative years of their existence, see Basil Zoustis, *O en Ameriki ellinismos kai i drasis aftou* (New York: D.C. Divry, 1954), esp. pp. 105-108 and 113-115.

[10] Charter of 1927, Art. 27.

[11] Counelis, James S., "Historical Reflections on the Constitutions of the Greek Orthodox Archdiocese of North and South America, 1922-1982," Workbook of the 26th Biennial Clergy-Laity Congress, San Francisco, 1982, p. 39.

[12] Counelis, p. 39. For a chronology of events leading up to the displacement of the two previous charters by the Charter of 1931, see Zoustis, pp. 193-207.

[13] Counelis, "Workbook", p. 40.

[14] The turbulence of this period is graphically portrayed in a lecture by Peter T. Kourides and published as a booklet bearing the title *The Evolution of the Greek Orthodox Church in America and Its Present Problems* (New York: Cos-

mos G/A Printing Co., 1959), see especially pp. 7-11. It is heartening to note an increasing interest in the early history of our Church's life in America. Such interest is reflected in a doctoral dissertation on the subject currently in preparation by the Rev. Thomas Fitzgerald, a member of the faculty of Hellenic College.

[15]Charter of 1931, Art. 4.
[16]Charter of 1931, Art. 12.
[17]Charter of 1931, Art. 13.
[18]Charter of 1931, Art. 14.
[19]Charter of 1931, Art. 15.
[20]Charter of 1931, Art. 16.
[21]Charter of 1931, Art. 19.
[22]Charter of 1931, Art. 20.
[23]Charter of 1931, Art. 22.
[24]Charter of 1931, Art. 6.
[25]Charter of 1977, Art. 4.
[26]Charter of 1977, Art. 1.
[27]Charter of 1931, Art. 7.
[28]Nikodemos Milasch, *To ekklisiastikon dikaion tis orthodoksou anatolikis Ekklisias* (Athens, 1906), pp. 419-421 and especially p. 420, footnote 15.
[29]Acts 1:15-26.
[30]Council of Laodicea, canon 13.

Metaxakis In Profile

The Controversial Visionary and His Contributions to Orthodoxy and the Founding of the Archdiocese

By
George Bebis

eletios Metaxakis, one of the most important ecclesiastical personalities of this century, has yet
to find his biographer. A full biography of this
great churchman of our times is still to be written.[1]
It is most fortunate, that Archbishop Methodius of
Thyateira and Great Britain, began the publication of Metaxakis' private *Diary*, which covers the period from January
30, 1925 to May 29, 1926.[2] However, a major biography will
have to study private and official papers and documents of his
long ecclesiastical career. Also, more information must be ex-
tracted from the archives of foreign governments and the
many Churches with which Metaxakis had contacts and relations. This short profile will focus on the activities of Metaxakis concerning the Greek Orthodox Church in America
and his special relations with the Anglican Church.

Metaxakis (whose baptismal name was ˙Emmanuel) was
born in Parsas, a village of Lasethis province of Crete, on September 21, 1871, one of four children born to Maria and Nicholas Metaxakis. After studying for the priesthood at the
Orthodox Seminary in Jerusalem, he was ordained to the dia-
conate in 1892 and, according to the monastic custom, he assumed the name Meletios. He continued his studies at the famous Theological School of Holy Cross in Jerusalem and immediately became a member of the Brotherhood of the "Holy
Sepulchre". He was appointed chief secretary to Damian,
the respected Patriarch of Jerusalem and served in that capacity from 1903 to 1909. His main interests were focused on
education, finance, publications and diplomacy. He was involved with the problems of the Patriarchate of Jerusalem in
relation to Arab Christians, as well as with the internal difficulties of the Church of Cyprus. While in Constantinople, in

1910, he was elected Metropolitan of Kition (Cyprus). He served there for eight years and in 1918 he was elected Metropolitan of Athens and titular head of the Church of Greece. He remained in this position until November 1920, when he was forced to resign because of the victory of the Royalist Party in Greece. Metaxakis was a great admirer of Eleftherios Venizelos and took side with that great Cretan politician and charismatic leader. As prime minister, Venizelos contributed to the territorial expansion of Greece during the Balkan War of 1912, but later, together with King Constantine, led the nation into factionalism and political strife which divided the Greek people.[3]

As Metropolitan of Athens and head of the Greek Church, Metaxakis left an indelible mark on the Church. Although his enemies[4] accused him of "satrapism", there is no doubt that he was interested in restoring the early practice of the Church in the ecclesiastical election of the clergy, especially the bishops. He also fought for the protection of the Church from the abuses by the State and from politicians who were not mindful of the canons of the Church. The Very Rev. Evangelos Mantzouneas, in his excellent study on Metaxakis, states that upon his enthronement the new Metropolitan realized that the Church of Greece had become merely a part of the Greek civil service.[5] Mantzouneas asserts all newly-elected bishops had to take the oath of allegiance in the presence of the King of Greece, "without even mentioning their faith and respect to the divine and sacred canons of the Church".[6] For Metaxakis, this absolute submission of the Church to the authority of the State was canonically and theologically unacceptable.

In 1919 he submitted his new "Constitutional Charter" concerning the administration of the Church of Greece. In it he introduced, as the supreme administrative body of the Church the Holy Synod of *all* the bishops of Greece (and not only the "residing" synod of few bishops). This has become, ever since, the ruling principle, of all the "Constitutional Charters" of the Church of Greece,[7] which is in accordance with the sacred canons of the Church. (5th canon of the 1st Ecumenical Council of Nicea; the 19th canon of the 4th Ecumenical Council of Chalcedon and the 37th Apostolic canon.) In addition Metaxakis issued a decree unheard of until this time that the Church has the right to legislate on matters which concern her own life, without the approval of the State.[8] Metaxakis also promulgated decrees which eliminated the parti-

Archbishop of Athens Meletios (Metaxakis), who later became Ecumenical Patriarch of Constantinople, arrived in America in 1918, accompanied by Bishop Alexandros; also organized and presided over the 1st Clergy-Laity Congress in New York, September 13 - 15, 1921.

95

cipation of the State in the election of new bishops and instituted the formation of an official "list" of those to be elected
to the office of the bishop. This "list" was to be voted upon,
by the bishops themselves through a secret balloting.[9] This
new practice was based on the 3rd canon of the 7th Ecumenical Council, which states very clearly, that an election of a
bishop by civil authorities is invalid. Moreover, Metaxakis introduced a new provision for the establishment of a "Diocesan Council" composed of both clergy and laity, which would
be responsible for the property and all the finances of the
local dioceses.[10]

A man of vision, abounding energy and well-meant ambition could not but come to the United States of America, where,
since the end of the nineteenth century, Greek Orthodox communities began to be established throughout the country. He
had grasped the great religious and "ethnic" importance of
the Greek Orthodox Church in the Diaspora. He also sought
contacts with the other Christian churches, whose clergy commanded respect and love by the American people. Saloutos
quotes Garret Droppers, the United States minister in Athens,
who telegraphed the State Department as follows: "Meletios,
Metropolitan, formerly Archbishop of Cyprus is much respected. He is a judicious but vigorous reformer of abuses in
the church. He occasionally exercises in preaching function,
a practice almost unknown in the Greek Church. In politics
he is a liberal and Venizelist. His intention, I believe, in going
to America is to render the Greek Church there a better instrument of religious work; also, if possible, to appoint a bishop;
also said to be interested in closer union of Christian churches."[11]
Droppers' comments were correct. But the situation which
Metaxakis faced in America was not pleasant. The Greeks in
America were divided into two great factions along the lines
of politics in Greece: the Royalists and the Venizelists. The
clergy and press took sides, but apparently, Metaxakis, with his
great respect for American institutions, and his adherence to
the sacred canons of the Church, tried to organize the Church
in America with a strong central authority but also with the
local democratic participation of clergy and laity. Metaxakis
brought with him a learned theologian and clergyman, Chrysostomos Papadopoulos, Professor of the University of Athens,
to whom he offered the post of Archbishop of the Greek Orthodox Church in America. He refused. Finally, he appointed
his auxiliary bishop Alexander of Rodostolou, as the first Bish-

op in America.[12] Alexander faced great obstacles in pursuing a canonical unifying policy, but his political affiliation with the Venizelist party and his personality did not contribute to the pacification of political and ecclesiastical conflict and passions. However, Metaxakis had planted deep roots in the American soil and he laid down the foundation for the Greek Orthodox Church in America as a strong and promising religious institution.[13]

In November 1920, Metaxakis was deposed from his Athens see by the Royalist government of prime minister Rallis. (Venizelos had lost the elections.) In February 1921 he arrived, for the second time, in the United States. He stayed in this country for 10 months. Through his extensive travels to all corners of this nation, his excellent organizational capacities and his persuasive power, Metaxakis promoted, along with bishop Alexander, unity among the Greek-Americans. He made clear, to friends and foes alike, that the Church in America had to become independent of the political fortunes and changes of Greece.[14]

Here it must be stated that Metaxakis, although a friend of Venizelos, was not an instrument of the Venizelist party and its politicians. Spanoudis has well-documented the fact that, even as Archbishop of Athens, he remained above all a genunine churchman, a man who respected the canons of the Church, and one who was mainly interested in the welfare and independence of the Church.[15] Many times he disagreed with Venizelos. One wonders whether he would have remained in the Metropolitan see of Athens if Venizelos had refused Metaxakis' ecclesiastical reforms. Be that as it may, the fact is that he offered amnesty to 500 clergymen who participated in the shameful and anti-canonical act of "anathematizing" Venizelos. He made appointments disregarding their political affiliations and even three out the five new assistants in his Metropolitan office in Athens were anti-Venizelists (Gerakakis, Papanastasiou and Athenagoras, who became later Archbishop of North and South America and Patriarch of Constantinople). Also, his attitude towards his predecessor of the Metropolitan see of Athens (Theoclitos), as well as to many bishops of the opposite faction, was generous, polite and exemplary.[16] When Venizelist members of the Government and members of the party in the Parliament attacked him publicly, because he refused to accept innovation and changes on the status of marriage, he remained unyielding in sustaining the theological and canonical tradition of the Church. Venizelos realized that he faced a man whose ecclesiastical integrity was beyond reproach.[17]

He followed the same policy upon his second arrival in the United States. He made it very clear that he was not fighting the new Royalist regime in Athens. That was not his main concern. In one of his encyclical letters to his flock in America, a copy of which was sent to the Greek Embassy in Washington, D.C., he wrote: "We have never thought of denying the legal basis of the Authorities who govern currently the Greek State". He continued: 'We deny the right of these legal authorities, i.e. the State, to violate sacred canons, to transcend ecclesiastical privileges and to confound the rights of God with the rights of Caesar".[18] Indeed this was a statement, of such candor, that only a great churchman could make. Metaxakis declared everywhere that the Church in America should not accept the meddling and interferences of the Greek Government in its internal affairs, and also made that very clear to the diplomatic representatives of Greece. He felt the clergy should pay their allegiance only to their ecclesiastical authorities. Saloutos notes, "Most of the disputes within the church in the United States revolved around ecclesiastical government and the struggle for power; differences over dogma and doctrine seldom became issues".[19] This is partially true. But behind this power-struggle the Church in America was trying to find a balance, or rather to free itself from the unhealthy Caesaropapism of the Greek Church and the secularism of the Protestant churches. Metaxakis, having the experience of both systems and adhering to the ancient canons of the Church, established the Greek Orthodox Archdiocese of North and South America, with a brilliant combination of ecclesiastical traditionalism and a healthy democratic process, which makes the Archdiocese, today, one of the most vibrant Orthodox churches in the world. A strong Archbishop, bishops in the large cities, an educated clergy, the participation of the laity in the financial affairs of the church, the establishment of the Clergy-Laity Conferences, all these were the products of Metaxakis' vision and labors. Alexander was the first to follow the great canonical scheme and in reality, he became the first Archbishop of North and South America. One should keep in mind, for a period the Greek Orthodox Church in America was under the jurisdiction of the Church of Greece. That was, of course, an anomaly.

Historians and canonists agree, that the 28th canon of the Fourth Ecumenical Council of Chalcedon (451 A.D.) granted ecclesiastical jurisdiction to the Ecumenical Patriarchate of Constantinople over the "barbaric nations", that is to say over all the churches, which were outside the geographical boundaries of the Roman Empire. The Ecumenical Patriarchate kept that privilege and jurisdiction by ordaining priests and appointing bishops and caring for all the spiritual needs of the churches established "abroad", that is, in "foreign territories".[20] The Ecumenical Patriarchate of Constantinople as "primus inter pares", and as the Church's "senior bishop, her representative and overseer",[21] had the right to exercise its authority over the churches in the Diaspora. Because of the difficult circumstances in which the Ecumenical Patriarchate found itself at the beginning of the 20th century, it empowered the Church of Greece "to look after and supervise the Orthodox communities of the Diaspora in Europe and America". This is the famous Patriarchal and Synodical Tome (No. 2388) signed by Patriarch Joachim III in March, 1908. Jerome Cotsonis, the former Archbishop of Athens, correctly stated that this was "a temporary measure"[22] and Michael Galanos, one of the great lay theologians of the early 20th century, in an excellent analysis, showed that the tome of 1908, first, was temporary; secondly, was not canonical (for only an Ecumenical Council could take such a step); thirdly, that the Ecumenical Patriarchate never relinquished its ultimate prerogatives over the churches in the Diaspora; and, fourthly, the Church of Greece herself had violated the rules of the tome by failing to appoint an archbishop for the period of ten years![23] Meletios Metaxakis was fully aware of all these grave repercussions of the Patriarchal Tome and realized that the Church of America dependence on the Church of Greece would have catastrophic results, precisely because of the Greek clergy's association with the Greek State and Greek political parties.

Metaxakis was still in America, practically a political exile, when on November 25, 1921 he was elected Ecumenical Patriarch of Constantinople.[24] Although his adversaries questioned the validity of his election, the great majority of the Greek Orthodox people in America—as well as the political and religious authorities and officials in America—bestowed upon him the greatest of honors. President Harding received him in the White House (as President Wilson had few years earlier) and an Ecumenical Service took place at the Cathedral of St. John the Divine, in New York City honoring the new Patriarch. Representatives of the Russian, Syrian and Armenian churches attended as well as representatives of the Episcopal and Protestant churches. One can understand the importance of this service if he keeps in mind the divisions, the hatred and the disarray of the Greek Orthodox communities at that time in the New World.

On January 1, 1922, he left New York and, after visiting England and France,[25] arrived in Constantinople on January 24, 1922. Professor Stavridis in his excellent biography on Metaxakis makes the point that Metaxakis was the last Ecumenical Patriarch under the rule of the Ottoman Empire. His failure to proceed after his arrival in Constantinople to his political installation, in accordance with Ottoman laws and regulations, as well as his attitude towards Turkish authorities and Ottoman sensitivities, were major tactical mistakes.[26] Consistent with his policies of reconciliation, in the first meeting of the Holy Synod under his presidency, on January 26, 1922, he granted complete amnesty to all those who resisted or opposed his election to the Patriarchal throne.[27] Of course, he did not forget his beloved Orthodox Churches in the Diaspora. In his enthronement address delivered in the Patriarchal church on January 24, 1922, he stated that although there are many problems which the hierarchy of the Ecumenical Patriarchate would face, he would mention by name the problem of the governance of the Orthodox church in the Diaspora. He said the arrangement made by the Patriarchate in 1908, for rather political reasons, had been proven not to be the proper one. Thus the Great Church of Christ had immediately, in accordance with canons, acted to take care of those churches and assist them in observing the unity of the spirit in the bond of peace. He also said he could speak from his personal experience in the United States and foretell that the name of Orthodoxy would be exalted, if more than two million American Orthodox were organized in one united ecclesiastical organization under the title, "American Orthodox Church". Here is the exact quotation. "I saw the largest and best part of the Orthodox Church in the Diaspora, and I understood how exalted the name of Orthodoxy could be, especially in the United States of America, if more than two million Orthodox people there were united into one church organization, an American Orthodox Church."[28] This final remark of Metaxakis was characterized as being above nationalistic limitations and proved

100

to be "an inspiring and broad vision of Orthodoxy"[29] which Metaxakis cultivated during his short Patriarchate. This may be true, but one must also remember that he remained a Greek until the end of his life. But because he had lived a great part of his life outside Greece, he had developed a balanced picture for the future of Orthodoxy. He worked hard for the convening of a Pan-Orthodox Synod; his relations with other autocephalous churches were correct; and in his enthronement address, he spoke with warmth and affection for all the Orthodox Patriarchates and the Autocephalous Churches.[30] In the same address, he spoke about the "democratic precepts" of the Church and the "recognized right of the people in the administration of the Church through elected representatives".[31] The wholesome education of the clergy, as well as the religious education of the people were his outmost concerns and it is well-known, that he founded in New York (October 1921) the Seminary of St. Athanasios.[32]

Consistent with the ideas and policy of Metaxakis, on May 17, 1922, the Ecumenical Patriarchate, under the leadership and direction of Patriarch Meletios IV, issued the celebrated Founding Tomos of the "Orthodox Archdiocese of North and South America".[33] The language and the spirit of Metaxakis is obvious throughout the text. This Tomos was issued in accordance with canonical order and the long practice of the Church. The previous status was contrary to the canonical order delivered to us by the Fathers therefore, canonically, all the Orthodox communities in America should be united under the "Orthodox Archdiocese of North and South America". That organization would be canonically dependent upon the Holy Patriarchal and Apostolic Throne as one of its eparchies.[34] It is number 15 in the list of the "Syntagmation" (Constitutional Charter) of the Ecumenical Patriarchate. It provided for the election and appointment of an Archbishop (who would be also responsible for Washington, D.C.) and for three bishops, those of Chicago, Boston and California. Finally, the document exhorted the people to respect their bishops as their own spiritual fathers, so that they might perform their duty with joy and not with groans.[35] It was a splendid document and it had far-reaching results.

Archbishop Iakovos, commenting on the sixtieth anniversary of the Greek Archdiocese (1982), wrote: "The amazing fact is that we are united as a Church and an Archdiocese, and this in contrast to other Orthodox jurisdictions and ethnic groups in America that are divided among themselves. We are grateful for and appreciative of the various achievements of the past. We are also hopeful for the future. . ."[36] Indeed Meletios Metaxakis had laid down a concrete and strong foundation. In August 1922, Archbishop Alexander called the second Clergy-Laity Conference, which drafted a constitution for the new Archdiocese, which he submitted to the Patriarchate. "The constitution provided for bishops in New York City, Chicago, Boston and San Francisco; the condition under which the delegates for the special ecclesiastical assemblies were to be selected in each diocese; and the procedure to be used in electing bishops".[37] In December 1922, the Ecumenical Patriarchate sent a telegram to Archbishop Alexander, informing him that the new Constitution of the new Archdiocese has been approved, in accordance, of course, with the Tomos, which stated that regulations would not become effective, unless approved by the Patriarch and his Synod.[38]

No one can claim that the history of the Greek Orthodox Church has been an easy one since Metaxakis. Many changes were inserted in the Constitution of the Archdiocese, all approved by the Ecumenical Patriarchate. The Archdiocese has adjusted itself to new conditions and new realities but the spirit of Meletios Metaxakis remains still alive and inspires American Orthodoxy.

Michael Galanos enumerates *thirteen* categories of people, who at that time were adversaries of the Greek Archdiocese, as it was founded by the Ecumenical Patriarchate: 1. The fanatical atheists; 2. those who were against the Church and religion, who wanted only schools and not the Church, or at least not a strong Church; 3. the converts to other religious groups; 4. those who considered the sacred canons irrelevant in our contemporary world; 5. the followers of John Metaxas in Greece; 6. the ambitious laymen, who believed that a strong Archdiocese would shake their influence in the communities; 7. those who believed that American civil laws were superior to the sacred canons of the Church; 8. those who claimed that each community had the right to select the ecclesiastical center to which they wanted to belong; 9. those who believed that only those who lived in the "unliberated" countries could be under the jurisdiction of the Ecumenical Patriarchate, whereas the Greeks from Greece should belong to the Church of Greece, the Cypriots to the Church of Cyprus, those from Egypt to the Patriarchate of Alexandria; 10. a number of priests who

Bishop Alexandros of Rodostolou and first Archbishop of the Greek Arch-diocese of North and South America.

were ordained without the proper qualifications; 11. those who believed that since there is a government in Greece, the Church of Greece is of higher status than the Ecumenical Patriarchate; 12. those who paid their allegiance to the political parties in Greece; 13. those who exploited the name of the Church of Greece in order to justify their disobedience to the Archdiocese.[39]

Under these circumstances, only a man with the determination of Meletios Metaxakis could have achieved the founding of the new Archdiocese. Correctly, Saloutos writes that "Metaxakis, whatever his limitations were, was an ecclesiastical leader of the first order who had a burning ambition to identify the Greek Church with the Western world. It was his misfortune that he became Metropolitan of Athens and Ecumenical Patriarch during the darkest hours of modern Orthodox history. One can only speculate what his accomplishments would have been had he appeared at a more tranquil time."[40]

During his short period of reign on the Patriarchal throne, Metaxakis remained extremely active. He instituted new Metropolitan sees in Asia Minor and Macedonia and Thrace. He recognized the Patriarchate of Serbia. He instituted (April 1922) the Metropolis of Thyateira, as an exarchy of Western and Central Europe and he granted autonomy to Finland, Czechoslovakia, Estonia, Latvia and Albania. He convened in Constantinople a "Panorthodox Conference", on May 10 - June 8, with participants from the Churches of Constantinople, Russia, Serbia, Cyprus, Greece, and Rumania. Apparently, not all the Orthodox Churches were represented, but Metaxakis with his vigorous determination accomplished results, unknown in the recent pages of church history. Decisions were taken on the following issues: 1. Correction of the Julian Calendar; 2. permission of marriage of deacons and presbyters after their ordination; 3. permission for deacons and priests to marry, if their wives died after cleric's ordination; 4. celebration of the 1600th anniversary of the First Ecumenical Council of Nicea. 5. Further study for the preparation of a Panorthodox Synod.[41] One year earlier (July 1922) Metaxakis recognized the validity of Anglican orders.[42]

But time was running short. On July 24, 1923, the Treaty of Lausanne was signed. The war between Greece and Turkey ended and the Ottoman Empire, which had granted so many privileges to the Ecumenical Patriarchate was eliminated. Also, in December 1922, Eleftherios Venizelos promised the Turkish delegation in Lausanne that he would extract the resignation of Metaxakis as Patriarch, because he was "persona non grata" to the Kemalic Turkish Government. The choice was between keeping in Constantinople the Patriarchate or Metaxakis. Venizelos made that very clear to Metaxakis, who, on July 10, 1923 left Constantinople for the Holy Mountain for a three-month "recuperative" leave of absence.[43] On October 2, 1923, the Holy Synod of the Patriarchate declared the Patriarchal throne vacant and in November 10, 1923, Metaxakis sent his own personal resignation (dated September 20, 1923).[44]

Later, he moved to Kifisia, a suburb of Athens, where on May 20, 1926, he accepted his election to the Patriarchal throne of Alexandria. In his personal calendar, and in the entries, during his stay in Kifisia, Metaxakis showed his concern and affection for the Greek-Americans and their life in the United States. On February 11, 1925, for instance, he expressed his opposition to an argument in a publication of the Ku Klux Klan (*The Fellowship Forum*) which said the Greeks in America were of the most criminal nature.[45] On July 8, 1925, he wrote that he had invited for dinner Senator King of Utah, together with other American and Greek-American officials.[46] On July 17, 1925, he wrote that religious conditions in America favor the ethnic organization of the religious communities, and that a union with the Anglicans would preserve the Greek East, as a religious center, instead of Moscow. On July 30, 1925 he wrote that Archdeacon Nicols visited him on behalf of Bishop Darlington from America, and that his friends in America inquired about him and asked if he would visit America again.[47] On August 26, 1925 he was visited by Michael Galanos from Chicago, who brought him good news from the Church in America (Archbishop Alexander was doing fine; only Joakim, the Bishop of Boston was not successful in his diocese and wanted to return to Greece).[48]

Metaxakis' enthronement as Patriarch of Alexandria took place on June 13, 1926 and he added to his title "of all Africa". Typically, he worked hard to consolidate the position of the Alexandrian Patriarchate in canonical and spiritual matters. He strengthened the missionary activities in the African countries and paid much attention to education, preaching, philanthropy and publications. Specifically, he sent representatives to the Pan-Orthodox Conference in Mount Athos (1930), participated at the Lambeth Conference in 1931 and, as Patriarch of Alexandria, recognized the validity of Anglican orders.

He died on Saturday night on July 27, 1935. In his funeral address, Metropolitan Parthenios of Pelusium praised Metaxakis as an "outstanding ecclesiastical personality" as "a star of great magnitude", as a man of unceasing labor, of profound philanthropy and finally as a "kindled candle", which was spent for the benefit of his people.[49]

Much has been written and said about the relationship of Metaxakis and the Anglican Church. Meletios was an Anglophile as was his friend and mentor Eleftherios Venizelos. Metaxakis was born in a geographical area where English influence was more than apparent. Both Venizelos and Metaxakis were convinced that the future and the interests of Greece, as well as Orthodoxy, were associated with England and America.[50] Metaxakis was fascinated with the honesty and openness of the English and American Anglican clergymen. He believed a union between Orthodox and Anglicans could eventually lead to better understanding and finally to unity between the East and the West. The first contacts of Metaxakis and the Anglicans took place in the United States of America. As Metropolitan of Athens and all Greece, Metaxakis was received by the Anglicans in America with great honors. He was invited, together with his assistants, to numerous receptions and banquets. They attended a special service at the Episcopal Church of St. Michael in the presence of Bishop Darlington and Bishop Parker. Meletios prayed and preached in this church and later invited his American friends to attend the Divine Liturgy at the Holy Trinity Greek Orthodox Cathedral on September 23, 1918. One of the accusations of his adversaries was that he prayed with the Anglicans, kneeling in front of their altar.[51] They forgot to mention that in 1920, Dorotheos, Metropolitan of Proussa and Locum Tenens of the Ecumenical Patriarchate, had invited Bishop Darlington to the Patriarchal church of St. George and after putting on his shoulders an Orthodox bishop's "omophorium" (pallium)—asked him to jointly distribute the "antidoron" (the special blessed bread which is given to the lay people at the end of the Liturgy).

The most important meeting took place at the General Theological Seminary in New York. Metaxakis made it clear to the Anglicans that he did not have any official authorization to proceed with formal negotiations with the Anglicans in America or in England. Everything had an unofficial and academic character. The Anglicans, however, submitted many questions to Metaxakis and to his committee in written form so that they might find out where they stood on contemporary church problems. For example, when in a small community there is only one priest, Orthodox or Anglican, was it possible for one of them to serve the needs of both churches? Was it difficult for one of their Churches to recognize marriages performed by either Church? In case of sacramental inter-communion, should it be mandatory for the faithful of both Churches to receive both the elements of the Holy Eucharist? What about the validity of Anglican Orders?[52] Meletios replied that both the Greeks and the Russians recognized the validity of Parker's ordination. He asked whether the Anglicans were ready to accept ordination as a sacrament of the Church and whether the Ecumenical Councils were infallible. The answers were positive. Professor Hall made clear that the 30 articles of the Church of England were not considered, any more, to be official dogmatical documents of the Anglican Communion.[53]

Meletios visited England twice before his enthronement as Bishop of the Patriarchal see in Constantinople. He was very popular in London. The newspapers followed his activities closely and the English faithful used to kneel in order to receive his blessing.[54] In 1922, Metaxakis received an honorary Doctor's Degree from highly-respected Oxford University, a rare distinction bestowed upon a foreign clergyman. But in that same year, Prime Minister Lloyd George received Metaxakis in a private audience and told him openly that the Greek people in Asia Minor should not expect British protection and support as long as Constantine remained King of the Hellenes. The revelation and publication of this statement caused an increase in the hostile feelings of his adversaries and his Anglophile attitude made his position in Constantinople precarious.

In any case, Metaxakis took three basic steps concerning the Anglican Church, upon his ascending to the Patriarchal Throne of Constantinople. First, he instituted the Metropolitan see of Thyateira in London and made the Metropolitan of Thyateira the official representative of the Ecumenical Patriarchate of Constantinople to the Church of England and to the Archbishop of Canterbury. His first appointment was Germanos Strinopoulos, one of the most outstanding churchmen of our times. Germanos, upon his arrival in London, was officially welcomed at Westminster, and during Vespers stood with his pastoral staff at the right side of the Westminster dean.[55] Secondly, Metaxakis proceeded with the official recognition of the Anglican Orders. In July 1922, he sent a letter to Dr. Randolph Davidson, the Archbishop of Canterbury. He called him "Brother" and "beloved" and without any hesitation made the following statement: "Accordingly, the Holy Synod has concluded that, as with the Catholic Church, the ordination of the Anglican Confession of bishops, priests, and deacons, possesses the same validity as those of the Roman, Old Catholic, and Armenian Churches, inasmuch as all essentials are found in them which are held indispensible from the Orthodox point of view for the recognition of the 'charisma' of the priesthood derived from Apostolic succession".[56] A second letter was sent to the heads of all the Orthodox Autocephalous Churches, in which he explained why he and his Synod accepted the validity of the Anglican Orders. He asked them, at the same time, to send back the results of their own Synodical discussions.[57] As we know the Churches of Jerusalem, Cyprus and Romania, agreed with Metaxakis, whereas the Holy Synod of the Church of Greece restated its position that "it intends as before to follow in each individual case that may arise of the adherence of an Anglican cleric to Orthodoxy the practice of the Church. . . in so far as she considers it proper and useful, in particular cases, after previous investigation of the current circumstances, to recognize, by Economy, the Ordination of those who come to Orthodoxy. . ."[58]

The third step, which Metaxakis took was to receive the Rt. Rev. Gore, the former Bishop of Oxford (appointed by the Archbishop of Canterbury as chairman of the Anglican Commission on the relations with the Orthodox Church), in full session of the Pan-Orthodox Conference held in Constantinople. The Patriarch, in a warm welcome address expressed the happiness of all the participants of the Pan-Orthodox Synod for the visit of Bishop Gore. Metaxakis assured him that one of the most serious problems which the Conference had on its agenda was the possible union with the Anglican Church. Then he thanked the clergy and the laity of the Anglican Church for the assistance and support offered to the Greek people in the East. Bishop Gore replied saying that he was happy to be at the Pan-Orthodox Conference and that he brought with him a document with the signatures of 5,000 Anglican clergymen, who claimed that they found no difficulty in accepting full union with the Orthodox Church. The Patriarch, surprisingly enough, did not answer Bishop Gore's statement. He only asked him to inform the Archbishop of Canterbury that the Orthodox Church would accept any revised Calendar the West proposed. The question of union was not brought up again at the Conference. A month later Metaxakis was forced to resign from this throne.[59]

However, Metaxakis continued to show the same interest in promoting good relations with the Anglicans after he became Patriarch of Alexandria. He himself headed the Orthodox delegation to the Lambeth Conference of 1930 and throughout the meetings he offered his experience with prudence, understanding and great eloquence. The Orthodox delegation, under his personal leadership, agreed that a joint commission of Orthodox and Anglican theologians should be appointed for the study of questions of doctrine. Also in the report of the Lambeth Conference, "The Orthodox Delegation stated that they were satisfied with regard to the maintenance of the Apostolic Succession in the Anglican Church in so far as the Anglican Bishops have already accepted Ordination as a *mysterion,* and have declared that the Doctrine of the Anglican Church is authoritatively expressed in the Book of Common Prayer, and that the meaning of the XXXIX Articles must be interpreted in accordance with the Book of Common Prayer."[60] Further, "It was stated by the Orthodox Delegation with regard to the Holy Eucharist that, pending a formal decision by the whole Orthodox Church and therefore without giving the practice official sanction, for which it has no authority, it is of the opinion that the practice of the Orthodox receiving Holy Communion from Anglican priests in case of need and where no Orthodox priest was available, might continue, provided that an Orthodox authority did not prohibit such a practice."[61]

The report included, indeed, some astonishing positions from both sides. There is no doubt that Metaxakis had made serious inroads towards union with the Anglican Church. Later, as Patriarch of Alexandria and on Christmas Day of 1930, Metaxakis wrote to Dr. Cosmo Lang, Lord Archbishop of Canterbury, announcing to him the resolution of the Holy Synod of the Patriarchate of Alexandria. The text is as follows: "The Holy Synod recognizes that the declarations of the Orthodox, quoted in the Summary, were made according to the spirit of Orthodox teaching. Inasmuch as the Lambeth Conference approved the declarations of the Anglican Bishops as a genuine account of the teaching and practice of the Church of England and the Churches in communion with it, it welcomes them as a notable step towards the Union of the two Churches. And since in these declarations, which were endorsed by the Lambeth Conference complete and satisfying assurance is found as to the Apostolic Succession, as to a real reception of the Lord's Body and Blood, as to the Eucharist being *thusia hilasterios* (Sacrifice), and as to Ordination being a Mystery, the Church of Alexandria withdraws its precautionary negative to the acceptance of the validity of Anglican Ordinations, and, adhering to the decision of the Ecumenical Patriarchate of July 28, 1922, pronounces that if priests, ordained by Anglican Bishops, accede to Orthodoxy, they should not be re-ordained, as persons baptized by Anglicans are not rebaptized."[62]

It was a bold statement. Indeed, Metaxakis never ceased to believe that a union with the Anglicans was possible, although, he recognized that a fuller discussion was needed on the more difficult doctrinal questions which separated both Churches.[63] But, apparently, Metaxakis was not interested in any serious discussion of theological problems. Finally, he realized, that a union should be based on concrete theological agreement and that a full intercommunion should be the last step of a full union.[64]

Chrysostomos Papadopoulos was convinced that the attitude of Metaxakis was influenced by the fact that until the sixth decade of the 19th century, there was no serious doubt about Parker's election and ordination. Thus, Orthodox bishops freely associated with Anglican clergymen.[65] Professor John Karmires seemed hesitant to accept all these statements of Metaxakis as having Pan-Orthodox recognition and validity and suggested that more serious discussions lay ahead.[66] Moreover, Ieronymos Kotsonis (following Prof. Bratsiotis) seemed to agree that Meletios' approach was unacceptable, because it obscured the deep-rooted problems, which separated the two Churches and offered to the Anglicans vain hopes for a full union.[67]

In retrospect, one may say, that Metaxakis went too fast, despite his deep faith and broad and resilient spirit, his impatience made him insensitive, many times, to the deep-rooted divisions between East and West. But he must be admired for his sincerity and good intentions. He realized that a union between the two Churches should not only be "an agreement between Hierarchs, but a union of the faith and the hearts of the people."[68] To be sure, he worked for that purpose with vigor and determination.

Professor Andreas Fytrakis in commenting on Meletios Metaxakis wrote he was a "strong personality with intense activism" in difficult and troubled times, and therefore, he has become controversial. However, Fytrakis concluded that Metaxakis was the greatest ecclesiastical personality of the first half of the 20th century.[69] Dr. Pantelos added, that Metaxakis was one of the greatest men of later times of our Church history.[70] He was accused as being too innovative. His proposition for the formation of an Autocephalous Pan-Orthodox Church in America and his strange suggestion of transferring the Ecumenical Patriarchate from Constantinople to Greece, did not endear him to many of his adversaries. But great men are bound to be controversial. Metaxakis' foresight and his vision in establishing the Greek Orthodox Archdiocese in America, proves the genuiness of his convictions and the extraordinary power of his spirit.

FOOTNOTES

[1]A good bibliographical note has been written by Dr. John Constantinides in the "Threskeftiki kai Ethnike Enkyklopaideia", Vol. 8 (Athens, 1966), pp. 965-969.

[2]Meletios Metaxakis (On his 100th Anniversary), by Methodius, Metropolitan of Axum. Reprint from *Ekklesiastikos Pharos,* Vol. 53, 1971 (Athens, 1971).

[3]An excellent account of the political and religious strife of those times. See Theodore Saloutos, *The Greeks in the United States* (Cambridge: Harvard University Press, 1964), pp. 281ff.

[4]Dem. Mavropoulos, *Patriarchikai Selides* (Athens, 1960), pp. 154-198.

[5]Preb. Evangelos Mantzouneas, *Meletios Metaxakis* (Athens, 1972), p. 5.

[6]Mantzouneas, p. 5.

[7]Mavropoulos, p. 11.

[8]Mavropoulos, p. 13.

[9]Mavropoulos, p. 15ff.

[10]Mavropoulos, p. 22ff.

[11]Saloutos, p. 282.

[12]Among those who came with Metaxakis to America, were Hamilcar Alivizatos of the Ministry of Education and Religious Affairs (and later Professor of Canon Law at the University of Athens) and deacon Chrysostomos Agelidakis. The Committee of ecclesiastical personalities left Athens on July 19 and returned to Greece on December 27, 1918.

[13]Of course, the publication of all documents relating to this mission of Metaxakis and his Committee, will show how great was the task and the achievement of this great churchman.

111

[14]Constantine Spanoudi, *Meletiou IV, Erga kai Ideai* (Athens, 1925), pp. 35 ff.
[15]Spanoudi, pp. 17ff.
[16]Spanoudi, pp. 20-22.
[17]Spanoudi, p. 18.
[18]Spanoudi, 1p. 34. Saloutos, p. 285.
[19]Spanoudi, p. 286.
[20]The granting of "Autocephaly" to the "Metropolia", by the Patriarchate of Moscow on April 10, 1970 provoked a heated discussion on the rights of the Ecumenical Patriarchate over the churches in the Diaspora. See the following literature: *Autocephaly, The Orthodox Church in America (New York: St. Vladimir's Seminary Press, 1971). Russian Autocephaly and Orthodoxy in America* (New York: The Orthodox Observer Press, 1972). P.N. Trempelas, *The Autocephaly of the Metropolia in America.* Translated and edited by G.S. Bebis, R.G. Stephanopoulos and N.M. Vaporis. (Brookline, Mass.: Holy Cross Theological School Press, 1973). The Ecumenical Patriarchate, however, produced the best of the arguments. Only a Pan-Orthodox Council could discuss and grant Autocehpaly to the Churches in the Diaspora. See the letter of the late Patriarch Athenagoras to Metropolitan Pimen, Locum Tenens of the Patriarchal Throne of Moscow, in *Russian Autocephaly and Orthodoxy in America,* pp. 34ff.
[21]See letter of Ieronymos, Archbishop of Athens and all Greece to Metropolitan of Krutitsa and Kolomna, Locum Tenens of the Holy Patriarchal Throne of Moscow and Russia, p. 59.
[22]Ieronymous, p. 59.
[23]Michael Galanos, *To Ekklesiastikon Zitima tou Ellenismou tis Amerikis* (Chicago, Ill.: 1924), p. 39ff.
[24]For the details of Metaxakis' election see Prof. Vasilios Stavridis, *Oi Oikoumenikoi Patriarchai 1860-Simeron* (Thessaloniki, 1977), Vol. A. p. 443ff.
[25]For details of this trip see: C. Kallimachos, *With Patriarch Meletios IV from London to Bosporus* (New York, 1922).
[26]Stavridis, p. 447.
[27]Stavridis, p. 445.
[28]Stavridis. Also Dmitry Grigorieff, *The Orthodox Church in America, From the Alaska Mission to Autocephaly,* in St. Vladimir's Theological Quarterly, Vol. 14, No. 4 (1970), p. 206.
[29]Grigorieff.
[30]Stavridis, p. 464ff.
[31]Stavridis, p. 466.
[32]Saloutos, p. 293ff. Only nine students were accepted, but they did not finish their studies. In 1923 (with only 6 students) the Seminary closed. None of these students were American-born. Undoubtedly, it was not supported by the Greek Orthodox people.
[33]See excerpts of the document in M. Galanos, pp. 54-56.
[34]Galanos, pp. 54-56.
[35]*Year Book 1982,* Greek Orthodox Archdiocese of North and South America. p. 8.
[36]*Year Book, 1982,* p. 7.
[37]Saloutos, p. 289.
[38]*Ecclesiastical Herald* (July 1, 1922), p. 442.
[39]Galanos, pp. 71-72.
[40]Saloutos, p. 293.
[41]Stavridis, p. 446.
[42]Stavridis, p. 446.
[43]Stavridis, p. 448.

[44]Methodius, p. 406ff.

[45]Methodius, p. 476.

[46]Methodius, pp. 480-481.

[47]Methodius, pp. 485.

[48]Methodius, pp. 489.

[49]Stavridis, p. 479.

[50]Methodius, pp. 480-481. One can see here that Metaxakis in his personal writing, expresses his faith and that a union with the Anglican Church would benefit the interests of the Greek Orthodox peoples.

[51]Mavropoulos, pp. 175-176.

[52]Chrysostomos Papadopoulos, *To Zitima peri tou Kyrous ton Anglikanikon Heirotonion*, reprint, *Nea Sion, 1925*, pp. 52-53. Ieronymos Kotsonis, *I Kanoniki Apopsis peri tis Epikoinonias meta ton Enterodoxon* (Intercommunion), (Athens, 1957), p. 33ff.

[53]Papadopoulos, pp. 52-53

[54]Kotsonis, p. 252ff.

[55]Kotsonis, p. 111.

[56]E.R. Hardy, ed., *Orthodox Statements on Anglican Orders* (New York, 1946), pp. 1-2.

[57]Hardy, p. 5.

[58]Hardy, p. 18.

[59]Methodius, p. 517. Stavridis, p. 446ff.

[60]Hardy, p. 20ff.

[61]Hardy, p. 22.

[62]Hardy, p. 24.

[63]Hardy, pp. 11-12.

[64]Pantainos, 1930. pp. 816-817.

[65]Kotsonis, pp. 46-47.

[66]Kotsonis, p. 71.

[67]John Karmires, *Orthodoxia kai Protestantismos*, Vol. I (Athens, 1937), pp. 310ff.

[68]Kotsonis, pp. 45-46.

[69]Kotsonis, p. 63.

[70]Stavridis, p. 451.

ΥΠΟΥΡΓΕΙΟΝ ΕΞΩΤΕΡΙΚΩΝ

Κρυπτογραφικὸν Γραφεῖον

Ἀντίγραφον

Ἀριθ. Πρωτ. 5514.

Ἐξοχώτατε,

Εἰς ἀπάντησιν τῆς ἀπὸ θ΄. τοῦ τρέχον-
τος μηνὸς ἀριθ. 5765 ἐπιστολῆς τῆς περισπου-
δάστου ἡμῖν Ἐξοχότητος Αὐτῆς περὶ τῆς ἐν Ἀμερικῇ Ἐκκλησίας, λαμβάνομεν τὴν τιμὴν συνοδικῇ διαγνώμῃ, ἵνα δηλώσωμεν Αὐτῇ μετὰ τῆς
παρακλήσεως, ὅπως διαβιβάσῃ ταῦτα ἁρμοδίως ἀπὸ μέρους τῆς ἡμετέρας Ἐκκλησίας, ὅτι
ὑφίσταται καὶ ἐξακολουθεῖ ἰσχύων ὁ Πατριαρχικὸς καὶ Συνοδικὸς Τόμος τοῦ 1908 "περὶ
χειραφετήσεως τῶν ἐν διασπορᾷ Ὀρθοδόξων
Ἑλληνικῶν Ἐκκλησιῶν εἰς τὴν Ἱερὰν Σύνοδον
τῆς Ἐκκλησίας τῆς Ἑλλάδος" καὶ ὅτι τὸ Οἰκουμενικὸν Πατριαρχεῖον οὐδέποτε μέχρι
σήμερον ἐπενέβη εἰς τὰ τῆς ἐν Ἀμερικῇ Ἐκκλησίας πράγματα, οὐδὲ ἔδωκέ ποτε ἐντο-
λὴν εἴς τινα κληρικὸν ἢ λαϊκόν, ὅπως ἀπὸ
μέρους αὐτοῦ ἐνεργῇ καὶ ἀναμιγνύηται εἰς

τήν διοίκησιν καί διοργάνωσιν τῶν ἐκεῖ
Ὀρθοδόξων Ἑλληνικῶν Ἐκκλησιῶν καί Κοινοτήτων.

Ἐπί δέ τούτοις διατελοῦμεν μετ' εὐχῶν
ἐγκαρδίων καί βαθείας τιμῆς.

͵αϡκα!, Αὐγούστου κ!

Ὁ
Τοποτηρητής τοῦ Οἰκουμενικοῦ Θρόνου
Διάπυρος ἐν Χριστῷ εὐχέτης.
† Ὁ Καισαρείας Νικόλαος.

Ἀκριβές ἀντίγραφον
Ἐν Ἀθήναις τῇ 16 Ὀκτωβρίου 1921

Ὁ Ἀ. Γραμματεύς

[signature]

ΒΑΣΙΛΕΙΟΝ ΤΗΣ ΕΛΛΑΔΟΣ

Η ΙΕΡΑ ΣΥΝΟΔΟΣ ΤΗΣ ΕΚΚΛΗΣΙΑΣ ΤΗΣ ΕΛΛΑΔΟΣ

Πρός

Τόν Σεβασμιώτατον Ἐπίσκοπον Μονεμβασίας καί Λακεδαίμονος κ.Γερμανόν
Συνοδικόν Ἔξαρχον Ἀμερικῆς καί Καναδᾶ

Ἀποστέλλεται Ὑμῖν ὧδε περικλείστως ἀντίγραφον τῆς ὑπό τοῦ Ὑ-
πουργείου τῶν Ἐξωτερικῶν διά τοῦ ὑπ'ἀριθμ.9901 ἐ.ἔ.ἐγγράφου αὐτοῦ
διαβιβασθείσης τῇ Ἱ.Συνόδῳ ἐν ἀντιγράφῳ ἀπαντήσεως τοῦ Οἰκουμενικοῦ
Πατριαρχείου περί τοῦ ἐν ἰσχύει ἐξακολουθοῦντος νά ὑφίσταται Πατριαρχικοῦ καί Συνοδικοῦ Τόμου τοῦ 1908 περί ἐκχωρήσεως τῶν ἐπί τῶν ἐν τῇ
Διασπορᾷ Ὀρθοδόξων Ἑλληνικῶν Ἐκκλησιῶν εἰς τήν Ἱ.Σύνοδον τῆς Ἐκκλησίας τῆς Ἑλλάδος,ἵνα λάβητε γνῶσιν.

*Document to Germanos, the Synodical Exarch of the Church in America and Canada, con-
firming the Patriarchal decision found in the Tomos of 1908, signed October 18, 1921.*

Certificate-of-Incorporation

-of-

GREEK ARCHDIOCESE OF NORTH AND SOUTH AMERICA

We, the undersigned, MELETIOS METAXAKIS, ARCHBISHOP OF
ATHENS, residing at the Hotel Majestic in the Borough of Manhattan in the City of New York, and GERMANOS POLYZOIDES, ARCH⁺
DEACON, residing at 339 East 88th Street in the Borough of Manhattan in the City of New York, being respectively the presiding
officer, namely, the president, and the clerk of Greek Archdiocese of North and South America and members thereof, for the
purpose of procuring the incorporation of the same pursuant to
the provisions of section 15 of the Religious Corporations Law
of the State of New York, execute and acknowledge this certificate as follows:

1st- The said Greek Archdiocese of North and South America
is an unincorporated governing and advisory body of the Greek
Orthodox Church having jurisdiction over and relations with
several Greek Orthodox Churches, some of which are located in
the State of New York.

2nd- At a meeting of the said body, at which a quorum was
present, duly held in the Holy Trinity Hellenic Orthodox Church
located at 153 East 72nd Street in the Borough of Manhattan in
the City of New York on the 15th day of September, 1921, it,
by a resolution duly adopted by it, determined to become incorporated under the laws of the State of New York under the name,
GREEK ARCHDIOCESE OF NORTH AND SOUTH AMERICA, Incorporated.

3d - At the said meeting the said body, by a plurality vote
of its members, also elected nine persons, whose names and post-
office addresses are hereinafter stated, to be the first trustees of such corporation.

4th- The name by which such corporation is to be known is:
GREEK ARCHDIOCESE OF NORTH AND SOUTH AMERICA
Incorporated

5th - The objects for which such corporation is to be formed are:

To edify the religious and moral life of the Greek Orthodox Christians in North and South America on the basis of the Holy Scriptures , the rules and canons of the Holy Aposteles and of the seven Oecumenical Councils of the ancient undivided church as they are or shall be actually interpreted by the Great Church of Christ in Constantinople and to exercise governing authority over and to maintain advisory relations with Greek Orthodox Churches throughout North and South America and to maintain spiritual and advisory relations with synods and other governing authorities of the said church located elsewhere.

6th - The number of trustees of such corporation shall be nine.

7th- The names and post office addresses of the first trustees of such corporation, as elected by the said body, as aforesaid , are as follows:

Names	Post-Office Addresses
ALEXANDER, BISHOP OF RODOSTOLOU,	140 East 72nd Street New York City;
REV. METIODIOS KOURKOULIS,	1030 Beverly Road, Brooklyn, N.Y.;
REV. DEMETRIOS KALLIMACHOS,	64 Schermerhorn Street, Brooklyn, N. Y.;
REV. STEPHANOS MAKAFONIS,	259 West 94th Street, New York City;
REV. GERMANOS POLYZOIDES,	339 East 86th Street, New York City;
LEONIDES CALVOKORESSI,	11 William Street, New York City;
PANAGIOTIS PANTEAS,	3904 - 3d Avenue, Brooklyn, N.Y.;
GEORGE KONTOMANOLIS.	100 West 36th Street, New York City;
ALEXANDER ALEXION	152 West 82nd Street, New York City.

8th - The principal office of such corporation is to be located in New York County in the City of New York in the State of New York; but branch offices thereof may be maintained and its functions may be exercised in any part of North or South America.

I n t e s t i m o n e w h e r e o f, we have heretoo set our hands and affixed our seals in the City of New York on thi s 17th day of September, 1921.

MELETIOS METOXAKIS, ARCHBISHOP OF ATHENS

President

GERMANOS POLYZOIDES, ARCHDEACON

Clerk

State of New York

New York County;

On this 17th day of September 1921, personally appeared

GERMANOS

to be

nstrument,

same.

County

March 30,

of New York,
of New York, } ss.:

I, WILLIAM F. SCHNEIDER Clerk of the said County and Clerk of the Supreme Court of said State for said County, Do CERTIFY, That I have compared the preceding with the original certificate of Incorporation of

Greek Archdiocese of North and South America

on file in my office, and that the same is a correct therefrom, and of the whole of such original.

Indorsed Filed. Recorded Sept 19 - 1921 3.43 P.M.

IN WITNESS WHEREOF, I have hereunto subscribed my name and affixed my official seal, this 9 day of April 1924

James Donegan Clerk.

Certificate of incorporation of the Greek Orthodox Archdiocese of North and South America, establishing the Archdiocese in New York State as a recognized religious institution, September 17, 1921.

Τῶν ὀρθοδόξων παροικιῶν τῶν ἔξω τῶν κανονικῶν ὁρίων ἑκάστης τῶν ἐπὶ μέρους Ἁγίων τοῦ Θεοῦ Ἐκκλησιῶν εὑρισκομένων τὴν ποιμαντικὴν διακυβέρνησιν τῷ Ἁγιωτάτῳ Ἀποστολικῷ Πατριαρχικῷ Οἰκουμενικῷ θρόνῳ αἱ κανονικαὶ διατάξεις καὶ ἡ διὰ τῶν αἰώνων κρᾶσις τῆς Ἐκκλησίας ἀναγράφουσιν.

Ἐπειδὴ δὲ ἀνὰ τὴν Ἀμερικὴν Βόρειόν τε καὶ Νότιον ὀρθόδοξοι παροικίαι, ἀσύντακτοι τέως καὶ ἀσυνάρτητοι, ἔστι δ' ὅπου καὶ ἐν ἀκωλύσει ἀπὸ τῆς παραδεδομένης ὑπὸ τῶν Πατέρων κανονικῆς τάξεως διατελοῦσαι, ἐκρίθησαν δέον γενεσι ἐκκλησιαστικῆς νομοθετικῆς χρονείας, πρὸς τὸ "πάντα εὐσχημόνως καὶ κατὰ τάξιν γίγνεσθαι", κατὰ τὴν τοῦ μακαρίου Παύλου ἐντολήν, ἡ μετριότης ἡμῶν μετὰ τῶν περὶ ἡμᾶς Ἱερωτάτων Μητροπολιτῶν καὶ ὑπερτίμων, τῶν ἐν Ἁγίῳ Πνεύματι ἀγαπητῶν ἡμῖν ἀδελφῶν καὶ συλλειτουργῶν, συνοδικῶς περὶ τῆς ἀνάγκης διασκεψάμενοι, ἔγνωμεν τὰς ὀρθοδόξους ταύτας παροικίας εἰς μίαν Ἐκκλησιαστικὴν Ἐπαρχίαν συναγαγεῖν, τίτλον μὲν Ἀρχιεπισκοπῆς φέρουσαν, εἰς ἐπισκοπικὰς δὲ δικαιοδοσίας ὑποτεμνομένην, ἐν ἐξαρτήσει δὲ κανονικῇ ἀπὸ τοῦ Ἁγιωτάτου Ἀποστολικοῦ καὶ Πατριαρχικοῦ Οἰκουμενικοῦ θρόνου ὡς μία τῶν αὐτοῦ Μητροπόλεων διατελοῦσαν.

Διὰ τοι τοῦτο καὶ ἐν Ἁγίῳ Πνεύματι συνοδικῶς ἀποφαινόμενοι ὁρίζομεν ὅπως ἀπὸ τοῦ νῦν πᾶσαι αἱ ἐν Ἀμερικῇ Βορείῳ καὶ Νοτίῳ εὑρισκόμεναι ὀρθόδοξοι παροικίαι καὶ οἱαιδήποτε ὀρθόδοξοι ἐκκλησιαστικαὶ συσσωματώσεις αἵ τε νῦν ὑφιστάμεναι καὶ αἱ ἐν τῷ ἐφεξῆς χρόνῳ ἱδρυθησόμεναι γνωρίζωνται ὡς μέλη σώματος ἑνός, ὅπερ ἐστὶν ἡ "Ὀρθόδοξος Ἀρχιεπισκοπὴ Ἀμερικῆς Βορείου καὶ Νοτίου".

Ἡ Ἀρχιεπισκοπὴ δὲ αὕτη, μία οὖσα τῶν Μητροπόλεων τοῦ Ἁγιωτάτου Ἀποστολικοῦ καὶ Πατριαρχικοῦ Οἰκουμενικοῦ θρόνου Κωνσταντινουπόλεως, τὴν δεκάτην πέμπτην θέσιν ἐν τῷ Συντάγματι κατέχουσα, θεμελιοῦται μὲν ἐπὶ τῶν δογμάτων καὶ τῶν ἱερῶν Ἀποστολικῶν καὶ Συνοδικῶν Κανόνων τῆς μιᾶς Ἁγίας Ὀρθοδόξου Καθολικῆς καὶ Ἀποστολικῆς Ἐκκλησίας, διασῴζει δὲ στενὸν τὸν μετὰ τοῦ Οἰκουμενικοῦ θρόνου, ὡς καὶ αἱ λοιπαὶ αὐτοῦ Μητροπόλεις, σύνδεσμον, ἡρμοσμένη πρὸς αὐτὸν ὡς μέλος ζῶν πρὸς κεφαλὴν σώματος ζῶντος.

Διὰ τὴν ἀνάγκην δὲ πληρεστέρας ποιμαντορίας ὑπὲρ τοῦ περιουσίου λαοῦ χρονείας καὶ εὐκαρπεστέρας ἐπιτοπίου ἐκκλησιαστικῆς διοικήσεως ἡ ὅλη Ἐκκλησιαστικὴ Ἐπαρχία, ταύτην εἶναι ἡ τῆς ὅλης Ἀρχιεπισκοπῆς περιφέρεια, ὑποτέμνεται εἰς ἐπισκοπικὰς δικαιοδοσίας ἐχούσας τὸν ἴδιον αὐτῆς ἑκάστην ἐπίσκοπον, τοιαύτας δὲ συνιστῶμεν ἐν τῷ παρόντι τέσσαρας, ἤτοι τὴν τοῦ Ἀρχιεπισκόπου αὐτοῦ ὡς ἑνὸς τῶν ἐπισκόπων θεωρουμένου, περιλαμβάνουσαν ἀπαραιτήτως τὰς πόλεις Φιλαδελφείαν τῆς Πανσυλβανίας καὶ Οὐάσιγκτωνα τὴν ἕδραν τῆς Κυβερνήσεως τῶν Ἡνωμένων Πολιτειῶν, μετὰ ταύτην δὲ τὰς ἐπισκοπὰς Σινάλου, Βοστώνης καὶ Καλλιφορνίας.

Ὁ Ἀρχιεπίσκοπος μετὰ τῶν ἐπισκόπων συγκροτεῖ Σύνοδον Ἐπαρχίας συνερχομένην ἀπαραιτήτως δὶς τοῦ ἔτους πρὸ τοῦ Πάσχα καὶ κατὰ τὸ φθινόπωρον, κατὰ τὴν διάταξιν τῶν σχετικῶν κανόνων, ὅπου ἂν ὁρίσῃ ὁ Ἀρχιεπίσκοπος. Ἔχει δὲ ἡ Σύνοδος αὕτη πάσας τὰς ἐξουσίας καὶ τὰς εὐθύνας, ὅσας οἱ Ἱεροὶ Κανόνες τῇ "Συνόδῳ τῆς Ἐπαρχίας" ἀναγράφουσιν, βοηθεῖ δὲ σὺνόλογος ὅθεν ἐπὶ μὲν τῇ περὶ τὸν Πατριάρχην Συνόδῳ τῶν Μητροπολιτῶν τοῦ Οἰκουμενικοῦ θρόνου. Ἐκ τῶν δικαιωμάτων καὶ καθηκόντων αὐτῆς ἐστι τὸ ψηφίζειν τε τὸν Ἀρχιεπίσκοπον καὶ τοὺς ἐπισκόπους ἐν περιπτώσεσι χηρείας τῶν θρόνων, μετὰ δὲ τὴν ἀποκατάστασιν εὐθύνει ὅτι τῶν πρώτων κανονικῶν κοιμωμένων ἐξ ἀποφάσεως τῆς περὶ ἡμᾶς Ἁγίας Συνόδου, ὑποβάλλουσι δὲ τὴν ψῆφον πρὸς ἔγκρισιν τῷ Πατριάρχῃ, ὃς χειροτονεῖ μὲν τὸν Ἀρχιεπίσκοπον, ἐκδίδωσι δὲ ἄδειαν πρὸς χειροτονίαν τῶν ἐπισκόπων. Προβάλλονται δὲ τῇ ψήφῳ τῆς Συνόδου οἱ διὰ τοὺς χηρεύοντας θρόνους κρινόμενοι ἐπὶ τοῖς πονίμοις ἐπὶ τῶν κληρικῶν καὶ λαϊκῶν Ἀποστολικῶς συνελέχθησαν Ἐκκλησίας, λαμβανόμενοι ἐκ καταλόγου τοῦ δὶς κατὰ τὸ ἔτος ἡ Σύνοδος ἐκδίδωσιν προεγκεκριμένου ὑπὸ τῆς περὶ τὸν Πατριάρχην Ἁγίας Συνόδου. Εἰ δὲ καὶ μετὰ τὴν πρώτην ἐξ ἀποφάσεως τῆς περὶ ἡμᾶς Ἁγίας Συνόδου κληρωσι τῶν ἐπισκοπικῶν θρόνων συμβαίνει ἐλλιγμένους εἶναι τῶν τριῶν τοὺς ἐπισκόπους, αἱ κανονικαὶ ψῆφοι ἐπὶ ἐκλογὴ Ἐπισκόπου γενέσθωσαν ἐν τῇ περὶ τὸν Πατριάρχην Συνόδῳ, συμψήφων γιγνομένων καὶ τῶν ἐν Ἀμερικῇ Ἐπισκόπων.

Ἕκαστος τῶν Ἐπισκόπων κοιμαίνων τῶν ἑαυτοῦ παροικίαν ἔχει τὰς ἐξουσίας καὶ τὰς εὐθύνας, ὡς οἱ θεῖοι καὶ ἱεροὶ Κανόνες καὶ ἡ μακραίων τῆς Ἐκκλησίας πρᾶξις τῷ ἐπισκοπικῷ ἀξιώματι ἀναγράφουσιν μετὰ καὶ τοῦ ἐν τῷ Ἱερῷ Συνθρόνῳ ἐγκαθιδρύεσθαι. Εἰδικώτερον δὲ μνημονευτέον τῶν δικαιωμάτων καὶ καθηκόντων ἑκάστου Ἐπισκόπου τοῦ ἁγιάζειν καὶ τῇ λατρείᾳ καθιεροῦν τοὺς ναοὺς καὶ εὐκτηρίους οἴκους, τοῦ ἀποκαθιστᾶν ἐν αὐτοῖς ὅθεν λειτουργοῦντας καὶ ψάλλοντας ἱερεῖς, διακόνους καὶ λοιπὸν κληρικούς, τοῦ ἐπιμελεῖσθαι τοῦ εὐθηγησσῶνα καὶ κατὰ τάξιν καὶ πρὸς τὸ κοινὸν ὄφελος συμφέρον διοικήσεως πάντων τῶν ἐν αὐτοῖς, τοῦ τελεῖν τε τὸ μυστήριον τοῦ γάμου καὶ τὰ διαζευκτήρια γράμματα τοῖς ἐκ τούτων ἁρμοδίως καὶ διαλυθέντων κηρυσσθέντων, τοῦ διανέμειν τοῖς ἱερεῦσι τὸ Μύρον τοῦ Ἁγίου Χρίσματος λαμβανόμενον διὰ τοῦ Ἀρχιεπισκόπου παρὰ τοῦ Οἰκουμενικοῦ Πατριάρχου.

Ἴστω δὲ παρ' ἑκάστῳ Ἐπισκόπῳ Δικαστήριον Πνευματικὸν ἐκ δύο γε βαθμούχων ἱερέων καὶ τοῦ Ἐπισκόπου αὐτοῦ ἢ τοῦ νομίμου ἀναπληρωτοῦ αὐτοῦ συγκροτούμενον ἐπὶ πρωτοδίκῃ ἐκδικήσει πάντων τῶν κανονικῶν παρεκτραπήσων τοῦ κλήρου καὶ τοῦ λαοῦ, ἐν οἷς καὶ τὰ συντρέχοντα εἰς λύσιν τῶν γαμικῶν δεσμῶν κατὰ τοὺς ἐν τῷ Οἰκουμενικῷ Πατριαρχείῳ ἰσχύοντας λόγους διαζυγίου. Τῶν πρωτοδίκων δὲ τούτων Δικαστηρίων αἱ ἀποφάσεις ἐκκαλοῦνται ὡς εἰς δευτέρειον νευματικῶν πρὸς τὴν περὶ Ἀρχιεπίσκοπον Σύνοδον τῆς Ἐπαρχίας καὶ ἀπὸ ταύτης πρὸς τὴν περὶ τὸν Πατριάρχην Σύνοδον τῶν Μητροπολιτῶν ὡς πρὸς ἀνωτέριον.

Ὀφείλουσι δὲ πάντες τὰς τελετὰς οἱ μὲν ἱερεῖς καὶ διακόνου τοῦ Ἐπισκόπου αὐτῶν, οἱ δὲ Ἐπίσκοποι τοῦ Ἀρχιεπισκόπου, ὁ δὲ Ἀρχιεπίσκοπος τοῦ Πατριάρχου, δι' οὗ ἡ κοινωνία μετὰ πάσης Ἐπισκόπου Ὀρθοδόξου τῶν ὁρθοτομούντων τὸν λόγον τῆς τοῦ Χριστοῦ Ἀληθείας.

Τῶν διατάξεων τούτων ὡς βάσεων ἀκαρασαλεύτων φυλαττομένων, ὁ Ἀρχιεπίσκοπος καὶ οἱ Ἐπίσκοποι μετὰ κλήρου καὶ τοῦ λαοῦ ἔχουσι τὸ ἐλεύθερον συσσωματῶσαι τὴν ἐν Ἀμερικῇ Ὀρθόδοξον Ἐκκλησίαν κατὰ τοὺς νόμους ἑκάστης τῶν Πολιτειῶν ὑπὸ τὴν ἀπαραίτητον ὅρον ὅτι οὐδεμία διάταξις τοῦ πολιτικῶν τῶν πολιτικῶν κανονισμῶν ἔσται ἀντικειμένη ᾗ διδασκαλίᾳ ἢ τοῖς ἱεροῖς Κανόσι τῆς Ἁγίας Ὀρθοδόξου Ἐκκλησίας ὡς οὗτοι εἰσιν ἡρμηνευμένοι ἐν τῇ πράξει τοῦ Οἰκουμενικοῦ Πατριαρχείου, οὗ ἕνεκεν καὶ οὐδεὶς μὲν ἐνοριακὸς κανονισμὸς ἔσται ἔγκυρος ἐὰν μὴ φέρῃ τὴν ἔγκρισιν τοῦ ἁρμοδίου Ἐπισκόπου ἢ τῆς Ἐπαρχιακῆς Συνόδου, οὐδεὶς δὲ κανονισμὸς Ἐπισκοπῆς ἢ τῆς ὅλης Ἀρχιεπισκοπῆς τεθήσεται ἐν ἰσχύϊ ἐὰν μὴ κρατερὸν κεκυρωμένος ὑπὸ τοῦ Πατριάρχου.

Παραγγέλλομεν δὲ πατρικῶς ἐν τῷ Κυρίῳ καὶ διὰ λόγῳ τιμᾶν τοὺς Ἐπισκόπους αὐτῶν καὶ ὡς ἡγουμένους καὶ πνευματικοὺς πατέρας λόγον μέλλουσιν ἀποδοῦναι ὑπὲρ τῶν ψυχῶν αὐτῶν "πείθεσθαι καὶ ὑπείκειν", κατὰ τὴν τοῦ Ἀποστόλου ἐντολήν, ἵνα μετὰ χαρᾶς τὸ ἔργον τῆς διακονίας ποιῶσι καὶ μὴ στενάζοντες.

Ταῦτα οὕτω δόξαντα καὶ κριθέντα, ἐκυρώθησαν συνοδικῶς, εἰς μόνιμον δὲ αὐτῶν παραφυλακὴν ἐκδίδοται ὁ Πατριαρχικὸς καὶ Συνοδικὸς οὗτος Τόμος, καταστρωθεὶς καὶ ἐν τῷ Κώδικι τῆς καθ' ἡμᾶς Ἁγίας Μεγάλης τοῦ Χριστοῦ Ἐκκλησίας.

Ἐν τοῖς Πατριαρχείοις τῇ 17 Μαΐου 1922.

† ὁ Πατριάρχης Κωνσταντίνος Μελέτης ἀποφαίνεται

† ὁ Πισιδίας Βασίλειος.

† ὁ Σερρῶν Λεόνδιος   † ὁ Σαμψουντος Γρηγόριος

† ὁ Ἀμασείας Γερμανός   ὁ Χαλδίας-Κερασοῦντος Λαυρέντιος   † ὁ Βιζύης Ἀνθιμος

† ὁ Ἡλιουπόλεως Ἑμμανουήλ

ὁ Ἀρχιγραφεὺς

† ὁ Κυζίκου Ἀθανάσιος

Damaskinos of Corinthos

His Contributions to the Development of the Archdiocese

By
George Papaioannou

A. INTRODUCTION

he life of the first ten years of the Greek Archdiocese of North and South America resembles a Greek tragedy. In this modern Greek tragedy, the protagonist and victim is the Greek Orthodox population of America. Among its other characters are the hierarchy, the Ecumenical Patriarchate, the Church of Greece, the Greek government and the Greek-American press. Finally, the role of the saving catalyst is played by an outstanding personality, the Metropolitan of Corinth Damaskinos, who was dispatched to America as the Exarch of the Ecumenical Patriarchate to restore ecclesiastical order.

What were the reasons for this turmoil in the young Archdiocese that necessitated outside interference for the restoration of its peace? Other contributions in this volume deal with major developments that preceded the era covered by this paper—an analysis of the problem Metropolitan Damaskinos was asked to solve.

Although the first Greek Orthodox Church in the Americas was founded in 1864, and 29 church communities had been established by 1906, for more than forty years there was no episcopal authority in America. During that period of accelerating growth, those Greek communities tried to maintain direct ties with the mother churches but had to so without any real organization apart from the parish itself. Also adding to the potential for confusion was the fact that the priests who served the Greek-American parishes had come from Greece, yet they were canonically under the jurisdiction of the Ecumenical Patriarchate of Constantinople.

In 1908, the Ecumenical Patriarchate recognized the need

121

to establish episcopal authority, and its inability to provide it due to the difficult political conditions at the time. Through the promulgation of the Tomos of 1908, the Ecumenical Patriarchate transferred its rights to, and responsibilities for, the Greek-American community to the Church of Greece. After this act, it was expected the Greek Church would immediately start organizing the immigrant faithful into a unified body under its jurisdiction by first appointing a qualified bishop to the Greek Diaspora in America. But months and years went by and this basic duty of the Greek Synod was not fulfilled. The delay in the appointment of a bishop continued to keep the Church without a leader. Continued pleas for the assignment of a guiding head went unanswered.

This was the ecclesiastical situation in America when the King Constantine - Eleftherios Venizelos controversy (which brought the latter to power) erupted in Greece. The controversy was political; nevertheless, the Greek Church, being a state church, was hopelessly involved. The conflict brought tragic consequences not only upon the faithful in Greece but also upon the Greek Diaspora, especially in the United States. Ironically, however, the political turmoil in Greece had some beneficial effects on the Greek Orthodox Church in America. The new pro-Venizelos Metropolitan of Athens, Meletios Metaxakis, in the midst of his problems in Greece showed great concern for the Greek Diaspora in America. He visited the United States twice. In his first visit, August, 1918, he spent 82 days and brought some order out of the existing chaos. He installed Bishop Alexandros of Rodostolou as the "Synodikos Epitropos" to govern the Church.

In Metaxakis' second visit, August, 1921, after he was ousted by the Royalist Government as Metropolitan of Athens, he initiated the work for the establishment of the Greek Archdiocese of North and South America.[1] By a twist of fate, while still in America, Metaxakis was elected Ecumenical Patriarch.[2] In his new position, Patriarch Meletios reversed the Tomos of 1908 by which the Greek-American Church was placed under the jurisdiction of the Holy Synod of Greece, because its stipulations had not been fulfilled.[3] On May 11, 1922, the Greek Archdiocese of North and South America was canonically created and its legal establishment formally ratified by the Ecumenical Patriarchate. The Synodikos Epitropos, Bishop Alexander, was chosen as its first Archbishop.[4]

While this was taking place in America and Constantinople,

Archbishop Damaskinos of Corinthos, attending a service at Canterbury Cathedral in March 1945.

conflicting events occurred in Greece. The promonarchist Government of Greece and the Royalist Holy Synod under Metropolitan Theoklitos, reacted rather violently: First, by ordering Bishop Alexander to return to Greece, and second, by sending the Metropolitan of Monemvasia and Lacedaemon, Germanos Troianos to the United States to take over those communities that remained loyal to the King and the Church of Greece. Although the majority of the parishes followed Alexander, a great number of parishes and individuals remained fanatically devoted to the Royalist cause. The two factions struggled constantly to sway the communities to their side even through court proceedings. Peter Kourides described this period in very somber tones: "During this interval our Churches and communities had become divided battlegrounds in which Venizelists and Royalists fought even within the sanctuary of the Holy Altar. Police were stationed at strategic positions within some of the Churches to actually prevent bloodshed and of course these shameful and disgusting incidents were duly reported on the front pages of the American press to the awful humiliation and irreparable damage of the Greek people throughout the country."[5]

B. A MOVEMENT TOWARD PEACE

In 1923, Chrysostomos Papadopoulos, a respected hierarch of moderate political views and a renowned scholar, was elected Metropolitan of Athens. Chrysostomos immediately began working on the restoration of the relations between the Mother Church (the Ecumenical Patriarchate of Constantinople) and the Church of Greece. Although the climate between Constantinople and Athens improved, the American situation worsened. The royalist Metropolitan Germanos Troianos returned to Greece but another controversial personality, Vasilios, Metropolitan of Chaldea, a fanatic Royalist, made his appearance on the American scene. On February 13, 1924, Archbishop Alexander telegraphed the Ecumenical Patriarchate asking for the severest punishment of Vasilios. On May 10, 1924, the Holy Synod of Constantinople responded by reducing Vasilios to the rank of the laity.[6] Vasilios, however, instead of losing authority after his down grading, became the hero of the Royalists. He proclaimed himself as the head of the Greek Orthodox Churches in the United States and Canada.

The opposition to Alexander grew. Not only Royalists but many of his former Venizelist friends turned against him. An unusual "holy alliance" led by the two dailies, the pro-Royalist "Atlantis" and the pro-Venizelist "National Herald", demanded nothing less than the replacement of Alexander.[7]

The situation became so deplorable that it demanded urgent solution. In response to the cries of despair from all sides, Church leaders in Athens and Constantinople, joined forces with the political leaders of both political parties in Greece to give top priority to the solution of the Greek-American problem. It was this joint effort that set the stage for the restoration of harmony in the Greek-American community. Patriarch Photios, in agreement with Athens and Constantinople, designated Metropolitan Damaskinos of Corinth, one of the most capable and respected hierarchs of Greece, as his Exarch to the United States.

C. ABOUT THE MAN

The man who was destined to bring peace to American-Hellenism and later to serve the people of Hellas both as head of the Church and during the most difficult days of the Civil War, following the German occupation, as the Regent, was born in 1890 in the village of Dorvitsia of the Province of Nafpatia. He came from an obscure and poor family. His uncle, Christophoros Papandreou, who was then the abbot of the Monastery of Koroni, placed the young Damaskinos under his patronage and took care of his education.

Following his graduation from high school, he was admitted to the University of Athens where he studied theology and law concurrently, completing both with very high honors.

In 1917 he was ordained deacon and received his assignment as managing secretary of the Offices of the Athenian Archdiocese, a position that gave him opportunity to demonstrate his tremendous organizational ability and administrative acumen. This was an exceptionally abnormal period for Greece pregnant with political and ecclesiastical upheavals. Damaskinos managed to remain above the divisions and factions, earning the respect and admiration of both ecclesiastical and political sides, the Venizelists and the Royalists.

During the time he served in the Athenian Archdiocese, he organized the "Pan-Clerical Union" that brought together the best elements the Greek clergy had to offer.[8] In 1918 he was ordained archimandrite. He was assigned by Metropolitan Meletios Metaxakis as abbot of the two most prestigious monasteries, Petraki and Penteli, and helped restore their organizational and financial integrity.

By this time the integrity, intellectual acumen and electrifying personality of the young cleric had been recognized and acknowledged by both the church and state authorities in Greece. When the Government, the Church of Greece, and the Ecumenical Patriarchate responded to the call of the monastic community of Mount Athos for a change of its administrative structure, by mutual agreement, they appointed Damaskinos to study the situation and write the new Charter.[9] Having accomplished so much in such a brief span of service to the Church, it was to no one's surprise when the hierarchs of Greece elected Damaskinos Metropolitan to fill the vacancy in the historic and apostolic Metropolis of Corinth, on December 16, 1922, at the age of 32; he was one of the youngest clergymen ever to be elected Bishop of such an important diocese.[10]

In Corinth he labored as no other churchman had, since the days of St. Paul. He built a majestic church dedicated to St. Paul, patron saint of the city and established an ecclesiastical school for the training of the clergy. What raised his reputation to the greatest heights however, was his compassion and love for the victims of the most terrible earthquake that occurred on the night of April 22, 1928 and his superhuman efforts to rebuild the city. Damaskinos was faced then with a tragic situation. The city was literally demolished; the young bishop saw the fruits of his labors destroyed. The task he was facing in reconstructing the city would have discouraged any lesser man, but Damaskinos proved to be the right man for the seemingly impossible task that lay ahead. His words of encouragement and sympathy to the victims soon translated into works and deeds. He began a crusade to solicit funds on an international scale.

He was presented by the press of Greece as an example of a true shepherd, a hierarch completely dedicated to the welfare of his flock. His efforts were crowned with success. The international community responded to his plea; people helped. Thanks to him the city of Paul the Apostle, was rebuilt. The Greek-American community participated in Corinth's reconstruction. That assistance would be repaid later by Damaskinos with very high dividends. It was his expedition to rebuild Corinth that first brought Damaskinos to America. His visit to the U.S. for the solicitation of funds in 1928 brought him to every major city and gave him first-hand knowledge of the Greek-American problem. His majestic physical appearance and his other qualities, especially his reputation as a good shepherd, attracted the attention of many leaders in the Greek-American community. Many appealed to him to help bring the American church out of its chaos.

The two leading daily newspapers "Atlantis" and "National Herald" promised to put their antagonisms aside and give him their unqualified support if he would undertake the mission of reuniting the Greek-American community. The publisher of the "National Herald" talked about these overtures: "In the presence of Damaskinos, I telephoned my colleague, the editor-in-chief of the other daily ("Atlantis") and said: 'You will agree with me that we are hit by a terrible curse. The responsibility of the two dailies is awesome. . . . .The young generation is being turned away from both ethnic heritage and Orthodox faith. . . . .we are fading away ingloriously. . . .Do your cross and come to unite our efforts and in brotherly fashion, and in the name of God and love, let us save our martyred Mother Church.' "[11]

Damaskinos also visited the publisher of the other daily "Atlantis", who pleaded with him to return to America and try to save Hellenism.[12] The Greek hierarch was moved deeply by the pleas of the two publishers and other similar appeals by other Greek-American leaders. Upon his return to Greece he reported to the Greek Government, the Greek Church, and the Ecumenical Patriarchate of the deplorable situation in America. The three groups consulted and agreed to persuade Damaskinos to undertake this important and difficult task.

His mission was to take sole authority over the Church in America and submit a report to the Holy Synod of the Patriarchate with recommendations for the final solution of the problem.

D. THE ARRIVAL OF THE EXARCH AND THE REACTION OF THE HIERARCHY TO HIM

Damaskinos arrived in New York on May 20, 1930. His difficult task was somewhat simplified by the immediate endorsement of his mission by the two Greek-American dailies, "Atlantis" and "National Herald". Both expressed editorially their sympathy and trust in him and promised to assist in every way they could to bring love, peace and harmony to the Greek-

American community. "The American Omogeneia was waiting in agony the olive branch, and the Greek-American press enthusiastically saluted the dawn of a new period in the life and progress of American Hellenism."[13]

One has to remember that the Greek Archdiocese of North and South America was governed then by the Charter of 1922 that gave the Greek-American Church some form of autonomy. The governing body of the Church was the Synod of Bishops presided over by the Archbishop. No solution could be given to the problem without the cooperation of the hierarchy. Metropolitan Damaskinos, being a lawyer himself, was careful not to rush and force a solution that could entangle the Church in long court proceedings. A master tactician, Damaskinos visited first with embattled Archbishop Alexander who had never been receptive to the mission of the Exarch. What transpired between the two hierarchs? Damaskinos gives an account of the meeting which he describes as polite. "During my meeting with his Eminence Alexander, I used the best possible diplomacy and *persuasion* to make it understood that the Mother Church was determined to lift the ecclesiastical divisions in America and that no one could stop her from making this decision a reality."[14]

Damaskinos advised Alexander to rise above human weaknesses and like a true soldier of the faith, to give "an example of self-sacrifice".[15] He let it be known to Alexander that the Church would not let his (Alexander's) generous gesture go unnoticed and would reward it accordingly.[16] The opposite would happen if he was to refuse to give his unqualified support and cooperation to the Exarch.[17] Alexander listened and gave the impression that he clearly understood the message. Damaskinos felt that the Archbishop would eventually succumb to the inevitable: resign and thus open the way for the pacification of the Greek-American community.[18]

The events that followed did not prove Damaskinos right. Although Archbishop Alexander had promised the Exarch that he would respond to his recommendations as soon as he had the chance to consult his Bishops, he forgot his promise. In fact, after the meeting of the Synod of Bishops he issued an encyclical to the pastors and parish councils of the Archdiocese in which he defied the Exarch. Alexander declared that he remained the canonical Archbishop and would accept no interference from the outside. He accused Damaskinos of taking a dictatorial attitude and claimed that the Exarch had no authority to interfere with the administration of the Archdiocese. His mission was only to study the situation and then report the findings to the Ecumenical Patriarchate. In the same encyclical, Alexander pleaded with the parishes to show their support for him by sending telegrams of protest for the Exarch's actions to the Ecumenical Patriarchate, the Church of Greece and the Greek Government.[19]

It was rumored, that wishing to surprise Damaskinos with a *de facto* solution of the Greek-American problem, Alexander met secretly with his arch rival, the defrocked Bishop Vasilios of Lowell, leader of the Royalist Churches. According to the same rumors, the two rivals had exchanged the kiss of peace and agreed on a plan that would allow them and the other bishops to remain in America to compose the leadership of a united Archdiocese.[20]

In response to Archbishop Alexander's actions, Damaskinos on May 31, 1930 issued his first encyclical to the priests and the parish councils, stating the objectives of his mission and the authority vested in him by the Ecumenical Patriarchate as its Exarch to administer the affairs of the American Church.[21] The encyclical included three important documents. One was the patriarchal letter designating Damaskinos as his Exarch, the second was a letter from Patriarch Photios to the Greek people of the United States admonishing them: "As of this day you have endured the hardships of partisanship and disunion. Now try the goodliness of peace and you shall observe the difference between peace and partisanship, love and hate, brotherly affection and fratricidal strife."[22]

The third was a letter from Archbishop Chrysostomos of Athens to Damaskinos in which he indicated his fullest cooperation in his efforts to bring peace and tranquility to the faithful in the U.S.[23] Damaskinos then called upon the other three members of the Holy Synod, Philaretos of Chicago, Kallistos of San Francisco and Joakeim of Boston to separate their positions from Archbishop Alexander and submit to the wishes of the Mother Church.

E. THE COALITION CRACKS

On June 6, 1930 Bishop Joakeim of Boston, a moderate and farsighted hierarch, separated his position from Alexander and the other bishops in a dramatic fashion. In a letter to the Archbishop, that was released to the press and officially announced to his parishes, he stated: "I believe that it is my duty, instead of continuing to involve myself in fruitless discussions on the ecclesiastical problem, to submit to the decisions of the Exarch (the sole responsible representative of the Ecumenical Patriarchate) whatever they may be, even if they are unfavorable to me personally."[24]

In the same latter Joakeim strongly advised the Archbishop that he too, (Alexander), give the Exarch the same obedient cooperation and help bring all the forces of the Church in one unified Archdiocese that would eventually shelter not only the Greeks but all the Orthodox faithful on the American continent.[25] The other two bishops, Philaretos and Kallistos, behind the scenes advised Alexander to be more flexible and bargain with the Exarch for the best possible compromise. In public, however, they demonstrated their loyalty to the Archbishop, who being encouraged by a fanatic following desperately hoped that the ultimate victory was to be his. In a joint letter to Damaskinos, the two bishops wrote that although they were in sympathy with the Exarch's most difficult task and ready to cooperate with him for the peaceful solution of the ecclesiastical problem, they felt that the discussions should be held within the confines of the Charter of the Archdiocese; the responsible head of which was Archbishop Alexander.[26]

F. THE DISMISSAL OF ARCHBISHOP ALEXANDER

The joint letter of bishops Kallistos and Philaretos convinced Metropolitan Damaskinos that only the dismissal of Alexander could open the way for the normalization of the ecclesiastical situation. Having failed in all his previous attempts to persuade the embattled Archbishop to resign voluntarily, Damaskinos was forced to ask the Patriarchate to dismiss the physically and emotionally demolished but amazingly stubborn churchman. The unaninous decision of the Holy Synod was announced to Alexander through its Exarch in a telegram on June 18, 1930. The dismissal of Alexander brought to a conclusion the first part of the mission of the Exarch.

G. THE REPLACEMENT OF ALEXANDER AND THE PLAN FOR THE REPATRIATION AND REASSIGNMENT OF ALL BISHOPS

Although stripped of his ecclesiastical authority, Alexander refused to accept the decision of his dismissal as final. He was still hoping to prevail.

Damaskinos acted in a masterly fashion, however, to shatter this last hope of Alexander by recommending the swift appointment of a new Archbishop. "It was necessary to be made clear to everyone, especially to His Eminence Alexander, that the decision of the dismissal was irrevocable and that the Church was prepared to face with determination even the most extreme consequences of that decision. In other words, it was imperative to close the road leading to any hope of Alexander regaining the Archbishop's throne. This could not be accomplished in any other way except by the appointment of a permanent Archbishop."[27]

The swift appointment of the new Archbishop was necessitated also by the attempts of Alexander and his friends to discredit the mission of the Exarch by circulating false rumors that he sought the dismissal of the Archbishop for personal gains. Damaskinos wished also, with the appointment of the new Archbishop, to discourage his friends who were pressing him to remain in America. "The swift resolution of this problem was necessitated also by the fact that the true intentions of the Exarch be known to all because even in this instance there were those who attributed the dismissal of Alexander in his (Exarch's) selfish desire to remain in America. . . I had to confirm with decisive deeds here what I had summoned to the authorities in Constantinople that only if I reached the lowest ebb of morality could I even discuss my staying in America."[28] On the 8th of August, 1930, Damaskinos telegraphed Patriarch Photios recommending the appointment of Metropolitan Athenagoras of Corfu and Paxoi as Archbishop of North and South America. On the 13th of August of the same year the Patriarchal Synod responded favorably to the Exarch's request by unanimously electing Athenagoras.

The new Archbishop was a very dynamic hierarch who had already distinguished himself in the service of the Church but was little known outside of Greece, especially in America. For this reason there were many who expressed their displeasure for the appointment especially among the friends and admirers of Damaskinos who were hoping that he would remain permanently as the head of the Archdiocese. Damaskinos wrote about the dissatisfaction that the appointment of Athenagoras caused his friends and of his own efforts to erase any doubts about the ability of the man whom he had so strongly recommended: "At this point I must add that this appointment at the beginning did not sound very well in the soul of the Omogeneia which was hoping that the Exarch would reconsider his original decision and remain permanently in America. Nevertheless, every effort was made by the Exarch to convince the Greek-American community of the extraordinary qualities of the new Archbishop and the bright future that his election symbolized for Greek Orthodoxy in America."[29]

The disappointment of the Greek-Americans on the election of Athenagoras is confirmed by Demetrios Kallimachos, the venerable editor-in-chief of the "National Herald": "I went to bid farewell to Damaskinos at his hotel. I must confess that I challenged angrily his decision to leave so soon. But the Patriarchal Exarch calmed me down. He told me to write in my memoirs these encouraging words: 'The peace that we achieved together will never be in jeopardy again. The man whom I recommend as your Archbishop is superior to me. Archbishop Athenagoras will be able to reorganize the Archdiocese, and his accomplishment will make you very proud. And you will also bless me for recommending him.' "[30]

The plan of Damaskinos was working perfectly. As far as Alexander was concerned, the Exarch's thinking was right. The election of Athenagoras erased all hopes for him to regain his throne. Now it was the time for the Exarch and the Mother Church to show magnanimity to him and the other bishops by reassigning them to vacant dioceses in Greece. Damaskinos offered Alexander, and he gladly accepted, the appointment to the Metropolis of Corfu and Paxoi which became vacant with the election of Athenagoras as Archbishop of North and South America.

On September 12, 1930, Alexander visited Damaskinos and in a dramatic fashion surrendered everything that had been left of his authority as Archbishop. The visit concluded with a joint communique by the two men urging the faithful to accept the decision of the Mother Church and work together for peace and harmony under the leadership of Archbishop Athenagoras. Alexander signed the communique as former Archbishop of North and South America.

Having solved Alexander's problem with his assignment to the Metropolis of Corfu, Damaskinos proceeded with the reassignment of his bishops. Bishop Philaretos of Chicago became Metropolitan of the Islands of Syros and Tinos; Joakeim of Boston was assigned to the Metropolis of Fokis. At the request of Archbishop-Elect Athenagoras, Bishop Kallistos of San Francisco was permitted to remain in America. Damaskinos felt that the solution to the ecclesiastical problem would give satisfaction to all sides, including the Royalist segment, by restoring their defrocked hero, Metropolitan Vasilios. At his (Damaskinos') recommendation, Vasilios was restored and assigned to the Metropolis of Drama. The mission of the Exarch seemed to be coming to a successful end; yet Damaskinos was not so comfortable with the ecclesiastical situation. He believed that in order to prevent a new fragmentation of the Church in the future, some additional drastic changes had to be made.

H. THE LIFTING OF THE AUTONOMY AND THE PROPOSED NEW CHARTER

Historians have failed to credit Exarch Damaskinos with one of his greatest accomplishments—restructuring the governance of the Greek Archdiocese. Damaskinos was convinced that the chaotic situation of the young American Church was a by-product of the political divisions and the system of government. The elimination of the political divisions did not mean a permanent solution to the ecclesiastical problem. The system of government also had to be changed. The autonomy granted by the Charter of 1922 had not been proven beneficial. The Synodic system and the independent dioceses had to be eliminated because the variety of the administrative systems created administrative anarchy.[31]

Damaskinos saw another benefit in again making the Greek Archdiocese of the Americas a province of the Ecumenical Patriarchate: The Archdiocese would be first in prestige and honor among the other Orthodox branches in the New World. Damaskinos envisioned the day when the other Orthodox branches would see the benefits and place themselves under the wings of this great Patriarchal Province. Also the new status of the Greek Archdiocese would raise the honor and prestige of the Ecumenical Patriarchate in the Western Hemisphere.[32] Damaskinos wrote a draft for the proposed new Charter and submitted it to the Ecumenical Patriarchate and the Greek Government. The revised Charter of 1931 is based on this draft.[33]

Thus Damaskinos went beyond his original mission. Not only did he succeed in restoring peace in the Greek American

133

community but supplied the Church with a centralized system of government that suited magnificiently the man he chose to lead the Greek-American faithful, in harmony, to great heights of success.

The service of Damaskinos to the Greek-American Archdiocese was the briefest of all major hierarchal personalities; yet the value of his contribution is unique and indispensable. Not only did he bring the young, tormented, Church out of chaos and restore peace and tranquility to American Hellenism, but also set the foundations upon which Athenagoras, Michael, and Iakovos built today's magnificent structure. Without the careful preparation of the ground by Damaskinos and his brilliant recommendations to the Ecumenical Patriarchate on the appointment of Athenagoras as Archbishop and the drastic changes in the Charter that abolished autonomy and brought the Archdiocese under the direct jurisdiction of the Mother Church of Constantinople, it would have been impossible for us to have achieved the progress we enjoy today. Damaskinos departed for Greece on the 9th of February, 1931. He took with him the gratitude of the Greek-American faithful for bridging the chasm that had divided Hellenism and moreover for setting the pace for the continuous advancement of the Greek Orthodox faith on the American Continent. His American mission was not the end of his career or ministry— other great episodes were to follow during the war years and the civil war period of Greece. However, Damaskinos will forever be remembered in the United States for carrying out a seemingly impossible task—the reconciliation of two great opposing forces that threatened to extinguish the development of Greek Orthodoxy in the New World.

FOOTNOTES

[1]Papaioannou, George, *From Mars Hill to Manhattan:* The Greek Orthodox in America under Athenagoras A (Light and Life Publishing Company, Minneapolis, MN 1976), p. 33.

[2]Papaioannou, pp. 33-34.

[3]The decision of the Patriarchal Synod was announced to Bishop Alexander by Patriarch Meletios in a letter dated March 1, 1922, Greek Archdiocese Archives.

[4]Patriarch Meletios: A letter to Bishop Alexander, May 17, 1922, Greek Archdiocese Archives.

[5]Kourides, Peter, *The Evolution of the Greek Archdiocese of North and South America* (New York: Cosmos Greek American Printing Co., 1959), p. 8.

[6]Patriarchal encyclical on Demotion of Metropolitan Vasilios, May 10, 1924, Protocol No. 20222, Greek Archdiocese Archives.

[7]Papaioannou, George, *"The Odyssey of American Hellenes"* (Unpublished Manuscript) pp. 124-129.

[8]Kollias, Sifis G., Arhiepiskopos-Antivasilefs Damaskinos o Apo Korinthias *(Archbishop-Regent Damaskinos—Formerly of Corinth)* Athens 1963, p. 32.

[9]The Charter makes Mount Athos a permanent part of Greece and ecclesiastically places it under the jurisdiction of the Ecumenical Patriarchate. See Kollias, Sifis, p. 33, and Konidaris, Gerasimos I., "Kathemerini" November 6, 1938.

[10]Erineos of Samos and Iaria, In "Ekklesia", November 5, 1949.

[11]Kallimachos, Demetrios, *"To orama tis frihtis anhonis"*, Vol. II (Argonaut, New York 1960-1962), p. 20.

[12]Kallimachos, p. 20.

[13]Damaskinos, Exarch, "Peri tis thothisis lyseos is to ekklesiastikon zitima tou ellinismou tis Amerikis" (Unpublished Report: Greek Archdiocese Archives), p. 1.

Elsewhere in the Report, Damaskinos calls the press his most valuable and indispensable ally in his quest for peace and reconciliation. p. 35.

[14]Damaskinos, p. 3.

[15]Damaskinos, p. 3.

[16]Damaskinos, p. 3.

[17]Damaskinos, p. 4.

[18]Damaskinos, p. 5.

[19]Alexander, Archbishop, An encyclical to the pastors and parish councils of the Greek Archdiocese of North and South America on the activities of the Exarch, May 26, 1930, Greek Archdiocese Archives.

[20]Zoustis, Vasilios, O en Ameriki Ellinismos ke e drasis aftou (New York: D.C. Divry, Inc. 1954), p. 205.

Damaskinos writes that the plan received initially the support of "Atlantis". Damaskinos "Peri tis thotheisis lyseos", p. 6.

[21]Damaskinos, May 31, 1930.

[22]Patriarch Photios, An encyclical to the Greek American faithful, April 9, 1930, Greek Archdiocese Archives.

[23]Chrysostomos, Archbishop of Athens, A letter to Metropolitan Damaskinos, May 3, 1930, Greek Archdiocese Archives.

[25]Joakeim, Bishop of Boston, A letter to Archbishop Alexander, June 6, 1930, Greek Archdiocese Archives.

[25]Joakeim, June 6, 1930.

[86]Bishop Kallistos of San Francisco and Philaretos of Chicago, A letter to Exarch Damaskinos, June 7, 1930, Greek Archdiocese Archives.

[27]Damaskinos, p. 16.

[28]Damaskinos, p. 17.

[29]Damaskinos, p. 18.

[30]Kallimachos, p. 21.

[31]Damaskinos, p. 27.

[32]Damaskinos, p. 32.

The same argument was used later by Athenagoras to convince the delegates of the Fourth Clergy-Laity Congress in New York, November 14, 1931, that eliminated the autonomy and made the Archdiocese of North and South America a province of the Ecumenical Patriarchate. See Papaioannou, pp. 60.62.

[33]Segments of the proposed Charter are to be found in Damaskinos, pp. 27-34.

Ἀριθμ. Πρωτ. 1022. Ἐν Ἀθήναις τῇ 19 Μαΐου 1924
Διεκπ. 414.

Ἡ
Ἱερὰ Σύνοδος τῆς Ἐκκλησίας τῆς Ἑλλάδος
Πρὸς
τὰς ἐν τῇ διασπορᾷ ὀρθοδόξους Ἑλληνικὰς Παροικίας.

Ἡ Ἱερὰ Σύνοδος τῆς Αὐτοκεφάλου Ἐκκλησίας τῆς Ἑλλάδος ἐκ γραμμάτων τοῦ Οἰκουμενικοῦ Πατριαρχείου ἐπληροφορήθη τὴν ὑπ᾽ αὐτοῦ ἵδρυσιν Ἀρχιεπισκοπῶν καὶ Μητροπόλεων ἐν ταῖς ἐν τῇ διασπορᾷ ὀρθοδόξοις Ἑλληνικαῖς παροικίαις ὑπὸ τὴν πνευματικὴν δικαιοδοσίαν τοῦ Πατριαρχείου ὑπαγομέναις, καὶ τὴν τούτου ἐπικαιρουμένην κατάρχων τοῦ Πατριαρχικοῦ Τόμου τοῦ 1908.

Ὅθεν ἐν τῇ συνεδρίᾳ τῆς 16ης Μαΐου ἐ. ἔ. ἡ Ἱερὰ Σύνοδος ἐξ ἀπείρου σεβασμοῦ πρὸς τὸ Οἰκουμενικὸν Πατριαρχεῖον ᾧ εὐπόθου τῆς ἠθικῆς αὐτοῦ ὑποστηρίξεως, συμφώνως ἄλλως τε καὶ πρὸς τὰς ἀρχαίας κανονικὰς διατάξεις, ἀπεφάσισε ν᾽ ἐπικυρώσῃ τὴν κατάρχων τοῦ Τόμου τοῦ 1908 καὶ τὴν ἐκ νέου ὑπαγωγὴν πασῶν τῶν ἐν τῇ διασπορᾷ ὀρθοδόξων Ἑλληνικῶν παροικιῶν ὑπὸ τὴν τοῦ Οἰκουμενικοῦ Πατριαρχείου δικαιοδοσίαν. Τὴν ἀπόφασιν δὲ αὐτῆς ταύτην ἐξαγγέλλουσα ἡ Ἱερὰ Σύνοδος, παραγγέλλει, ὅπως πάντες οἵ τε εὐλαβέστατοι κληρικοὶ καὶ οἱ εὐσεβεῖς λαϊκοὶ τῶν ἐν Εὐρώπῃ Ἀμερικῇ καὶ Καναδᾷ, ἐν Αὐστραλίᾳ καὶ ἀλλαχοῦ εὑρισκομένων ὀρθοδόξων παροικιῶν ἀναγνωρίζωσι τὰς πνευματικὴν αὐτῶν Ἀρχὴν τὸ Οἰκουμενικὸν Πατριαρχεῖον καὶ πρὸς αὐτὸ ἢ πρὸς τοὺς ὑπ᾽ αὐτοῦ καθισταμένους πνευματικοὺς ποιμένας ἀναφέρωνται, πειθόμενοι εὐπροθύμως ταῖς διατάξεσιν αὐτῶν.

Εἰδικώτερον δὲ ταῖς ἐν Ἀμερικῇ ὀρθοδόξοις Ἑλληνικαῖς παροικίαις θερμῶς συνιστᾷ ἡ Ἱερὰ Σύνοδος, ὅπως ἀναγνωρίζωσιν ὡς μόνην κανονικήν, τὴν ὑπὸ τοῦ Πατριαρχείου κατασταθεῖσαν Ἀρχιεπισκοπὴν καὶ τοὺς Ἐπισκόπους αὐτῆς, εὐπειθῶς καὶ ἐπι-

ξένωσι καὶ αἰτίας τῆς ἀξιοθρηνήτου διαιρέσεως αὐτῶν, ἥτις καὶ πρὸ τῆς
ὁριστικῆς τοῦ περὶ αὐτὰς πεπολιτισμένου καὶ θρησκευτικῶς ἀνε-
πτυγμένου κόσμου διαπομπεύσεως καὶ πεῖσαν πρόοδον αὐτῶν πα-
ρακωλύει. Ἄλλως τε οὐδὲν κῦρος δύνανται αἱ ἑκάστοτε πράξεις ἐκκλησιῶν ἀντικανονικαὶ ἄνευ ἀδελφῆς ἑτεροτοπικῆς εὐλογίας τελούμεναι.

Ταῦτα ἔχουσαι ὑπ' ὄψει αἱ εὐσεβεῖς ὀρθόδοξοι ἑλληνικαὶ παροικίαι τῆς διασπορᾶς, τοῦ λοιποῦ, δεῖ τῶν αὐτῶν ἐκκλησιασμικῶν
ζητημάτων, ὅσον ἡ εὐπερίστασις οὐχὶ πρὸς τὴν ἱερὰν Σύνοδον τῆς Ἐκκλησ-
σίας τῆς Ἑλλάδος, ἀλλὰ πρὸς τὸ Οἰκουμενικὸν Πατριαρχεῖον καὶ πρὸς
τοὺς ὑπ' αὐτοῦ καθισταμένους Ἀρχιερεῖς ἢ ἐξάρχους.

Ὁ δὲ Θεὸς τῆς εἰρήνης καὶ ἀγάπης δῴη ὑμῖν τὸ αὐτὸ φρονεῖν
καὶ ἑρμηνεύειν ἐν ἀλλήλοις εἰς ἐπίδοσιν τῆς Χριστιανικῆς πίστεως
πέλαγος ἀρετῆς καὶ εἰς δόξαν τοῦ Ἁγίου Αὐτοῦ ὀνόματος. Ἀμήν.

† Ὁ Ἀθηνῶν Χρυσόστομος Πρόεδρος.
† Ὁ Ζακύνθου Διονύσιος.
† Ὁ Φωκίδος Ἀμβρόσιος.
† Ὁ Κεφαλληνίας Δαμασκηνός.
† Ὁ Λαρίσης Ἀρσένιος.
† Ὁ Μεσσηνίας Μελέτιος.
† Ὁ Μαντινείας καὶ Κυνουρίας Γερμανός.
† Ὁ Πατρῶν Ἀντώνιος.
† Ὁ Σύρου, Τήνου κ.λ. Ἄνδρου Ἀθανάσιος.
† Ὁ Λευκάδος καὶ Ἰθείκης Δανιήλ.
† Ὁ Δημητριάδος Γερμανός.
† Ὁ Ὕδρας Προκόπιος.
† Ὁ Ναυπακτίας καὶ Εὐρυτανίας Ἀμβρόσιος.
† Ὁ Ὕδρας καὶ Σπετσῶν Προκόπιος.
† Ὁ Θηβῶν κ. Λεβαδείας Συνέσιος.
† Ὁ Καλαβρύτων κ. Αἰγιαλείας Τιμόθεος.
† Ὁ Ἄρτης Σπυρίδων.
† Ὁ Παρναξίας Ἱερόθεος.
† Ὁ Φθιώτιδος Ἰάκωβος.
† Ὁ Τρίκκης καὶ Σταγῶν Πολύκαρπος.
† Ὁ Μάνης καὶ Οἰτύλου Διονύσιος.
† Ὁ Κορινθίας Δαμασκηνός.
† Ὁ Κυθήρων Δωρόθεος.

+ ὁ Ἀκαρνανίας Κωνσταντῖνος.

+ ὁ Χελιδὼς Γρηγόριος.

+ ὁ Τρίκκης καὶ Ὀλύμπου Ἀνδρέας

+ ὁ Καρυστίας Παντελεήμων.

+ ὁ Ἤλιδος Ἀντώνιος.

+ ὁ Γόρτυνος καὶ Μεγαλοπόλεως Πολύκαρπος.

Ὁ Ἀ. Γραμματεὺς

Ἀρχιμ. Λεόντιος Παπαγιαννόπουλος.

Ἀκριβὲς ἀντίγραφον

Ἐν Ἀθήναις τῇ 25ῃ Ἰουλίου 1924.

+ ὁ Ἀθηνῶν Χρυσόστομος Παπαδόπουλος

ὁ Ἀ. Γραμματεὺς

Ἀρχιμ. Λεόντιος Παπαγιαννόπουλος

Μετὰ τῆς περὶ ἡμᾶς
Ἁγίας καὶ Ἱερᾶς Συνόδου
ἐγκρίναντες ἐπικυροῦμεν τὰ ἀκόλουθα:

ΣΥΝΤΑΓΜΑ

ΤΗΣ ΕΛΛΗΝΙΚΗΣ ΟΡΘΟΔΟΞΟΥ ΑΡΧΙΕΠΙΣΚΟΠΗΣ

ΑΜΕΡΙΚΗΣ ΒΟΡΕΙΟΥ ΚΑΙ ΝΟΤΙΟΥ

Ἄρθρον Α.

Ἡ Ἑλληνικὴ Ὀρθόδοξος Ἀρχιεπισκοπὴ Ἀμερικῆς Βορείου καὶ Νοτίου, ἀπο-
τελοῦσα Θρησκευτικὸν Σωματεῖον ὑπὸ τὸ ὄνομα ΕΛΛΗΝΙΚΗ ΑΡΧΙΕΠΙΣΚΟΠΗ ΑΜΕ-
ΡΙΚΗΣ ΒΟΡΕΙΟΥ ΚΑΙ ΝΟΤΙΟΥ, εἶνε Ἐπαρχία τοῦ Κλίματος τοῦ Ἁγιωτάτου Ἀποστολικοῦ καὶ Πατριαρχικοῦ Οἰκουμενικοῦ Θρόνου, ὅστις καὶ ἀποτελεῖ μέλος καὶ
τὴν Πρωτόθρονον Ἕδραν τοῦ σώματος τῆς Μιᾶς, Ἁγίας, Καθολικῆς καὶ Ἀποστολικῆς Ὀρθοδόξου Ἀνατολικῆς Ἐκκλησίας, ἧς Κεφαλὴ ὁ Χριστός, διοικεῖται δὲ
ἡ Ἀρχιεπισκοπὴ αὕτη ἐπὶ τῇ βάσει τοῦ παρόντος Συντάγματος καὶ τῶν ὑπ'
αὐτοῦ προβλεπομένων Κανονισμῶν.

Ἄρθρον Β.

Ἡ Ἑλληνικὴ Ἀρχιεπισκοπὴ Ἀμερικῆς περιλαμβάνει πάντας τοὺς Ὀρθοδόξους, τοὺς κατοικοῦντας ἐν τῇ Ἀμερικανικῇ Ἠπείρῳ καὶ ταῖς προσκειμέναις
αὐτῇ νήσοις, τοὺς ἔχοντας ὡς γλῶσσαν λειτουργικὴν ἀποκλειστικῶς ἢ πρωτευόντως τὴν Ἑλληνικήν, ἐν ᾗ ἐγράφησαν τὰ Ἅγια Εὐαγγέλια καὶ τὰ λοιπὰ
Ἱερὰ Βιβλία τῆς Καινῆς Διαθήκης.

Ὑπὸ τὴν Ἀρχιεπισκοπὴν ταύτην δύνανται νὰ ὑπαχθῶσι καὶ ἄλλαι, διαφό-
ρου φυλῆς, Ὀρθόδοξοι ἐν Ἀμερικῇ Κοινότητες ἐπὶ τῇ αἰτήσει αὐτῶν καὶ μετ'
ἔγκρισιν τοῦ ἀσκοῦντος, κανονικῷ καὶ ἱστορικῷ δικαιώματι, τὴν πνευματικὴν

δικαιοδοσίαν τῶν ἐν Διασπορᾳ Ὀρθοδόξων Κοινοτήτων Οἰκουμενικοῦ Πατριαρχείου, διατηροῦσαι τὴν ἥν ἔχουσι λειτουργικὴν γλῶσσαν.

Ἄρθρον Γ´.

Ἡ Ἑλληνικὴ Ἀρχιεπισκοπὴ Ἀμερικῆς Βορείου καὶ Νοτίου ὑπάγεται, κατὰ
τοὺς κανόνας καὶ τὴν πρᾶξιν τῆς Ἐκκλησίας, ὑπὸ τὴν πνευματικὴν καὶ ἐκκλησιαστικὴν ἐξάρτησιν, δικαιοδοσίαν καὶ ἐποπτείαν τοῦ Οἰκουμενικοῦ Πατριαρχείου Κωνσταντινουπόλεως.

Ἄρθρον Δ´.

Σκοπὸς τῆς Ἀρχιεπισκοπῆς ταύτης εἶνε:

Α´.) Νὰ διατηρῇ καὶ νὰ διαδίδῃ τὴν Ὀρθόδοξον Χριστιανικὴν Πίστιν ἐπὶ
τῇ βάσει τῶν Ἁγίων Γραφῶν, τῆς Ἱερᾶς Παραδόσεως, τῶν Ὅρων καὶ Κανόνων τῶν
Ἁγίων Ἀποστόλων καὶ τῶν Ἑπτὰ Οἰκουμενικῶν Συνόδων τῆς Ἀρχαίας Ἀδιαιρέτου Ἐκκλησίας, ὡς ἑρμηνεύονται ἐν τῇ πράξει ὑπὸ τῆς ἐν Κωνσταντινουπόλει
Μεγάλης τοῦ Χριστοῦ Ἐκκλησίας.

Β´.) Νὰ οἰκοδομῇ τὸν θρησκευτικὸν καὶ ἠθικὸν βίον τῶν Ὀρθοδόξων Χριστιανῶν, συμφώνως πρὸς τὴν πίστιν καὶ τὰς παραδόσεις τῆς Ἐκκλησίας ταύτης.

Γ´.) Νὰ διδάσκῃ τὴν πρωτότυπον γλῶσσαν τοῦ Εὐαγγελίου.

Λεπτομερέστερον τὰ καθήκοντα ταῦτα κανονισθήσονται δι᾽ εἰδικοῦ Κανονισμοῦ.

Ἄρθρον Ε´.

Ἡ Ἑλληνικὴ Ἀρχιεπισκοπὴ Ἀμερικῆς Βορείου καὶ Νοτίου πρὸς ἐκπλήρωσιν
τοῦ ἐν ἄρθρῳ Δ´. ἀναγραφομένου σκοποῦ διαθέτει ἱεροὺς Ναούς, σχολικὰ καὶ
φιλανθρωπικὰ Ἱδρύματα, δημοσιεύματα καὶ οἰαδήποτε ἄλλα νόμιμα μέσα.

Δι᾽ ἕκαστον τῶν Ἱδρυμάτων τούτων καὶ μέσων θέλει ἐκπονηθῇ, εἰδικὸς Κανονισμός.

Ἄρθρον ΣΤ´.

Τὴν διοίκησιν ἔχει ὁ Ἀρχιεπίσκοπος, ἀσκῶν πάσας τὰς ἐξουσίας καὶ τὰ
καθήκοντα συμφώνως πρὸς τοὺς Ἱεροὺς Κανόνας καὶ τὴν πρᾶξιν τῆς Μεγάλης
Ἐκκλησίας καὶ τὰς εὐθύνας ὑπέχων ἐνώπιον τοῦ Οἰκουμενικοῦ Πατριάρχου
καὶ τῆς περὶ Αὐτὸν Ἁγίας καὶ Ἱερᾶς Συνόδου.

Παρὰ τῷ Ἀρχιεπισκόπῳ δύναται νὰ διορισθῇ καὶ εἷς Βοηθὸς Ἐπίσκοπος.

Ἄρθρον Ζ´.

Ὁ Ἀρχιεπίσκοπος ἐκλέγεται παρὰ τῆς Ἱερᾶς Συνόδου τοῦ Οἰκουμενικοῦ

Πατριαρχείου,ὁ δὲ Βοηθὸς Ἐπίσκοπος παρ'αὐτῆς μετὰ πρότασιν τοῦ Ἀρχιεπισκότου.

Οὐδεὶς δύναται προταθῆναι ἤ ἐκλεγῆναι Ἀρχιεπίσκοπος ἤ Βοηθὸς Ἐπίσκοπος,ἐὰν μὴ ᾖ κάτοχος διπλώματος ἀνεγνωρισμένης Ὀρθοδόξου Θεολογικῆς Σχολῆς,προδεδοκιμασμένος ἐν κατωτέροις ἱερατικοῖς βαθμοῖς ἐπὶ χρόνον οὐχ ἦσσονα τῶν πέντε ἐτῶν καὶ ἐν ἡλικίᾳ μὴ ὑποβαινούσῃ τὰ τριάκοντα ἔτη.

Ἄρθρον Η΄.

Παρὰ τῇ Ἀρχιεπισκοπῇ ὑφίσταται Γραφεῖον,οὗτινος ἡ λειτουργία θέλει ρυθμισθῇ ὑπὸ εἰδικοῦ Κανονισμοῦ.

Ἄρθρον Θ΄.

Πρὸς ἐξυπηρέτησιν τῶν διὰ τοῦ παρόντος τιθεμένων σκοπῶν τῆς Ἀρχιεπισκοπῆς Ἀμερικῆς καθορίζονται Ἐκκλησιαστικαὶ Συνελεύσεις,τῶν ὁποίων τὴν σύνθεσιν,τὸν χρόνον καὶ τὸν τρόπον τῆς συγκλήσεως,τὴν ἁρμοδιότητα,ὡς καὶ πᾶσαν ἄλλην σχέσιν καὶ λεπτομέρειαν τῆς λειτουργίας αὐτῶν,θέλει ὁρίσῃ εἰδικὸς Κανονισμός.

Ἄρθρον Ι΄.

Πρὸς ἐνίσχυσιν τῆς Ἀρχιεπισκοπῆς διὰ τὴν πραγμάτωσιν τῶν σκοπῶν αὐτῆς καὶ ἰδίᾳ διὰ τὴν διοίκησιν τῆς ἐκκλησιαστικῆς περιουσίας λειτουργεῖ,κατὰ τὰ ὁρισθησόμενα ὑπὸ εἰδικοῦ Κανονισμοῦ,παρὰ τῷ Ἀρχιεπισκόπῳ Μικτὸν Συμβούλιον.

Τὸ Μικτὸν Συμβούλιον θέλει μεριμνήσῃ διὰ τὴν ἵδρυσιν Γενικοῦ Ἐκκλησιαστικοῦ Ταμείου καὶ τὴν σύνταξιν τοῦ Ἱεροῦ Κλήρου.

Προσωρινῶς καὶ μέχρι τῆς ἐκδόσεως τοῦ Κανονισμοῦ τούτου θέλει συνεχίσῃ τὸ ἔργον αὐτοῦ τὸ νῦν ὑφιστάμενον Μικτὸν Συμβούλιον τῆς Ἀρχιεπισκοπῆς,κατὰ τὰς διατάξεις τῶν ἄρθρων ΚΑ΄. καὶ ΚΒ΄. τοῦ μέχρι τοῦδε ἰσχύοντος Καταστατικοῦ αὐτῆς.

Παραιτουμένου ἤ ὁπωσθήποτε ἄλλως ἐκλείποντος μέλους τινὸς τοῦ προσωρινοῦ τούτου Συμβουλίου,ὁ Ἀρχιεπίσκοπος διορίζει τὸν ἀντικαταστάτην αὐτοῦ.

Ἄρθρον ΙΑ΄.

Διὰ τὸν αὐτὸν ἐν τῷ ἀνωτέρω ἄρθρῳ ὁριζόμενον σκοπὸν λειτουργεῖ παρ' ἑκάστῃ Κοινότητι,ὑπαγομένῃ ὑπὸ τὴν Ἀρχιεπισκοπὴν ταύτην,Συμβούλιον,τὴν ἁρμοδιότητα,καθήκοντα καὶ λειτουργίαν καθόλου τοῦ ὁποίου θέλει καθορίσῃ εἰδικὸς Κανονισμός.

Ὅπου ὑπάρχουσιν ἐν τῇ αὐτῇ Παροικίᾳ πλείονες τοῦ ἑνὸς Ἱεροὶ Ναοὶ καὶ Κοινότητες δύναται δι'εἰδικοῦ Κανονισμοῦ νὰ διαιρεθῇ αὕτη εἰς ἐνοριακὰς περιφερείας,τὰς σχέσεις μεταξὺ τῶν ὁποίων καὶ τῆς Ἀρχιεπισκοπῆς θὰ ὁρίσῃ ὁ αὐτὸς Κανονισμός.

Ἄρθρον ΙΒ΄.

Πρὸς ἐκπλήρωσιν τῶν θρησκευτικῶν ἀναγκῶν τῶν Χριστιανῶν,τῶν μὴ ὠργανωμένων εἰς Θρησκευτικὰ Σωματεῖα,ἰδρύεται θεσμὸς Ἱεραποστολῆς,λειτουργῶν κατὰ τὰ εἰδικώτερον καθορισθησόμενα δι'ἰδιαιτέρου Κανονισμοῦ.

Ἄρθρον ΙΓ΄.

Διὰ τὴν προαγωγὴν καὶ συστηματικὴν ὀργάνωσιν τῶν Σχολείων λειτουργεῖ,καθ'ἃ θέλει ὁρίσῃ εἰδικὸς Κανονισμός,παρὰ τῇ Ἀρχιεπισκοπῇ,ὑπὸ τὴν προεδρείαν τοῦ Ἀρχιεπισκόπου Ἀνώτερον Ἐκπαιδευτικὸν Συμβούλιον.

Ἄρθρον ΙΔ΄.

Διὰ τὴν συστηματικὴν θρησκευτικὴν μόρφωσιν τῶν τέκνων τῆς Ἐκκλησίας λειτουργεῖ,καθ'ὃν τρόπον θέλει ὁρίσῃ Κανονισμὸς εἰδικός,παρὰ τῇ Ἀρχιεπισκοπῇ ὑπηρεσία Κατηχητικῶν Σχολείων μὲ διευθύνον αὐτὴν Συμβούλιον.

Ἄρθρον ΙΕ΄.

Συμφώνως πρὸς τοὺς Ἱεροὺς Κανόνας καὶ τὸ Δίκαιον τῆς Ἐκκλησίας λειτουργοῦσιν,ἐπὶ τῷ σκοπῷ τῆς τηρήσεως τῆς τάξεως καὶ πειθαρχίας ἐν αὐτῇ, παρὰ τῇ Ἀρχιεπισκοπῇ Ἀμερικῆς Πνευματικὰ Δικαστήρια,τῶν ὁποίων ἡ σύνθεσις,ἡ ἔκτασις τῆς ἁρμοδιότητος,ἡ διαδικασία,αἱ ποιναί,ἡ λειτουργία,ὡς καὶ πᾶσα ἄλλη λεπτομέρεια,θὰ καθορισθῶσιν ὑπὸ εἰδικοῦ Κανονισμοῦ.

Ἄρθρον ΙΣΤ΄.

Ὁ γάμος,ὡς καὶ τὸ διαζύγιον,κατὰ τὸ ἐκκλησιαστικὸν καὶ πνευματικὸν αὐτῶν μέρος,ὑπάγονται εἰς τὰς ἐκκλησιαστικὰς Ἀρχάς.

Εἰδικὸς Κανονισμός,ἐκδοθησόμενος ὑπὸ τῆς Ἀρχιεπισκοπῆς,θέλει κανονίσῃ λεπτομερῶς τὰ τῶν γάμων καὶ διαζυγίων.

Ἄρθρον ΙΖ΄.

Γενικαὶ Διατάξεις

Οἱ ὑπὸ τῶν ἄρθρων Δ΄.,Ε΄.,Η΄.,Θ΄.,Ι΄.,ΙΑ΄.,ΙΒ΄.,ΙΓ΄.,ΙΔ.,ΙΕ΄.,ΙΣΤ΄.,προβλεπόμενοι Κανονισμοὶ θέλουσι συνταχθῇ ὑπὸ Ἐπιτροπῶν,ὁρισθησομένων ὑπὸ τοῦ Ἀρχιεπισκόπου Ἀμερικῆς καὶ Προεδρευομένων ὑπ'Αὐτοῦ,θέλουσι δὲ ἀποτελέ-

ση ἀναπόσπαστον μέρος τοῦ παρόντος Συντάγματος,μετὰ τὴν ἔγκρισιν αὐτῶν
ὑπὸ τοῦ Οἰκουμενικοῦ Πατριαρχείου.Οἱ Κανονισμοὶ οὗτοι θέλουσιν ὁρίσῃ
τὴν σύνθεσιν τῶν ἁρμοδίων Συμβουλίων,τὰ καθήκοντα καὶ ἁρμοδιότητα αὐτῶν
καὶ πᾶσαν ἐν γένει λεπτομέρειαν τῆς λειτουργίας τῶν εἰς οὓς ἀναφέρονται
'Οργανισμῶν.

"Αρθρον ΙΗ.'

Πᾶσα ἀναγκαία συμπλήρωσις καὶ λεπτομέρεια τοῦ παρόντος Συντάγματος
θέλει καθορίζεσθαι ἑκάστοτε δι'ἀποφάσεως τοῦ ..ιχτοῦ Συμβουλίου τῆς'Αρχιεπισκοπῆς.

Πᾶσα διάταξις τῶν εἰδικῶν Κανονισμῶν ἤ ἀπόφασις τοῦ Μιχτοῦ Συμβουλίου ἀντικειμένη εἰς τὸ παρὸν Σύνταγμα εἶνε ἄκυρος.

"Αρθρον ΙΘ.'

Οἱ κληρικοὶ τῆς'Αρχιεπισκοπῆς'Αμερικῆς καθίστανται εἰς τὰς διακονίας αὐτῶν διὰ πράξεως τῆς κανονικῆς καὶ νομίμου'Εκκλησιαστικῆς'Αρχῆς,νό-
μος δὲ χειραγωγὸς εἰς ὅλην τὴν διοίκησιν τῆς'Αρχιεπισκοπῆς'Αμερικῆς θὰ
εἶνε οἱ'Ιεροὶ'Αποστολικοὶ καὶ Συνοδικοὶ Κανόνες καὶ οἱ κατ'αὐτοὺς καὶ
τὴν πρᾶξιν τῆς'Εκκλησίας,ὡς ἑρμηνεύεται ἑκάστοτε ὑπὸ τοῦ Οἰκουμενικοῦ
Πατριαρχείου,τιθέμενοι εἰς ἐνέργειαν Κανονισμοί.

"Αρθρον Κ.'

Οὐδεὶς κληρικὸς ἤ λαϊκὸς δύναται νὰ εἶνε ἀξιωματοῦχος ἤ καὶ ἁπλοῦν
μέλος τῆς'Ελληνικῆς'Ορθοδόξου'Αρχιεπισκοπῆς τῆς'Αμερικῆς,ἐὰν δὲν ἀνή-
κῃ εἰς τὴν'Ορθόδοξον τοῦ Χριστοῦ'Εκκλησίαν,καὶ οὐδεὶς δύναται νὰ μείνῃ
εἰς τὸ ἀξίωμά του ἤ καὶ ὡς ἁπλοῦν μέλος τῆς ἐν λόγῳ'Αρχιεπισκοπῆς,ἐὰν
παύσῃ νὰ εἶνε ἐν τάξει πρὸς Αὐτήν.

"Αρθρον ΚΑ.'

Πᾶσα διάταξις οἱουδήποτε 'Εκκλησιαστικοῦ 'Οργανισμοῦ ἀντιβαίνουσα εἴτε πρὸς τοὺς νόμους τῶν 'Ηνωμένων Πολιτειῶν εἴτε πρὸς τοὺς
'Ιεροὺς Κανόνας τῆς 'Ορθοδόξου 'Ανατολικῆς 'Εκκλησίας εἶνε αὐτοδικαίως ἄκυρος.

"Αρθρον ΚΒ.'

Τὸ παρὸν Σύνταγμα,συγκείμενον ἐξ ἄρθρων εἴκοσι δύο καὶ συνταχθὲν ἐ-
πὶ τῇ βάσει καὶ εἰσηγήσεως ἁρμοδίας εἰδικῆς'Επιτροπῆς,κατὰ τὸ ΚΗ.'ἄρθρον
τοῦ μέχρι τοῦδε ἐν ἰσχύϊ Καταστατικοῦ τῆς'Αρχιεπισκοπῆς'Αμερικῆς Βορείου

καὶ Νοτίου,ἐγκριθὲν δὲ καὶ ἐπικυρωθὲν συνοδικῶς ὑπὸ τῆς Αὐτοῦ Θειοτάτης Παναγιότητος τοῦ Οἰκουμενικοῦ Πατριάρχου,τίθεται εἰς ἐφαρμογὴν,δύναται δὲ νὰ τροποποιηθῇ εἰς τὰς μὴ θεμελιώδεις διατάξεις αὐτοῦ ὑπὸ εἰδικῆς Ἐπιτροπῆς,παρὰ τοῦ Ἀρχιεπισκόπου διοριζομένης,πάσης τροποποιήσεως οὔσης ἐγκύρου καὶ ἐφαρμοσίμου μετὰ τὴν ἔγκρισιν αὐτῆς ὑπὸ τοῦ Οἰκουμενικοῦ Πατριαρχείου.

Ἐν μηνὶ Ι Α Ν Ο Υ Α Ρ Ι ΩΙ (ι.) , Ι Ν Δ Ι Κ Τ Ι Ω Ν Ο Σ   Ι Δ .(ᾳϡλα.)

The first Constitution (Syntagma) organizing the Archdiocese in the Americas, issued by the
Ecumenical Patriarchate, January 10, 1931.

The First Charter of the Archdiocese

Τό πρῶτον Καταστατικόν τῆς Ἱερᾶς Ἀρχιεπισκοπῆς
Ἀμερικῆς καί ὁ πρῶτος Ἀρχιεπίσκοπος κατασταθείς
ὑπό τοῦ Οἰκουμενικοῦ Πατριάρχου Μελετίου Δ´
τοῦ Μεταξάκη

ὑπό
Ἐπισκόπου Μελόης Φιλοθέου

ποτελεῖ παράδοσιν μακραίωνα ἡ διαποίμανσις
τῶν ἐν τῇ διασπορᾷ Ὀρθοδόξων Ἐκκλησιῶν ὑπό
τοῦ Οἰκουμενικοῦ Πατριαρχείου, ἐφ᾽ ὅσον αὗται
κεῖνται ἐκτός τῶν ὁρίων τῶν καθορισθέντων ὑπό
τῶν Οἰκουμενικῶν Συνόδων ἤ ἐκτός τῆς δικαιοδοσίας τῶν Πατριαρχικῶν ἤ Αὐτοκεφάλων Ἐκκλησιῶν[1]. Ἐρί-
ζεται δέ ἡ παράδοσις αὕτη κυρίως ἐπί τοῦ κη´ Κανόνος τῆς Δ´
Οἰκουμενικῆς Συνόδου (Χαλκηδών 451)[2]. Τοιουτοτρόπως ἐπι-
τυγχάνεται ἡ ἑνότης τῆς ἐν τῇ διασπορᾷ ἐκκλησιαστικῆς ζωῆς
περί ἕν κέντρον[3]. Οὕτω πᾶσαι ἐν Εὐρώπῃ, Ἀμερικῇ καί Αὐστραλίᾳ συσταθεῖσαι Ὀρθόδοξοι Ἑλληνικαί Κοινότητες ὑπή-
γοντο ὑπό τόν Οἰκουμενικόν Πατριάρχην, ὁ ὁποῖος εἶχε τήν
διακυβέρνησιν καί τήν πνευματικήν ἐποπτείαν τῶν Ἐκκλησιῶν τούτων, ἀρχῆς γενομένης ἀπό τῆς ἱδρύσεως ἐν ἔτει 1577
ἐν Εὐρώπῃ καί δή ἐν Βενετίᾳ τῆς Μητροπόλεως Φιλαδελ-
φείας[4].

Editor's Note: One of the most important documents of the early organization of the Greek Orthodox Archdiocese of North and South America was
the first official canonical Constitution, created and approved by the Ecumenical Patriarchate, which assigned His Eminence Alexander as the first
Archbishop of the newly proclaimed Greek Orthodox Archdiocese of North
and South America.
Because of the many intricate and specific issues which the young church
had to deal with at the time, and because the language spoken and written
was Greek, it was decided to print this article by His Grace Bishop Philotheos of Meloa as it was written in order to convey the proper and specific
tenets of this first Constitution which in essence reversed the Tomos Decla-
ration of 1908 by which the American Church was placed under the jurisdiction of the Holy Synod of Greece. The importance of this article lies in
the fact that with the official proclamation of the Greek Orthodox Archdiocese and the selection of Alexander as Archbishop, the Ecumenical Patriar-

Ἡ ἐκ τοῦ κανονικοῦ τούτου δικαιώματος τοῦ Οἰκουμενικοῦ θρόνου διαμορφωθεῖσα κατά τά ἀνωτέρω τάξις ἐμφανίζεται διαταρασσομένη ἀπό τῆς συστάσεως τοῦ Ἑλληνικοῦ Βασιλείου καί τῆς ἀνεξαρτοποιήσεως τῆς ἐν Ἑλλάδι Ἐκκλησίας, καί μάλιστα λήγοντος τοῦ δεκάτου ἐνάτου αἰῶνος καί ἀρχομένου τοῦ εἰκοστοῦ, ὅτε παρατηρεῖται ἐπαύξησις τῆς μεταναστευτικῆς κινήσεως πρός τήν Ἀμερικήν κυρίως⁵, τῶν μεταναστῶν διατηρούντων στενόν δεσμόν μετά τῆς Ἐκκλησίας τῆς γενετείρας των⁶. Διά τοῦτο καί ἡ Μεγάλη τοῦ Χριστοῦ Ἐκκλησία ἀπεφάσισε πατριαρχεύοντος Ἰωακείμ τοῦ Γ΄ τήν ἐκχώρησιν πρός τήν Ἐκκλησίαν τῆς Ἑλλάδος «τό κανονικόν κυριαρχικόν τῆς πνευματικῆς προστασίας καί ἐποπτείας δικαίωμα ἐπί πασῶν τῶν ἐν τῇ διασπορᾷ, ἔν τε τῇ Εὐρώπῃ καί Ἀμερικῇ καί ταῖς λοιπαῖς χώραις, ὀρθοδόξων ἑλληνικῶν Ἐκκλησιῶν . . .» πλήν ἐκείνης τῆς Βενετίας, ὅπως διαλαμβάνεται εἰς τόν ἀπό τῆς 18 Μαρτίου 1908 πατριαρχικόν καί συνοδικόν τόμον⁷. Ἡ τοιαύτη ἀπόφασις ἀνεκοινώθη πρός τάς ἐν διασπορᾷ Ὀρθοδόξους Ἑλληνικάς Κοινότητας διά τῆς ὑπ᾽ ἀριθμ. πρωτ. 3498 ἀπό 21ης Ἀπριλίου 1908 πατριαρχικῆς Ἐγκυκλίου. Ἐν τῷ μεταξύ εἰς τόν Ἑλλαδικόν Χῶρον ἀρχομένου τοῦ εἰκοστοῦ αἰῶνος καί μάλιστα ἀμέσως μετά τήν ἔκδοσιν τοῦ ὡς ἄνω τόμου ἀλλεπάλληλοι πολιτικαί ἐξελίξεις συνετέλεσαν ὄχι μόνον εἰς τήν ἀπελευθέρωσιν μέρους τῆς Ἠπείρου, Θράκης, Κρήτης καί Μακεδονίας, ἀλλά καί εἰς τήν διαμόρφωσιν κλίματος πολιτικοῦ διχασμοῦ, αἱ συνέπειαι τοῦ ὁποίου ἐπί τι διάστημα ἐπιφαίνονται εἰς τήν Ἐκκλησίαν τῆς Ἑλλάδος καί ἐπεκτεί-

chate set a new plan of government in the Greek Orthodox Church of America. This plan came to be known as the Synodical System.

The official document cited by Bishop Philotheos in his article as well as the corresponding encyclical letters of the Ecumenical Patriarch was a plan that was to be adopted by the New York Clergy-Laity Congress on August 8, 1922, at which time two Bishops were elected and ordained: Bishop Philaretos for Chicago and Bishop Joachim for Boston. The Archdiocese was to be divided into four Dioceses: New York, Chicago, Boston and San Francisco. New York was to be the Archdiocesan See, the headquarters of the Archdiocese. Each Diocese would have its own annual conference of the elected clergy and laity representatives. The plan also provided for a biennial conference of the entire Archdiocese, and for at least two meetings of the Bishops with the Archbishop annually. Bishops were to be nominated at the annual Diocesan conferences and approved by the Ecumenical Patriarchate.

The importance of the specific points cited by Bishop Philotheos show that indeed, the "Katastatikon" was a landmark document of the Greek-American Church, for it marked the beginning of the Greek Orthodox Archdiocese of North and South America.

νονται καί εἰς τόν ἀπόδημον Ἑλληνισμόν. Καθίσταται οὕτως εὐνόητον ὅτι ὑπό τοιαύτας συνθήκας ἡ δυνατότης τῆς Ἐκκλησίας τῆς Ἑλλάδος, ὅπως ποιμάνῃ τάς ἀποδήμους Ὀρθοδόξους Ἑλληνικάς Ἐκκλησίας δοκιμάζεται οὐκ ὀλίγον[8]. Ἡ προσπάθεια ἐξ ἄλλου τοῦ Ἀθηνῶν Μελετίου Μεταξάκη, ὑποχρεωθέντος κατά τό ἔτος 1920 εἰς φυγήν ἐξ Ἀθηνῶν, ἅμα τῇ ἀνατροπῇ τῆς Κυβερνήσεως Βενιζέλου, ὅπως ποδηγετήσῃ κατά τήν ἐξορίαν του τά ἐν τῇ Ἀμερικῇ ἀποδημικά, διά τῆς ἀπομονώσεως τούτων ἐκ τοῦ συνόλου τῶν ἐν τῇ διασπορᾷ ὀρθοδόξων Ἐκκλησιῶν, ἀσφαλῶς δέν ἐλέγχεται κανονικῶς, ἀλλ᾽ οὔτε καί εὑρίσκει σύμφωνον τήν τότε ἑλληνικήν Ἐκκλησιαστικήν ἀρχήν.

Ὁ παρεπιδημῶν εἰς Ἀμερικήν Ἀθηνῶν Μελέτιος Μεταξάκης ὅτε ἐκλήθη τήν 25ην Νοεμβρίου τοῦ ἔτους 1921 εἰς τόν ἁγιώτατον ἀποστολικόν θρόνον Κωνσταντινουπόλεως θά πρέπει νά εἶχε συσσωρεύσει ἱκανήν ἐμπειρίαν τῆς διανυούσης τότε τήν δευτέραν δεκαετίαν ἐκχωρήσεως τῶν ὀρθοδόξων ἑλληνικῶν ἐκκλησιῶν τῆς διασπορᾶς ὑπό τοῦ Οἰκουμενικοῦ Πατριαρχείου εἰς τήν Ἐκκλησίαν τῆς Ἑλλάδος, δι᾽ ὅ καί ἅμα τῇ ἐνθρονίσει του τόν Ἰανουάριον τοῦ ἐπομένου ἔτους, ἐφρόντισε διά τήν ὑπαγωγήν τῆς Ἐκκλησίας τῆς Ἀμερικῆς ὑπό τό Οἰκουμενικόν Πατριαρχεῖον ἐν τῷ πλαισίῳ τῆς ἐπαναφορᾶς τοῦ συνόλου τῶν ὀρθοδόξων ἑλληνικῶν Ἐκκλησιῶν τῆς διασπορᾶς ὑπό τήν Μητέρα Ἐκκλησίαν καί τῆς ἀναδιαρθρώσεως αὐτῶν[9]. Οὕτως ἀπεφάσισεν ἡ Ἁγία καί Ἱερά Σύνοδος τοῦ Οἰκουμενικοῦ Πατριαρχείου κατά τήν συνεδρίαν αὐτῆς τῆς 1ης Μαρτίου 1922[10]:

«Ἐπειδή ὁ σκοπός τῆς εἰς τήν Ἐκκλησίαν τῆς Ἑλλάδος ἐκχωρήσεως κατ᾽ οἰκονομίαν καί ὑπό τύπον ἐντολῆς τοῦ δικαιώματος τῆς διοικήσεως τῶν ἐν τῇ διασπορᾷ Ὀρθοδόξων Ἑλληνικῶν Παροικιῶν οὐκ εὐωδώθη· ἐπειδή οὐκ ἐτηρήθησαν οὐδέ ἐξετελέσθησαν οἱ ὅροι τῆς ἐκχωρήσεως, οἱ διαλαμβανόμενοι ἐν τῷ Πατριαρχικῷ καί Συνοδικῷ Τόμῳ, τῷ ἀπολυθέντι ὑπό ἡμερομηνίαν η΄ Μαρτίου αϡθη΄· ἐπειδή, ἐκλιπουσῶν ἤδη τῶν ἀπό τῶν καιρικῶν περιστάσεων ἀφορμῶν, συνεξέλιπε καί ὁ ἀπ᾽ αὐτῶν σοβαρός λόγος, ὁ εἰς τήν κατ᾽ οἰκονομίαν διευθέτησιν ἐκείνην ἀγαγών· καί τό δή σπουδαιότερον, ἐπειδή ἐκ τῆς ἐκχωρήσεως ποικίλη προέκυψε κανονική ἀνωμαλία, τήν ἑνότητα διαταράττουσα τῆς Ἐκκλησιαστικῆς διοικήσεως· διά ταῦτα ἡ καθ᾽ ἡμᾶς Ἁγία Μεγάλη τοῦ Χριστοῦ Ἐκκλησία, ἐν τῷ ἀπαραγράπτῳ δικαιώματι αὐτῆς τοῦ διέπειν καί διαχειρίζεσθαι αὐτεξουσίως τήν ἀπό τῶν ἱε-

147

ρῶν κανόνων καί τῆς Ἐκκλησιαστικῆς τάξεως ἀνήκουσαν αὐτῇ κανονικήν ἐξουσίαν, ἐν ᾗ περιλαμβάνεται καί
ἡ ἐπί τῶν ἔξω καί ἐν τῇ διασπορᾷ Ὀρθοδόξων Παροικιῶν ἐκκλησιαστική ἐποπτεία, καί ἐκ καθήκοντος ὀφειλετικῆς προνοίας αἴρει μέν καί ἀκυροῖ τήν ἐκδεδομένην
καί ἐν τῷ διαληφθέντι Πατριαρχικῷ καί Συνοδικῷ ὑπό
ἡμερομηνίαν η΄ Μαρτίου αϡιη΄ Τόμῳ, τῷ ἐν σελίδι 108 τοῦ
εἰδικοῦ Κώδικος ἐν τοῖς Πατριαρχείοις κατεστρωμένῳ,
περιλαμβανομένην ἀπόφασιν αὐτῆς, τήν δι᾽ Ἐγκυκλίου
Πατριαρχικῆς ὑπό ἡμερομηνίαν κα΄ Ἀπριλίου αϡιη΄ καί
ἀριθμ. Πρωτ. 3498 ἀνακοινωθεῖσαν ταῖς Κοινότησι, περί
ἐκχωρήσεως τῇ Ἐκκλησίᾳ τῆς Ἑλλάδος τοῦ δικαιώματος τῆς διακυβερνήσεως τῶν ἐν τῇ διασπορᾷ Ὀρθοδόξων Ἑλληνικῶν Παροικιῶν μετά πασῶν τῶν συναφῶν
τῇ ἀποφάσει ταύτῃ παραχωρήσεων, τῶν ἀναγραφομένων
ἐν τῷ διαληφθέντι Τόμῳ, ἀποκαθίστησι δέ καί αὖθις πλήρη καί ἀκέραια τά κανονικά κυριαρχικά αὐτῆς δικαιώματα τῆς ἀμέσου ἐποπτείας καί διακυβερνήσεως ἐπί πασῶν ἀνεξαιρέτως τῶν ἔξω τῶν ὁρίων ἑκάστης ἐπί μέρους
Αὐτοκεφάλου Ἐκκλησίας, ἔν τε Εὐρώπῃ καί ἐν Ἀμερικῇ καί ἀλλαχοῦ εὑρισκομένων Ὀρθοδόξων Παροικιῶν,
ὑπάγουσα καί αὖθις αὐτάς ὑπό τήν ἄμεσον αὐτῆς Ἐκκλησιαστικήν ἐξάρτησιν καί χειραγωγίαν καί ὁρίζουσα,
ὅπως πρός αὐτήν μόνον ἔχωσιν αὗται ἐφεξῆς τήν ἀναφοράν αὐτῶν καί παρ᾽ αὐτῆς τό κῦρος τῆς Ἐκκλησιαστικῆς συγκροτήσεως καί ὑποστάσεως αὐτῶν, ὡς τέτακται,
αἰτῶνται καί ἀρύωνται, μνημονευομένου, κατά τήν τάξιν,
τοῦ Πατριαρχικοῦ ὀνόματος ἐν αὐταῖς.»

Ἡ ἀπόφασις αὕτη ἀνηγγέλθη τηλεγραφικῶς τήν 22 Μαρτίου τοῦ αὐτοῦ ἔτους τῷ Ἐπισκόπῳ Ῥοδοστόλου Ἀλεξάνδρῳ.
Διά τοῦ αὐτοῦ τηλεγραφήματος ὁρίζεται ὁ παραλήπτης Ἐπίσκοπος ὡς Πατριαρχικός Ἐπίτροπος[11]. Ὁ Ῥοδοστόλου Ἀλέξανδρος εἰς ἀπάντησιν ἀπό 12 Ἀπριλίου 1922 ἀναφέρει πρός
τό Σεπτόν Κέντρον τάς πρώτας του ἐνεργείας ὡς Πατριαρχικοῦ Ἐπιτρόπου[12].

Ἐν τῷ μεταξύ τήν 26ην Ἀπριλίου διά Πατριαρχικῆς καί
Συνοδικῆς Γράξεως ἱδρύεται ἡ Ὀρθόδοξος Ἀρχιεπισκοπή
Ἀμερικῆς Βορείου καί Νοτίου,[13] ἡ θέσις τῆς ὁποίας ὁρίζεται 15η ἐν τῷ Συνταγματίῳ, «μετά τῶν ὑπ᾽ αὐτήν ἐπισκόπων,
ἔν γε τῷ παρόντι Σικάγου, Βοστώνης καί Ἁγίου Φραγκίσκου»[14].

Ὁ εἰρημένος τόμος ὁρίζει τάς ἐπισκοπάς «θεμελιωδῶς»
εἰς τέσσαρας, «διά λόγους σχετιζομένους πρός τήν ἀνάγκην
αὐταρκείας πρός ἐπιτόπιον διοίκησιν, ἀλλά καί πρός τό σχέ-

148

διον ὀργανώσεως τῆς ὅλης ἐν Ἀμερικῇ Ὀρθοδοξίας εἰς ἕν ἑνιαῖον σύνολον», ὀνομάζων μόνον αὐτάς, καί μή καθορίζων τά ὅρια ἑκάστης. Σχεδιάζει δηλ. ἐν γενικότητι τήν ὅλην διάρθρωσιν τῆς Ἀρχιεπισκοπῆς Ἀμερικῆς προσβλέπων κυρίως εἰς τήν «συσσωμάτωσιν». Προβλέπει πρός τοῦτο ἐν ἑκάστῃ ἐπισκοπικῇ περιφερείᾳ ἰδίαν συνέλευσιν, καί Γενικήν Συνέλευσιν ὡς σῶμα τῆς ὅλης Ἀρχιεπισκοπῆς. «Ἀλλά τό νέον Καταστατικόν οὐκ ὀφείλει χαρακτηρισθῆναι ὁριστικόν, ἀλλά διατηρείσθω καί ἐν αὐτῷ ὁ τίτλος τοῦ προσωρινοῦ, μέχρις οὗ διά τῆς δοκιμαστικῆς ἐξελίξεως φθάσωμεν εἰς τό τέλειον». Ἰδιαίτερα δέ γίνεται προτροπή τοῦ καταρτισμοῦ «Συνόδου Ἀρχιερέων πρός ἐπιτόπιον Συνοδικήν διακυβέρνησιν».

Πρῶτος ποιμενάρχης τῆς νέας Ἀρχιεπισκοπῆς ψήφων κανονικῶν γενομένων ἐκλέγεται τῇ αὐτῇ ἡμέρᾳ κς΄ Ἀπριλίου 1922 ὁ ἀπό Ῥοδοστόλου Ἀλέξανδρος. Διά τήν πλαισίωσιν δέ τοῦ Ἀρχιεπισκόπου ἐκλέγεται ὑπό τῆς Ἁγίας καί Ἱερᾶς Συνόδου τοῦ Οἰκουμενικοῦ Πατριαρχείου ὁ ἀρχιμ. Φιλάρετος (Ἰωαννίδης) ὡς Ἐπίσκοπος Σικάγου τήν 1ην Μαΐου 1923, ἀναλαμβάνων συγχρόνως καί τά καθήκοντα τοῦ τοποτηρητοῦ Ἁγ. Φραγκίσκου[15] καί ὁ ἀρχιμ. Ἰωακείμ (Ἀλεξόπουλος) τήν 28ην Ἰουνίου ὡς Ἐπίσκοπος Βοστώνης, διαμορφουμένων οὕτω τῶν προϋποθέσεων διά τόν καταρτισμόν τό ταχύτερον Συνόδου Ἀρχιερέων «πρός ἐπιτόπιον Συνοδικήν τῆς Ἐκκλησίας διακυβέρνησιν».

Ἡ προσπάθεια αὕτη, παρά πᾶσαν προσδοκίαν κανονικήν, ἀντιμετώπισε τήν ἀμφισβήτησιν ὑπό τῆς Ἐκκλησίας τῆς Ἑλλάδος, συνεπικουρημένης ὑπό τοῦ Ἑλληνικοῦ Κράτους, καίτοι τό Οἰκουμενικόν Πατριαρχεῖον διέπεμψε τήν Πατριαρχικήν καί Συνοδικήν Πρᾶξιν περί ἄρσεως καί ἀκυρώσεως τοῦ Πατριαρχικοῦ καί Συνοδικοῦ Τόμου τοῦ ἔτους 1908, διά Πατριαρχικοῦ Γράμματος πρός τήν Ἱεράν Σύνοδον τῆς Ἐκκλησίας τῆς Ἑλλάδος ἐγκαίρως. Διό καί ἡ Μήτηρ Ἐκκλησία κατά Σεπτέμβριον τοῦ ἔτους 1923, μετά παρέλευσιν δηλ. ἑνός καί ἡμίσεος σχεδόν ἔτους, ἀπηύθυνε πρός τήν Ἐκκλησίαν τῆς Ἑλλάδος τό ὑπ᾽ ἀριθμ. πρωτοκόλλου 4534 Πατριαρχικόν Γράμμα, εἰς τό ὁποῖον ὑποδεικνύεται ὅπως «ὡς οἷόν τε τάχιον, τεθῇ τέρμα εἰς πᾶσαν περί τό σημεῖον τοῦτο ἐγερθεῖσαν ἀμφιβολίαν καί ἔτι μᾶλλον εἰς γενομένας ἤδη ἀντιθέτους ἐνεργείας»[16]. Οὐχ ἧττον ἡ Ἐκκλησία τῆς Ἑλλάδος ἐπηρεαζομένη ὑπό τῆς ὑφισταμένης τότε ἐν Ἑλλάδι πολιτικῆς διαιρέσως δέν ἀντιμετώπισεν ἐπισταμένως τό ὅλον θέμα, προκαλέσασα οὕτως νέαν Πατριαρχικήν ἐνέργειαν, ἐκδηλωθεῖσαν διά γράμματος τοῦ Πατριάρχου Γρηγορίου τοῦ Ζ΄, ὑπ᾽ ἀριθμ.

πρωτ. 688 ἀπό 23 Φεβρουαρίου 1924, εἰς τό ὁποῖον ἐκτιμου-
μένης τῆς ἐγκυμονούσης ζημίας ἐπαναλαμβάνεται ἡ ὑπόδει-
ξις τοῦ προηγουμένου γράμματος περί ἀπολύσεως ὑπό τῆς
Ἐκκλησίας τῆς Ἑλλάδος πρός τάς παροικίας τοῦ ἐξωτερι-
κοῦ σχετικῆς ἐγκυκλίου, διότι «οὕτως ἀσφαλέστερον μέλλουσιν διασκεδασθῆναι αἱ ὑπολειπόμεναι τυχόν ἀμφιβολίαι
καί ἐνισχυθήσεται τό κανονικῶς γενόμενον περί τάς ἐν διασπορᾷ παροικίας ὑπό τοῦ καθ᾽ ἡμᾶς Ἁγιωτάτου Πατριαρχικοῦ Θρόνου, παγιουμένης τῆς τάξεως καί τῆς ἡσυχίας ἔν τε
ταῖς παροικίαις αὐταῖς καί ἐν ταῖς περί αὐτάς προθέσεσι καί
ἐνεργείαις τῶν διαφόρων ἀδελφῶν Ὀρθοδόξων Ἐκκλησιῶν»[17].
Ἡ Σύνοδος τῆς Ἱεραρχίας τῆς Ἐκκλησίας τῆς Ἑλλάδος
κατά τήν Β´ ἔκτακτον σύνοδόν της κατά τόν Μάϊον τοῦ ἔτους
1924 συνοδικῇ ἀποφάσει ἀπεδέχθη τήν ἐκ νέου ὑπαγωγήν τῶν
ἐν διασπορᾷ Ὀρθοδόξων Ἑλληνικῶν παροικιῶν ὑπό τήν
πνευματικήν δικαιοδοσίαν τοῦ Οἰκουμενικοῦ Πατριαρχείου[18]
καί ἐξαπέλυσε πρός τάς εἰρημένας παροικίας τήν ὑπ᾽ ἀριθμ.
πρωτ. 1022 ἀπό 19 Μαΐου 1924 ἐγκύκλιον αὐτῆς[19], ἀντίγραφον τῆς ὁποίας ἐκοινοποίησε πρός τό Οἰκουμενικόν Πατριαρχεῖον, διά τοῦ ὑπ᾽ ἀριθμ. πρωτ. 830 ἀπό 22 Μαΐου 1924 ἀπαντητικοῦ αὐτῆς ἐγγράφου[20].
Κατόπιν τούτων ἀκολουθεῖ ἡ ἀπό 3ης Ἰουλίου 1924 Ἐγ-
κύκλιος τοῦ ἐν Οὐασιγκτῶνι Ἐπιτετραμμένου τῆς Πρεσβείας
τῆς Ἑλληνικῆς Δημοκρατίας Β. Μαμμωνᾶ πρός τάς ἐν Ἀμερικῇ Ἑλληνικάς Προξενικάς ἀρχάς περί ἀναγνωρίσεως ὡς
μόνης κανονικῆς ἀρχῆς τῆς ὑπό τοῦ Οἰκουμενικοῦ Πατριαρχείου καταστασθείσης ἐν Ἀμερικῇ Ἀρχιεπισκοπῆς[21].
Ἐν τέλει εἶναι ἀνάγκη ὅπως μνημονευθῇ τό συνῳδά τῷ Πατριαρχικῷ τόμῳ καταρτισθέν ἐν Νέᾳ Ὑόρκῃ Καταστατικόν
τῆς Ἑλληνικῆς Ἀρχιεπισκοπῆς Ἀμερικῆς Βορείου καί Νοτίου[22] ὑπό ἡμερομηνίαν 11 Αὐγούστου 1922, ἐπικυρωθέν ὑπό
τοῦ Οἰκουμενικοῦ πατριάρχου Μελετίου. Εἰς τοῦτο εἶναι σα-
φής ἡ ἐπίδρασις τοῦ προσωρινοῦ Καταστατικοῦ τῆς Ἀρχιεπισκοπῆς τοῦ συνταχθέντος ἐν Νέᾳ Ὑόρκῃ τῇ 24ῃ Σεπτεμ-
βρίου 1921 ὑπό τοῦ Ἀθηνῶν τότε Μελετίου καί τοῦ Ἐπισκόπου τότε Ροδοστόλου Ἀλεξάνδρου[23], ἐμφανιζομένη αὕτη
ἀκόμη καί εἰς τήν γλωσσικήν διατύπωσιν, ἀφοῦ ἀποτελεῖ ἐπε-
ξεργασμένην μορφήν τούτου ἐπί τῷ σκοπῷ τοῦ καταρτισμοῦ
τοῦ ἰδανικωτέρου, ὡς ἐπί λέξει σημειοῦται ἐν τῷ ἱδρυτικῷ τόμῳ τῆς Ἱερᾶς Ἀρχιεπισκοπῆς. Εἰς τό νέον καταστατικόν
διαγράφεται ἡ χαρακτηρίζουσα τήν ἐκκλησιαστικήν διακο-
νίαν τοῦ Μελετίου Μεταξάκη μεθοδολογία, ἥτις ἐκφράζεται
ὅπου οὗτος ὑπηρέτησε, διά τῆς ἐνισχύσεως τοῦ συνοδικοῦ

150

θεσμοῦ καί τῆς ἐκκλησιαστικῆς παιδείας[24]. Συναφής εἶναι ἄλλωστε ἡ προσπάθεια προσαρμογῆς τοῦ καταστατικοῦ εἰς τήν περί Ἐκκλησιαστικῶν ὀργανισμῶν νομοθεσίαν τῶν Η.Π.Α. Ἰδιαιτέρας προβολῆς ὅμως πρέπει νά τύχῃ τό ἄρθρον Ζ, ὑπό τόν τίτλον «Ἡ Σύνοδος τῆς Ἀρχιεπισκοπῆς» εἰς τό ὁποῖον στοιχειοθετοῦνται τό πρῶτον τά σχετικά τῆς πρώτης ὀρθοδόξου Συνόδου Ἐπαρχίας ἐν Ἀμερικῇ, διότι ἀποτελεῖ πρωτοπορίαν διά τήν ἐποχήν του καί καθιστᾷ τό Καταστατικόν πρόδρομον τοῦ νῦν ἰσχύοντος περί τῆς Ἱερᾶς Ἀρχιεπισκοπῆς.

ΠΑΡΑΡΤΗΜΑ

Ἐπίσημα ἔγγραφα

1. Πατριαρχική καί Συνοδική πρᾶξις περί ἄρσεως καί ἀκυρώσεως τοῦ προεκδεδομένου Πατριαρχικοῦ καί Συνοδικοῦ ὑπό ἡμερομηνίαν 8 Μαρτίου 1908 Τόμου περί τῶν ἐν τῇ διασπορᾷ Ἐκκλησιῶν.
(Τζωρτζάτου Β., Ἡ εἰς τήν Ἐκκλησίαν τῆς Ἑλλάδος ὑπαγωγή τῶν ἐν διασπορᾷ Ἑλληνικῶν Ἐκκλησιῶν καί ἀνάκλησις αὐτῆς. Ἐν Ἀθήναις 1977, σσ. 11 - 13).

ΠΑΤΡΙΑΡΧΙΚΗ ΚΑΙ ΣΥΝΟΔΙΚΗ ΠΡΑΞΙΣ

Ἄρσεως καί ἀκυρώσεως τοῦ προεκδεδομένου Πατριαρχικοῦ καί Συνοδικοῦ ὑπό ἡμερομηνίαν 8 Μαρτίου 1908 Τόμου περί τῶν ἐν τῇ διασπορᾷ Ἐκκλησιῶν.

† ΜΕΛΕΤΙΟΣ

ΕΛΕΩ ΘΕΩ ΑΡΧΙΕΠΙΣΚΟΠΟΣ ΚΩΝΣΤΑΝΤΙΝΟΥΠΟΛΕΩΣ ΝΕΑΣ ΡΩΜΗΣ ΚΑΙ ΟΙΚΟΥΜΕΝΙΚΟΣ ΠΑΤΡΙΑΡΧΗΣ

Τῶν περί τήν ἐγκόσμιον ἐκκλησιαστικήν διοίκησιν καί πειθαρχικήν τάξιν θεσμοθετημάτων, κατά τό χρεών καί πρός ὠφέλειαν γιγνομένων καί γίγνεσθαι ἀεί ὀφειλόντων, εὔδηλον ὅτι καί μεταβαλεῖν τι, ἄν γένοιτο, αὐτῶν καί ὅλως ποτέ ἆραι, ἐπάν ἡ χρεία καί ἡ ὠφέλεια τοῦτο ἐμφαίνωσι καί ὑπαγορεύωσιν.

Ἐπειδή τοίνυν καί ἡ καθ' ἡμᾶς Μεγάλη τοῦ Χριστοῦ Ἐκκλησία ἔφθασε μέν πρότερον ἐκ λόγων προνοίας διά τάς καιρικάς περιστάσεις καί ἐπ' ἐλπίδι μείζονος ὠφελείας πνευματικῆς ἐκχωρήσασα διά συνοδικῆς ἀποφάσεως καί πράξεως κατά τό ἔτος αϠη΄ τήν διαχείρισιν τοῦ κανονικοῦ αὐτῆς δικαιώματος τῆς ὑπάτης πνευματικῆς ἐξουσίας καί προστασίας ἐπί τῶν ἐν τῇ διασπορᾷ

151

καὶ ἔξω τῶν καθωρισμένων ὁρίων τῶν Ἁγιωτάτων Ὀρθοδόξων Αὐτοκεφάλων Ἐκκλησιῶν Ὀρθοδόξων Ἑλληνικῶν Παροικιῶν τῇ Ἁγιωτάτῃ Ἐκκλησίᾳ τοῦ Βασιλείου τῆς Ἑλλάδος ὡς ἀδελφῇ πεφιλημένῃ εὐχερέστερον, ὥς γε ὑπελαμβάνετο, δυναμένῃ φροντίζειν προσφόρως καὶ σωστικῶς περὶ τῆς πνευματικῆς διακυβερνήσεως τῶν εἰρημένων παροικιῶν, ἀλλ᾽ ἐκ τῆς πράξεως καὶ τῆς πείρας φανερὸν ἤδη ἐγένετο ὅτι ὁ τῆς καιρικῆς ἐκείνης οἰκονομίας σκοπός οὐκ εὐώδωται, οὐδ᾽ ἐπιτέτευκται, τοὐναντίον δὲ ἀνωμαλία τις ἤδη ἐξ αὐτῆς ἀνεφύη, βλάβην μὲν ταῖς Κοινότησι σύγχυσιν δὲ καὶ εἰς τὴν ὅλην κανονικὴν τάξιν τῆς Ἐκκλησίας δυναμένη ἐπενεγκεῖν, διὰ ταῦτα ἡ Μετριότης ἡμῶν μετὰ τῶν περὶ ἡμᾶς Ἱερωτάτων Μητροπολιτῶν καὶ ὑπερτίμων, τῶν ἐν Ἁγίῳ Πνεύματι ἀγαπητῶν ἡμῖν ἀδελφῶν καὶ συλλειτουργῶν, θέμα μελέτης νέας καὶ ἐξετάσεως συνοδικῆς ποιησάμενοι ὀφειλετικῶς τὸ ζήτημα τῆς ἐξαρτήσεως καὶ διοικήσεως τῶν ἐν τῇ διασπορᾷ Ὀρθοδόξων Παροικιῶν, ὡς Ἐκκλησιῶν κατὰ τοὺς κανόνας καὶ τὴν τάξιν τῆς Ἁγίας Ὀρθοδόξου Ἐκκλησίας ὑπὸ τὴν ὑπάτην δικαιοσίαν καὶ εὐθύνην ὑπαγομένων τοῦ καθ᾽ ἡμᾶς Ἁγιωτάτου Ἀποστολικοῦ καὶ Πατριαρχικοῦ Οἰκουμενικοῦ Θρόνου, διὰ λόγους δὲ καιρικῆς ἀνάγκης καὶ οἰκονομίας ὑπ᾽ αὐτοῦ τῶν γε ἑλληνοφώνων ἐξ αὐτῶν τῇ διοικήσει τῆς Ἱερᾶς Συνόδου τοῦ Βασιλείου τῆς Ἑλλάδος ὡς ἐντολοδόχου αὐτοῦ καὶ ἐπὶ ὡρισμένοις καθυπαχθεισῶν, ἔγνωμεν, διασκεψάμενοι συνοδικῶς, καὶ ἐν Ἁγίῳ Πνεύματι ἀποφαινόμενοι, ὁρίζομεν τὰ ἀκόλουθα: Ἐπειδὴ ὁ σκοπός τῆς εἰς τὴν Ἐκκλησίαν τῆς Ἑλλάδος ἐκχωρήσεως κατ᾽ οἰκονομίαν καὶ ὑπὸ τύπον ἐντολῆς τοῦ δικαιώματος τῆς διοικήσεως τῶν ἐν τῇ διασπορᾷ Ὀρθοδόξων Ἑλληνικῶν Παροικιῶν οὐκ εὐωδώθη· ἐπειδὴ οὐκ ἐτηρήθησαν οὐδὲ ἐξετελέσθησαν οἱ ὅροι τῆς ἐκχωρήσεως, οἱ διαλαμβανόμενοι ἐν τῷ Πατριαρχικῷ καὶ Συνοδικῷ Τόμῳ, τῷ ἀπολυθέντι ὑπὸ ἡμερομηνίαν η΄ Μαρτίου ͵αϡιη΄· ἐπειδή, ἐκλιπουσῶν ἤδη τῶν ἀπὸ τῶν καιρικῶν περιστάσεων ἀφορμῶν, συνεξέλιπε καὶ ὁ ἀπ᾽ αὐτῶν σοβαρός λόγος, ὁ εἰς τὴν κατ᾽ οἰκονομίαν διευθέτησιν ἐκείνην ἀγαγών· καὶ τὸ δὴ σπουδαιότερον, ἐπειδὴ ἐκ τῆς ἐκχωρήσεως ποικίλη προέκυψε κανονικὴ ἀνωμαλία, τὴν ἑνότητα διαταράττουσα τῆς Ἐκκλησιαστικῆς διοικήσεως· διὰ ταῦτα ἡ καθ᾽ ἡμᾶς Ἁγία Μεγάλη τοῦ Χριστοῦ Ἐκκλησία, ἐν τῷ ἀπαραγράπτῳ δικαιώματι αὐτῆς τοῦ διέπειν καὶ διαχειρίζεσθαι αὐτεξουσίως τὴν ὑπὸ τῶν ἱερῶν κανόνων καὶ τῆς Ἐκκλησιαστικῆς τάξεως ἀνήκουσαν αὐτῇ κανονικὴν ἐξουσίαν, ἐν ᾗ περιλαμβάνεται καὶ ἡ ἐπὶ τῶν ἔξω καὶ ἐν τῇ διασπορᾷ Ὀρθοδόξων Παροικιῶν ἐκκλησιαστικὴ ἐποπτεία, καὶ ἐκ καθήκοντος ὀφειλετικῆς προνοίας αἴρει μὲν καὶ ἀκυροῖ τὴν ἐκδεδομένην καὶ ἐν τῷ διαληφθέντι Πατριαρχικῷ καὶ Συνοδικῷ ὑπὸ ἡμερομηνίαν η΄ Μαρτίου ͵αϡιη΄ Τόμῳ,

152

τῷ ἐν σελίδι 108 τοῦ εἰδικοῦ Κώδικος ἐν τοῖς Πατριαρχείοις κατεστρωμένῳ, περιλαμβανομένην ἀπόφασιν αὐτῆς, τήν δι᾽ Ἐγκυκλίου Πατριαρχικῆς ὑπό ἡμερομηνίαν κα΄ Ἀπριλίου ᾳϡη΄ καί ἀριθμ. Πρωτ. 3498 ἀνακοινωθεῖσαν ταῖς Κοινότησι, περί ἐκχωρήσεως τῇ Ἐκκλησίᾳ τῆς Ἑλλάδος τοῦ δικαιώματος τῆς διακυβερνήσεως τῶν ἐν τῇ διασπορᾷ Ὀρθοδόξων Ἑλληνικῶν Παροικιῶν μετά πασῶν τῶν συναφῶν τῇ ἀποφάσει ταύτῃ παραχωρήσεων, τῶν ἀναγραφομένων ἐν τῷ διαληφθέντι Τόμῳ, ἀποκαθίστησι δέ καί αὖθις πλήρη καί ἀκέραια τά κανονικά κυριαρχικά αὐτῆς δικαιώματα τῆς ἀμέσου ἐποπτείας καί διακυβερνήσεως ἐπί πασῶν ἀνεξαιρέτως τῶν ἔξω τῶν ὁρίων ἑκάστης ἐπί μέρους Αὐτοκεφάλου Ἐκκλησίας, ἔν τε Εὐρώπῃ καί ἐν Ἀμερικῇ καί ἀλλαχοῦ εὑρισκομένων Ὀρθοδόξων Παροικιῶν, ὑπάγουσα καί αὖθις αὐτάς ὑπό τήν ἄμεσον αὐτῆς Ἐκκλησιαστικήν ἐξάρτησιν καί χειραγωγίαν καί ὁρίζουσα, ὅπως πρός αὐτήν μόνον ἔχωσιν αὗται ἐφεξῆς τήν ἀναφοράν αὐτῶν καί παρ᾽ αὐτῆς τό κῦρος τῆς Ἐκκλησιαστικῆς συγκροτήσεως καί ὑποστάσεως αὐτῶν, ὡς τέτακται, αἰτῶνται καί ἀρύωνται, μνημονευομένου, κατά τήν τάξιν, τοῦ Πατριαρχικοῦ ὀνόματος ἐν αὐταῖς.

Ἐπί τούτῳ δέ εἰς ἔνδειξιν καί βεβαίωσιν ἐγένετο ἡ παροῦσα Πατριαρχική καί Συνοδική Πρᾶξις, καταστρωθεῖσα ἐν τῷ Ἱερῷ Κώδικα τῆς τοῦ Χριστοῦ Μεγάλης Ἐκκλησίας καί ἐν ἰσχύι καί ἐνεργείᾳ ἀπό σήμερον ὑπάρχουσα, ἐντελλόμεθα δέ ἴσον καί ἀπαράλλακτον αὐτῆς δημοσιευθῆναι ἐν τῷ ἐπισήμῳ ὀργάνῳ τοῦ Οἰκουμενικοῦ Πατριαρχείου, τῇ «Ἐκκλησιαστικῇ Ἀληθείᾳ».

Ἐν ἔτῃ σωτηρίῳ ᾳϡκβ΄, μηνός Μαρτίου α΄, Ἐπινεμήσεως Ε΄.

† Ὁ Πατριάρχης Κωνσταντινουπόλεως ΜΕΛΕΤΙΟΣ ἀποφαίνεται
† Ὁ Καισαρείας ΝΙΚΟΛΑΟΣ
† Ὁ Νικαίας ΒΑΣΙΛΕΙΟΣ
† Ὁ Χαλκηδόνος ΓΡΗΓΟΡΙΟΣ
† Ὁ Ἀμασείας ΓΕΡΜΑΝΟΣ
† Ὁ Νεοκαισαρείας καί Κοτυώρων ΠΟΛΥΚΑΡΠΟΣ
† Ὁ Ρόδου ΑΠΟΣΤΟΛΟΣ
† Ὁ Ἀγκύρας ΓΕΡΒΑΣΙΟΣ
† Ὁ Βάρνης ΝΙΚΟΔΗΜΟΣ
† Ὁ Χαλδίας καί Κερασοῦντος ΛΑΥΡΕΝΤΙΟΣ
† Ὁ Ἡλιουπόλεως ΣΜΑΡΑΓΔΟΣ
† Ὁ Σαράντα Ἐκκλησιῶν ΑΓΑΘΑΓΓΕΛΟΣ
† Ὁ Μετρῶν καί Ἀθύρων ΙΩΑΚΕΙΜ

153

2. Τηλεγράφημα τοῦ Οἰκουμενικοῦ Πατριάρχου Μελετίου ἀπό 22 Μαρτίου 1922 πρός τόν Ἐπίσκοπον Ροδοστόλου Ἀλέξανδρον ἀγγελτήριον τῆς ἐπαναφορᾶς τῆς διασπορᾶς ὑπό τήν δικαιοδοσίαν τοῦ Οἰκουμενικοῦ Πατριαρχείου καί τοῦ διορισμοῦ αὐτοῦ ὡς Πατριαρχικοῦ Ἐπιτρόπου (ἐκ τοῦ ἀρχείου τῆς Ἱ. Ἀρχιεπισκοπῆς Ἀμερικῆς)

PARIS 43

BISHOP RODOSTOLOU ALEXANDER
140 East 72nd Street, New York City

The Tome regarding the Church in the Diaspora having been revoked by a Patriarchal and Synodical decision. They (the Churches of the Diaspora) are restored again to the Pastoral administration of Ecumenical Patriarchate and the spiritual protection of all Greek Churches in America is assigned to you as the Patriarchal Delegate, until a newer decision. Announcing these to the Clergy and Laity, through you, we invoke God's blessing on all.

† Meletios
Patriarch of Constantinople

3. Ἔγγραφον τοῦ Ροδοστόλου Ἀλεξάνδρου πρός τόν Οἰκουμενικόν Πατριάρχην Μελέτιον ἀπό 12 Ἀπριλίου 1922, ἀπαντητικόν πρός τό προηγούμενον τηλεγράφημα (ἐκ τοῦ ἀρχείου τῆς Ἱ. Ἀρχιεπισκοπῆς Ἀμερικῆς)

Ἀριθμ. Πρωτ. 1204
Διακ. 1292

Τῇ Α.Θ. Παναγιότητι τῷ Οἰκουμενικῷ Πατριάρχῃ
Κυρίῳ Κυρίῳ Μελετίῳ τῷ Δ΄

Παναγιώτατε Δέσποτα,

Πρό καιροῦ ἐγενόμην εὐσεβάστως κάτοχος τοῦ ἐκ Παρισίων διαβιβασθέντος μοι τηλεγραφήματος τῆς Ὑμετέρας Θειοτάτης Παναγιότητος τοῦ περί τῆς ἄρσεως τοῦ Πατριαρχικοῦ Τόμου, δι' οὗ αἱ ἐν Διασπορᾷ Ἐκκλησίαι ἐτίθεντο ὑπό τήν Ἐκκλησιαστικήν ποιμαντορίαν τῆς Ἐκκλησίας τῆς Ἑλλάδος καί ὁ διορισμός μου ὡς Πατριαρχικοῦ Ἐπιτρόπου. Κατά τήν ἄφιξιν τοῦ τηλεγραφήματος τούτου εὑρισκόμην εἰς Σικάγον διά τό ζήτημα τῆς Ἱερ. Σχολῆς τοῦ Ἁγίου Ἀθανασίου.

Ἅμα τῇ εἰς Νέαν Ὑόρκην ἐπανόδῳ μου συνεκάλεσα τό Διευθ. Συμβούλιον τῆς Ἀρχιεπισκοπῆς καί ἀνεκοίνωσα εἰς αὐτό τό

περιεχόμενον τοῦ ἱστορικοῦ τούτου τηλεγραφήματος. Τό Συμβούλιον λαβόν γνῶσιν αὐτοῦ, ὡς ἦτο ἑπόμενον εὐφροσύνως
ἐχαιρέτισε τό σπουδαῖον τοῦτο διά τήν Ἐκκλησίαν τῆς Ἀμερικῆς γεγονός.

Γενομένης μικρᾶς τινος συζητήσεως ἐξ ἀφορμῆς τοῦ τηλεγραφήματος τούτου, ἐθεωρήθη καλόν ὅπως ἀνακοινωθῇ μέν
τοῦτο δι᾽ Ἐγκυκλίου πρός ὅλους τούς Ἱερεῖς, τούς τε ὑπό τήν
ἐμήν ταπεινήν Ἐκκλησιαστικήν Δικαιοδοσίαν καί τούς μή, πα-
ραπεμφθῇ δέ εἰς τήν κατά Μάϊον συγκαλουμένην Γενικήν Συνέλευσιν, ἵνα αὕτη λαμβάνουσα ἐπισήμως γνῶσιν καί τοῦ τηλεγραφήματος τούτου καί τοῦ ἀναμενομένου ἐν τῷ μεταξύ σχετικοῦ
Πατριαρχικοῦ Γράμματος, προβῇ εἰς τήν τροποποίησιν τῶν
σχετικῶν ἄρθρων τοῦ Καταστατικοῦ τῆς Ἀρχιεπισκοπῆς.

Ἡ ἄρσις τοῦ Πατριαρχικοῦ Τόμου τοῦ 1908 ἐνεποίησεν ἀρί-
στην ἐντύπωσιν, ἐάν δέ δέν ἐξεδηλώθη αὕτη τόσον ἐμφανῶς,
τοῦτο ἀποδοτέον εἰς τό γεγονός, ὅτι ὁ κόσμος εἶναι τόσον ἀπη-
σχολημένος μέ τά Ἐθνικά ζητήματα, ὥστε ταῦτα καί μόνα ν᾽
ἀπορροφῶσι καί τήν ὅλην του προσοχήν.

Ἤδη ἀναμένω τό σχετικόν σεπτόν μοι Πατριαρχικόν Γράμμα
συμφώνως πρός τό ὁποῖον καί θά ἐνεργήσω.

Ἐπί τούτοις κατασπαζόμενος τήν σεπτήν μοι Παναγίαν δε-
ξιάν τῆς Ὑμετέρας Θειοτάτης Παναγιότητος διατελῶ

Ἐν Νέᾳ Ὑόρκῃ τῇ 12ῃ Ἀπριλίου 1922

† ὁ Ροδοστόλου Ἀλέξανδρος

4. Πατριαρχικός καί Συνοδικός Τόμος Ἱδρυτικός τῆς Ἀρχιεπισκοπῆς Ἀμερικῆς Βορείου καί Νοτίου (ἐκ τοῦ ἀρχείου τῆς Ἱ. Ἀρχιεπισκοπῆς Ἀμερικῆς)

† ΜΕΛΕΤΙΟΣ
ΕΛΕΩ ΘΕΟΥ ΑΡΧΙΕΠΙΣΚΟΠΟΣ ΚΩΝΣΤΑΝΤΙΝΟΥΠΟΛΕΩΣ,
ΝΕΑΣ ΡΩΜΗΣ ΚΑΙ ΟΙΚΟΥΜΕΝΙΚΟΣ ΠΑΤΡΙΑΡΧΗΣ

Ἀριθμ. Πρωτ. 2794

Ἱερώτατε Ἀρχιεπίσκοπε Ἀμερικῆς Βορείου καί Νοτίου, ἐν ἁγίῳ
Πνεύματι ἀγαπητέ ἀδελφέ καί συλλειτουργέ τῆς ἡμῶν Μετριότητος κύριε Ἀλέξανδρε, χάρις εἴη τῇ αὐτῆς Ἱερότητι καί εἰρήνη
παρά Θεοῦ.

Ὡς συνέπεια τῆς ἀνακλήσεως τοῦ περί Ἐκκλησιῶν τῆς Διασπορᾶς Πατριαρχικοῦ καί Συνοδικοῦ Τόμου καί ἀποκαταστά-
σεως αὐτῶν αὖθις ὑπό τήν ποιμαντορίαν τοῦ Οἰκουμενικοῦ Θρόνου ἦλθεν ἡ ἵδρυσις ἐν Εὐρώπῃ τῆς Μητροπόλεως Θυατείρων

καί Έξαρχίας Δυτικής καί Κεντρώας Ευρώπης, έν Άμερική δέ τής «Όρθοδόξου Αρχιεπισκοπής Αμερικής Βορείου καί Νοτίου», έν ή εύθύς μετά τήν περί ίδρύσεως Πατριαρχικήν καί Συνοδικήν Πράξιν, ψήφων κανονικών γενομένων τή κς΄ τού παρελθόντος μηνός Απριλίου, έξελέγη ή ύμετέρα λίαν αγαπητή Ιερότης.

Καί τήν μέν διά κανονικής αποφάσεως ίδρυσιν τής Αρχιεπισκοπής καί τήν έν αύτή κανονικήν εκλογήν τής ύμετέρας Ιερότητος ήγγείλαμεν ήδη τηλεγραφικώς, προσθέμενοι καί τά συγχαρητήρια ήμών τε καί τής Εκκλησίας διά τήν προαγωγήν αύτής αποτελούσαν άμοιβήν δικαίαν τής ύπέρ τριετίαν εύδοκίμου επιτελέσεως έν Αμερική τίτλω άλλω τού ποιμαντορικού καθήκοντος, προαγόμεθα δέ σήμερον αποστείλαι τή ύμετέρα Ιερότητι καί τό κείμενον αύτό τού Πατριαρχικού καί Συνοδικού Τόμου τού ίδρυτικού τής Αρχιεπισκοπής.

Τόν Τόμον τούτον αποτελούντα τήν κανονικήν βάσιν τής περαιτέρω ύπάρξεως τής έν Αμερική Όρθοδόξου Εκκλησίας στελλόμενον έν ίσω καί απαραλλάκτω τώ έν τώ Πατριαρχικώ Κώδικι κατεστρωμένω εύαρεστηθήσεται μέν ή ύμετέρα Ιερότης δημοσιεύσαι αύτόν τε καί τήν σχετικήν Εγκύκλιον διά τε τού επισήμου εκκλησιαστικού οργάνου καί δι' αναγνώσεως έπ' εκκλησίαις πρός γνώσιν τού χριστεπωνύμου πληρώματος, επιμεληθήναι δέ ώστε προσαρμοσθήναι τοίς έν αύτώ οριζομένοις τήν πάσαν εκκλησιαστικήν διοίκησιν.

Καί οίδαμεν μέν, έκ τού σύνεγγυς τά έν Αμερική εκκλησιαστικά πράγματα μεμελετηκότες, όσον έστίν εργώδες τό εντελλόμενον, άλλ' έν τούτω γνώσονται πάντες καί τόν ζήλον καί τήν σύνεσιν καί τήν ύπομονήν καί τήν επιμονήν τής ύμετέρας αγαπητής Ιερότητος, πολύτιμον ήδη κεκτημένης κεφάλαιον πείρας περί τών έν Αμερική προσώπων τε καί πραγμάτων.

Πρόδηλον δέ ότι πρό παντός ή ύφεστώσα ήδη τώ αύτώ ονόματι συσσωμάτωσις προσαρμοσθήσεται ταίς διατάξεσι τού Τόμου, μνημονεύουσα αύτού ρητώς ώς τής κανονικής τής οργανώσεως βάσεως. Καί έν πρώτοις συγκροτητέον τό σύνολον ούτως ώστε ύπάρχειν ίδίαν Συνέλευσιν διά μίαν έκάστην τών τεσσάρων Επισκοπών αποτελουμένην έκ πάντων τών κανονικών κληρικών αύτής καί έξ ένός αιρετού λαϊκού αντιπροσώπου έκάστου ωργανωμένου είς Σωματείον ιερού Ναού, εκλεγομένου ύπό τού ενοριακού Συμβουλίου. Ή Γενική δέ Συνέλευσις ώς σώμα τής όλης Αρχιεπισκοπής συγκροτείσθω έκ τού Αρχιεπισκόπου, τών Επισκόπων καί 24 αιρετών μελών, 12 κληρικών καί 12 λαϊκών, εκλεγομένων καί τούτων καί εκείνων ανά 6 ύφ' έκάστης τών τεσσάρων Εκκλησιαστικών Συνελεύσεων. Ούτος δοκεί ήμίν τά

ἐν Ἀμερικῇ μεμελετηκόσιν ὁ κρείσσων τρόπος τῆς συσσω-
ματώσεως.

Τό ἐν ἑκάστῃ Ἐπισκοπικῇ δικαιοδοσίᾳ διευθῦνον Συμβούλιον
ἔσται οἷον καί τό νῦν ὑφιστάμενον, τό ὁποῖον, παραμένον ὡς τό
εἰδικόν Συμβούλιον τῆς Ἐπισκοπῆς τοῦ Ἀρχιεπισκόπου, δύναται εἶναι συγχρόνως πρός ἀποφυγήν πολλῶν Σωμάτων καί τό
διευθῦνον Συμβούλιον τῆς ὅλης Ἀρχιεπισκοπῆς ἐκπροσωπουμέ-
νης ὑπό τῆς ὡς ἄνω περιγραφείσης ἐξ ἀρχιερέων καί 24 αἱρετῶν
μελῶν Γενικῆς Συνελεύσεως. Ἀλλά καί τό νέον Καταστατικόν
οὐκ ὀφείλει χαρακτηρισθῆναι ὁριστικόν, ἀλλά διατηρείσθω καί
ἐν αὐτῷ ὁ τίτλος τοῦ προσωρινοῦ, μέχρις οὗ διά τῆς δοκιμαστικῆς
ἐξελίξεως φθάσωμεν εἰς τό τέλειον.

Τάς ἐπισκοπάς ὡρίσαμεν θεμελειωδῶς εἰς τέσσαρας διά λόγους σχετιζομένους πρός τήν ἀνάγκην αὐταρκείας πρός ἐπιτόπιον διοίκησιν ἀλλά καί πρός τό σχέδιον ὀργανώσεως τῆς ὅλης
ἐν Ἀμερικῇ Ὀρθοδοξίας εἰς ἕν ἑνιαῖον σύνολον. Ἐάν ἡ Γενική
Συνέλευσις ἀποφασίσῃ τήν ἐν τῷ παρόντι πλήρωσιν καί τῶν τριῶν
Ἐπισκοπικῶν ἑδρῶν ἐξασφαλίζουσα τήν συντήρησιν τῶν Ἐπισκόπων καί τῶν Ἐπισκοπικῶν Γραφείων, ἔσται τοῦτο ἡ ἀρίστη
λύσις. Ἄλλως μετά τόν καθορισμόν τῶν ὁρίων ἑκάστης τῶν τεσσάρων Ἐπισκοπῶν ἐπί Γενικῆς Συνελεύσεως φροντίσει ἡ Ὑμε-
τέρα Ἱερότης ὅπως ὑπό τῆς εἰδικῆς Συνελεύσεως τῆς Ἐπισκοπῆς Σικάγου ἀποτελουμένης, ὡς εἴπομεν ἀνωτέρω, ὑπό πάντων
τῶν κανονικῶν κληρικῶν τῆς Ἐπισκοπῆς καί ὑπό ἑνός αἱρετοῦ
μέλους ἑκάστου ὠργανωμένου ἱεροῦ Ναοῦ, ὑπό τήν προεδρίαν τῆς
Ὑμετέρας Ἱερότητος συνερχομένης, ἐκλέξει τρεῖς ὑποψηφίους,
ἐφ' ὧν ἡ περί ἡμᾶς Ἁγία Σύνοδος ποιήσεται τάς κανονικάς ψήφους πρός ἐκλογήν τοῦ ἑνός. Ὑποτιθεμένου δέ ὅτι ἡ προτίμησις
τῆς Συνελεύσεως ἔσται διά κληρικούς θεολόγους ἐκ τῶν αὐτόθι
ὑπηρετούντων, δέον ἐγκαίρως προταθῆναι ὑπό τῆς ὑμετέρας Ἱερότητος πρός ἐγγραφήν ἐν τῷ καταλόγῳ τῶν δι' ἀρχιερατείαν ἐκ-
λεξίμων τούς ὑπ' αὐτῆς κρινομένους ἀξίους. Ὅταν δέ ἀποκα-
τασταθῇ εἰς τήν Ἐπισκοπήν Σικάγου Ἐπίσκοπος, καί ἄν ἐπί τινα
χρόνον ἀναβληθῇ ἡ πλήρωσις τῶν δύο ἑτέρων Ἐπισκοπῶν, οὐκ
ἔσται μεγάλη ζημία, καίτοι, ἐπαναλαμβάνομεν, θεωροῦμεν πρός
τό συμφέρον αὐτό τῆς ἐν Ἀμερικῇ Ἐκκλησίας τό καταρτισθῆναι
τό ταχύτερον Σύνοδον Ἀρχιερέων πρός ἐπιτόπιον Συνοδικήν τῆς
Ἐκκλησίας διακυβέρνησιν, οὐδέ βλέπομεν δυσυπέρβλητον τό οἰ-
κονομικόν ἐμπόδιον, ὅταν καταστῇ συνειδητή ἡ ἀνάγκη. Ὅπως
ποτ' ἄν ᾖ εὐθύς μετά τήν ἐκλογήν Ἐπισκόπου Σικάγου δυνατόν
μερισθῆναι μετ' αὐτοῦ τήν ποιμαντορικήν φροντίδα τῶν χειρευου-
σῶν ἐπαρχιῶν, τοῦ μέν Ἀρχιεπισκόπου τήν Ἐπισκοπήν Βοστώνης, τοῦ δέ Ἐπισκόπου Σικάγου τήν Ἐπισκοπήν Καλιφορνίας

τοποτηρητικῶς ποιμένοντος.

Ταῦτα τῇ ὑμετέρᾳ ἀγαπητῇ ἡμῖν Ἱερότητι ἐπεξηγηματικῶς ἐπιστέλλοντες καί τήν τούτων ἐφαρμογήν καί τήν ὅλην ἀποκατάστασιν παρά τοῦ ζήλου αὐτῆς ἀπεκδεχόμενοι, ἐξαιτούμεθα αὐτῇ ἀρωγόν τήν ἄνωθεν ἀντίληψιν καί διατελοῦμεν

ᾳ ꙽κβ´ Μαΐου ιη´

ἐν Χῷ ἀγαπητός ἀδελφός
† Ὁ Κωνσταντινουπόλεως Μελέτιος

5. Γράμμα τοῦ Πατριάρχου Κωνσταντινουπόλεως Μελετίου ἀπό 5 Μαΐου 1922 πρός τόν Ἀρχιεπίσκοπον Ἀμερικῆς Ἀλέξανδρον. (Ἐκ τοῦ ἀρχείου τῆς Ἱ. Ἀρχιεπισκοπῆς Ἀμερικῆς)

† ΜΕΛΕΤΙΟΣ

ΕΛΕΩ ΘΕΟΥ ΑΡΧΙΕΠΙΣΚΟΠΟΣ ΚΩΝΣΤΑΝΤΙΝΟΥΠΟΛΕΩΣ, ΝΕΑΣ ΡΩΜΗΣ ΚΑΙ ΟΙΚΟΥΜΕΝΙΚΟΣ ΠΑΤΡΙΑΡΧΗΣ

Ἀριθμ. Πρωτ. 2536

Ἱερώτατε Ἀρχιεπίσκοπε Ἀμερικῆς, ἐν ἁγίῳ Πνεύματι ἀγαπητέ ἀδελφέ καί συλλειτουργέ τῆς ἡμῶν Μετριότητος κύριε Ἀλέξανδρε, χάρις εἴη τῇ αὐτῆς Ἱερότητι καί εἰρήνη παρά Θεοῦ.

Διά τοῦ ἀποσταλέντος ἤδη τηλεγραφήματος ἐγνωστοποιήθη τῇ Ἱερότητι αὐτῆς ἡ ἵδρυσις τῆς ἰδιαιτέρας Ἀρχιεπισκοπῆς Ἀμερικῆς μετά τῶν ὑπ᾽ αὐτήν ἐπισκοπῶν ἔν γε τῷ παρόντι Σικάγου, Βοστώνης καί Ἁγίου Φραγκίσκου, καθώς καί ἡ διά κανονικῶν ψήφων ἐκλογή τῆς αὐτῆς Ἱερότητος ὡς πρώτου Ἀρχιεπισκόπου τῆς νέας ταύτης ἐκκλησιαστικῆς περιοχῆς. Σύν τῇ ἐκφράσει νῦν καί αὖθις τῶν συγχαρητηρίων ἡμῶν καί τῶν εὐχῶν ὑπέρ πάσης εὐδοκιμήσεως καί τῇ δηλώσει ὅτι τιμῆς ἕνεκα ἡ θέσις τῆς νέας ταύτης Ἀρχιεπισκοπῆς ὡρίσθη 15η ἐν τῷ Συνταγματίῳ, πληροφοροῦμεν τήν Ἱερότητα αὐτῆς ὅτι τά καθ᾽ ἕκαστα τῆς συγκροτήσεως τῆς Ἀρχιεπισκοπῆς ὁρίζονται ἐν τῷ ὁσονούπω ἐκδιδομένῳ καί ἀποσταλησομένῳ πρός αὐτήν Πατριαρχικῷ καί Συνοδικῷ Τόμῳ. Ἐπιβεβαιοῦντες δέ καί τήν παραλαβήν τοῦ σχετικοῦ ἀπαντητικοῦ αὐτῆς τηλεγραφήματος, αἰτούμεθα τά κράτιστα παρά Θεοῦ, οὗ ἡ χάρις καί τό ἄπειρον ἔλεος εἴη μετ᾽ αὐτῆς.

ᾳ ꙽ κβ´ Μαΐου ε´

158

6. Έγκύκλιος ἀπό 3 Ἰουλίου 1924 τῆς ἐν Οὐασιγκτῶνι Ἑλ- ληνικῆς Πρεσβείας περί τῆς ὑπαγωγῆς τῆς ἐν Ἀμερικῇ Ἐκκλησίας εἰς τό Οἰκουμενικόν Πατριαρχεῖον. (Ἐκ τοῦ ἀρχείου τῆς Ἰ. Ἀρχιεπισκοπῆς Ἀμερικῆς)

ΕΛΛΗΝΙΚΗ ΠΡΕΣΒΕΙΑ
ΕΝ ΟΥΑΣΙΓΚΤΩΝΙ

Τῇ 3ῃ Ἰουλίου 1924

Πρός τάς ἐν Ἀμερικῇ Προξενικάς Ἀρχάς

Ἔχω τήν τιμήν ν' ἀνακοινώσω ὑμῖν ὅτι τό Ὑπουργεῖον τῶν Ἐξωτερικῶν διά τοῦ ὑπ' ἀριθμ. 19758 ἀπό 7 παρελθόντος μηνός πρός ἡμᾶς ἐγγράφου του γνωρίζει ἡμῖν ὅτι ἡ Ἱερά Σύνοδος τῆς Ἐκκλησίας τῆς Ἑλλάδος, ἐν τῇ συνεδριάσει αὐτῆς τῆς 16ης Μαΐου ἐ. ἔ. ἀνεγνώρισε τήν ἄρσιν τοῦ τόμου τοῦ 1908 καί τήν ἐπί τῶν ἐν τῇ διασπορᾷ Ὀρθοδόξων Ἑλληνικῶν Παροικιῶν πνευματικήν δικαιοδοσίαν τοῦ Οἰκουμενικοῦ Πατριαρχείου καί ἐντέλλεται συγ- χρόνως ὅπως τό γεγονός τοῦτο ἀνακοινώσωμεν ἐπειγόντως εἰς ἁπάσας τάς ἐν Ἀμερικῇ Ἑλληνικάς Κοινότητας.

Μεταδίδων ὑμῖν τ' ἀνωτέρω, παρακαλῶ ὑμᾶς ὅπως εὐαρεστού- μενοι φέρητε ταῦτα ἐπειγόντως εἰς γνῶσιν ὅλων τῶν ἐντός τῆς Προξενικῆς ὑμῶν δικαιοδοσίας ὑφισταμένων Κοινοτήτων, ὑπο- μιμνήσκοντες ἅμα εἰς αὐτάς καί τήν σχετικήν ὑπ' ἀριθμ. 1022 τῆς 19 Μαΐου 1924 ἐγκυκλίου τῆς Ἱερᾶς Συνόδου τῆς Ἐκκλησίας τῆς Ἑλλάδος, διά τῆς ὁποίας αὕτη, ἀναγνωρίζουσα τήν κατάργησιν τοῦ Πατριαρχικοῦ Τόμου τοῦ 1908, συνιστᾷ εἰς τάς ἐν Ἀμερικῇ Ὀρθοδόξους Ἑλληνικάς παροικίας, ὅπως ἀναγνωρίζουσιν ἐφε- ξῆς ὡς μόνην κανονικήν τήν ὑπό τοῦ Πατριαρχείου κατασταθεῖσαν ἐν Ἀμερικῇ Ἀρχιεπισκοπήν, τούς Ἐπισκόπους αὐτῆς καί τούς ὑπ' αὐτῆς καθισταμένους πνευματικούς ποιμένας τῶν ὁποίων μό- νον αἱ πράξεις ἔχουσι κῦρος, διότι αἱ πράξεις κληρικῶν ἀντικανο- νικαί ἄνευ ἀδείας τῆς κανονικῆς πνευματικῆς Ἀρχῆς καί ἄνευ ἐπι- σκοπικῆς εὐλογίας τελούμεναι, οὐδέν κῦρος δύνανται νά ἔχωσιν.

Ἐπίσης παρακαλῶ ὑμᾶς, ὅπως εἰς τ' ἀνωτέρω δώσητε καί διά τῶν ἐγχωρίων ἐφημερίδων ὅσον τό δυνατόν εὐρυτέραν δημοσιό- τητα καί συγχρόνως μοί γνωρίσητε τήν λῆψιν καί ἐκτέλεσιν τῆς παρούσης.

Ὁ Ἐπιτετραμμένος
Β. Μαμμωνᾶς

7. Καταστατικόν τῆς Ἑλληνικῆς Ἀρχιεπισκοπῆς Ἀμερικῆς Βορείου καί Νοτίου.

Νέα Ύόρκη 1923. Σχ. 20Χ14, σσ. 12+1 ἄ.ἀ

† Ὁ Οἰκουμενικός Πατριάρχης Μελέτιος ἐπικυροῖ

ΚΑΤΑΣΤΑΤΙΚΟΝ
ΤΗΣ ΕΛΛΗΝΙΚΗΣ ΑΡΧΙΕΠΙΣΚΟΠΗΣ ΑΜΕΡΙΚΗΣ
ΒΟΡΕΙΟΥ ΚΑΙ ΝΟΤΙΟΥ

ΑΡΘΡΟΝ Α΄

Ἱδρύεται Θρησκευτικόν Σωματεῖον ὑπό τήν ἐπωνυμίαν ΕΛΛΗ-ΝΙΚΗ ΑΡΧΙΕΠΙΣΚΟΠΗ ΑΜΕΡΙΚΗΣ ΒΟΡΕΙΟΥ ΚΑΙ ΝΟΤΙΟΥ, χάριν τῶν αὐτόθι Χριστιανῶν τῶν ἀνηκόντων εἰς τήν Ἁγίαν Ὀρθό-δοξον Ἀνατολικήν Ἐκκλησίαν καί ἐχόντων ὡς γλῶσσαν λειτουργι-κήν ἀποκλειστικῶς ἤ πρωτευόντως τήν Ἑλληνικήν, ἐν ᾗ ἐγράφησαν τά Ἅγια Εὐαγγέλια καί τά λοιπά βιβλία τῆς Καινῆς Διαθήκης.

ΑΡΘΡΟΝ Β΄

Σκοπός

Σκοπός τῆς Ἐκκλησίας ταύτης εἶναι νά οἰκοδομῇ τόν θρησκευτι-κόν καί ἠθικόν βίον τῶν Ἑλλήνων καί τῶν ἐξ Ἑλληνικῆς καταγω-γῆς Ὀρθοδόξων Ἀμερικανῶν πολιτῶν, ἐπί τῇ βάσει τῶν Ἁγίων γραφῶν, τῶν ὅρων καί τῶν Κανόνων τῶν Ἁγίων Ἀποστόλων καί τῶν Ἑπτά Οἰκουμενικῶν Συνόδων τῆς Ἀρχαίας Ἀδιαιρέτου Ἐκκλη-σίας, ὡς οὗτοι ἑρμηνεύονται ἐν τῇ πράξει τῆς ἐν Κωνσταντινουπόλει Μεγάλης τοῦ Χριστοῦ Ἐκκλησίας.

ΑΡΘΡΟΝ Γ΄

Διοικητική ὑπαγωγή

Ἡ ΕΛΛΗΝΙΚΗ ΑΡΧΙΕΠΙΣΚΟΠΗ ΑΜΕΡΙΚΗΣ ΒΟΡΕΙΟΥ ΚΑΙ ΝΟΤΙΟΥ διατελεῖ κανονικῷ καί ἱστορικῷ δικαιώματι ὑπό τήν Ἀνω-τάτην Πνευματικήν καί Ἐκκλησιαστικήν Ἐποπτείαν τοῦ Οἰκουμε-νικοῦ Πατριαρχείου Κωνσταντινουπόλεως.

ΑΡΘΡΟΝ Δ΄

Διοικητική διαίρεσις

Ἡ ὅλη Ἀρχιεπισκοπή διαιρεῖται εἰς τέσσαρας Ἐπισκοπικάς πε-ριφερείας.

1ον. Τήν Νέας Ὑόρκης. Αὕτη περιλαμβάνει τάς Πολιτείας:
NEW YORK, μέ τάς ἐν τῇ πόλει τῆς Νέας Ὑόρκης καί Μπροῦκλυν
  Κοινότητας: Schenectady, Syracuse, Rochester, Buffalo, Endicott.
CONNECTICUT — Stamford, New Haven, Ansonia, Waterbury,
  New Britain, Danielson, Norwich, Thompsonville.
NEW JERSEY — Newark, Orange, New Brunswick, Trenton, Paterson,
  Passaic.
PENNSYLVANIA — Philadelphia, Reading, Bethlehem, Altoona,

Wilkes-Barre, Pittsburgh, Vandergrift, Erie, New Castle, Monessen, Woodlawn, Ambridge.
WEST VIRGINIA — Wheeling, Weirton, Clarksburg.
VIRGINIA — Norfolk, Richmond.
MARYLAND — Baltimore.
DISTRICT OF COLUMBIA — Washington.
KENTUCKY.
NORTH CAROLINA.
SOUTH CAROLINA — Charleston.
GEORGIA — Savannah, Atlanta, Augusta.
MISSISSIPPI.
FLORIDA — Jacksonville, Tarpon Springs, Pensacola.
Τήν Ἐπικράτειαν τοῦ Μεξικοῦ καί τάς χώρας τῆς Κεντρώας καί Νοτίου Ἀμερικῆς.

2ον. Τήν τῆς Βοστώνης. Αὕτη περιλαμβάνει τάς Πολιτείας:
MAINE, μέ Κοινότητας Biddeford, Lewiston.
NEW HAMPSHIRE — Manchester, Nashua, Dover, Somersworth.
VERMONT.
MASSACHUSETTS — Boston, Lynn, Peabody, Worcester, Fitchburg, Holyoke, Springfield, Webster, Clinton, Somerville, Cambridge, New Bedford, Lawrence, Brockton, Ipswich, Chicopee Falls, Marlboro, Southbridge.
RHODE ISLAND — Providence, Newport, Pawtucket.
COMMUNITIES IN CANADA — Montreal, Toronto, Fort William.

3ον. Τήν τοῦ Σικάγου. Αὕτη περιλαμβάνει τάς Πολιτείας:
MINNESOTA, μέ Κοινότητας Minneapolis, Duluth.
WISCONSIN — Milwaukee, Racine, Fond Du Lac, Sheboygan.
MICHIGAN — Detroit, Flint.
IOWA — Sioux City, Mason City, Waterloo.
ILLINOIS — Chicago, Chicago Heights, Moline, Joliet, Rockford.
IDIANA — Gary, Indianapolis.
MISSOURI — Saint Louis, Kansas City.
NORTH DAKOTA.
SOUTH DAKOTA.
KANSAS.
NEBRASKA — Omaha.
OKLAHOMA — Oklahoma City.
ARKANSAS — Little Rock.
LOUISIANA — New Orleans, Shreveport.
TEXAS — Dallas, Fort Worth, Houston.
OHIO — Cleveland, Toledo, Akron, Youngstown, Canton, Columbus, Cincinnati, Dayton, Warren, Martin's Ferry, Lorraine.
TENNESSEE — Memphis, Nashville.
ALABAMA — Birmingham, Mobile.
CANADA — Winnipeg.

4ον. Τήν τοῦ Ἁγίου Φραγκίσκου. Αὕτη περιλαμβάνει τάς Πολιτείας:
WASHINGTON — Seattle.
OREGON — Portland.
MONTANA — Great Falls.
NEVADA — McGill.
UTAH — Salt Lake City, Price.
COLORADO — Denver, Pueblo.
CALIFORNIA — San Francisco, Oakland, Los Angeles, Sacramento.
ARIZONA.
WYOMING.
NEW MEXICO.
ALASKA.
WEST CANADA — British Columbia.

Ἡ τῆς Νέας Ὑόρκης Ἐπισκοπική περιφέρεια κατέχει θέσιν Ἀρχιεπισκοπῆς καί ἔχει ἐπί κεφαλῆς αὐτῆς τόν Ἀρχιεπίσκοπον Ἀμερικῆς. Αἱ λοιπαί τρεῖς εἶναι Ἐπισκοπαί καί ἑκάστη ἐξ αὐτῶν διατελεῖ ὑπό τόν ἴδιόν της Ἐπίσκοπον, ἔχοντα ὡς ποιμαντορικόν του τίτλον τόν ἐκ τῆς ἕδρας τῆς Ἐπισκοπῆς του.

ΑΡΘΡΟΝ Ε΄
Ἀμετάθετον Ἀρχιεπισκόπου καί Ἐπισκόπων

Ὁ Ἀρχιεπίσκοπος καί οἱ Ἐπίσκοποι εἶναι ἀμετάθετοι, κατά τούς θείους καί Ἱερούς Κανόνας. Ἐν περιπτώσει χηρείας τοῦ Θρόνου τῆς Ἀρχιεπισκοπῆς, δύναται νά ἐκλεγῇ Ἀρχιεπίσκοπος εἷς ἐκ τῶν τριῶν Ἐπισκόπων, τηρουμένων τῶν σχετικῶν περί ἐκλογῆς Ἐπισκόπου διατάξεων τοῦ παρόντος Καταστατικοῦ.

ΑΡΘΡΟΝ ΣΤ΄
Ἄδεια ἀπουσίας Ἐπισκόπων

Ἕκαστος τῶν Ἐπισκόπων δικαιοῦται κατ᾽ ἔτος εἰς μηνιαίαν ἀπουσίαν, ἀπό τῶν καθηκόντων του. Διά περιπλέον χρονικόν διάστημα, ὄχι πέραν τοῦ τετραμήνου, ἀπαιτεῖται ἄδεια τῆς Ἱ. Συνόδου τῆς Ἀρχιεπισκοπῆς.

ΑΡΘΡΟΝ Ζ΄
Ἡ Σύνοδος τῆς Ἀρχιεπισκοπῆς

Ὁ Ἀρχιεπίσκοπος, μετά τῶν τριῶν Ἐπισκόπων, συγκροτοῦσι τήν Σύνοδον τῆς Ἑλληνικῆς Ἀρχιεπισκοπῆς Ἀμερικῆς Βορείου καί Νοτίου, συνερχομένην ἀπαραιτήτως κατά τούς σχετικούς ὁρισμούς τῶν Ἱ. Κανόνων τῆς Ὀρθοδόξου Ἀνατολικῆς Ἐκκλησίας δίς τοῦ ἔτους, πρό τοῦ Πάσχα καί κατά τό Φθινόπωρον, ὅπου ἄν ὁρίσῃ ὁ Ἀρχιεπίσκοπος.

Ἡ Σύνοδος αὕτη ἔχει πάσας τάς ἐξουσίας καί τάς εὐθύνας, ὅσας οἱ Ἱ. Κανόνες ἀναγράφουσι διά τήν «Σύνοδον Ἐπαρχίας», εὐθύνεται δέ διά τήν ἀπαρέγκλιτον τήρησιν τῶν θείων Δογμάτων καί τῶν Ἱερῶν

162

Κανόνων τῆς Ὀρθοδόξου Ἀνατολικῆς Ἐκκλησίας, ἐνώπιον τῆς περί τόν Πατριάρχην Συνόδου τῶν Μητροπολιτῶν τοῦ Οἰκουμενικοῦ Θρόνου.

Ἐν περιπτώσει ἰσοψηφίας ἐν τῇ Συνόδῳ ἐκνικᾷ ἡ γνώμη ὑπέρ ἧς ἐτάχθη ὁ Ἀρχιεπίσκοπος, ὅστις εἶναι καί ὁ Πρόεδρος.

Τόν Ἀρχιεπίσκοπον ἀπόντα ἤ κωλυόμενον ἀντικαθιστᾷ ἐν τῇ προεδρίᾳ ὁ ἔχων τά πρεσβεῖα τῆς χειροτονίας Ἐπίσκοπος.

ΑΡΘΡΟΝ Η΄

Δικαιοδοσία καί ἁρμοδιότης τῶν Ἐπισκόπων

Ἕκαστος τῶν Ἐπισκόπων ἔχει παρά τῇ ἑαυτοῦ Ἐπισκοπῇ τήν ἐξουσίαν καί τάς εὐθύνας, ἅς οἱ θεῖοι καί Ἱεροί Κανόνες καί ἡ μακραίων τῆς Ἐκκλησίας πρᾶξις καθορίζουσι διά τό Ἐπισκοπικόν ἀξίωμα, μετά καί τῆς ἐν τῷ Ἱερῷ Συνθρόνῳ ἐγκαθιδρύσεως.

Ἐκ τῶν δικαιωμάτων καί τῶν καθηκόντων τούτων εἶναι καί τό ἁγιάζειν καί καθιεροῦν τῇ λατρείᾳ τούς Ναούς καί τούς εὐκτηρίους οἴκους, τό ἀποκαθιστᾷν ἐν αὐτοῖς τούς λειτουργοῦντας καί ψάλλοντας Ἱερεῖς, διακόνους καί λοιπούς κληρικούς· τό ἐπιμελεῖσθαι τῆς ἐν εὐσχημοσύνῃ καί τάξει καί πρός τό κοινόν συμφέρον διοικήσεως πάντων τῶν ἐν αὐτοῖς· τό ἐκδιδόναι τάς ἀδείας πρός τέλεσιν τοῦ Μυστηρίου τοῦ Γάμου καί τά Διαζευκτήρια Γράμματα τῶν ἐκ τούτων ἁρμοδίως ὡς διαλυθέντων κηρυχθέντων· τό διανέμειν τοῖς ἱερεῦσι τό Μῦρον τοῦ Ἁγίου Χρίσματος, λαμβανομένου διά τοῦ Ἀρχιεπισκόπου παρά τοῦ Οἰκουμενικοῦ Πατριάρχου.

ΑΡΘΡΟΝ Θ΄

Τάξις Μνημονεύσεως

Ἐν τοῖς μυστηρίοις καί ταῖς τελεταῖς ὀφείλουσι νά μνημονεύωσιν οἱ μέν ἱερεῖς καί διάκονοι τοῦ Κανονικοῦ Ἐπισκόπου αὐτῶν, οἱ δέ Ἐπίσκοποι τοῦ Ἀρχιεπισκόπου, ὁ δέ Ἀρχιεπίσκοπος τοῦ Οἰκουμενικοῦ Πατριάρχου.

ΑΡΘΡΟΝ Ι΄

Πνευματικόν Δικαστήριον

Παρ' ἑκάστῳ Ἐπισκόπῳ λειτουργεῖ Πνευματικόν Δικαστήριον, ἀποτελούμενον ἐκ δύο, τοὐλάχιστον, βαθμούχων ἱερέων καί τοῦ Ἐπισκόπου αὐτοῦ, ὡς Προέδρου, ἤ τοῦ κατά τήν Ἐκκλησιαστικήν τάξιν ἀντιπροσώπου καί ἀναπληρωτοῦ αὐτοῦ. Τό Δικαστήριον τοῦτο ἐκδικάζει πρωτοδίκως, πλήν τῶν ἐπαγομένων καθαίρεσιν, πάντα τά λοιπά κανονικά παραπτώματα τοῦ Κλήρου.

ΑΡΘΡΟΝ ΙΑ΄

Τά Πνευματικά Δικαστήρια τῶν Ἐπισκόπων δέν δικάζουν ὑποθέσεις, περί ὧν ἤθελον μορφώσει γνώμην ἐκ τῶν γενομένων προανακρίσεων, ὅτι δύναται νά ἐπιβληθῇ εἰς τόν κατηγορούμενον ἡ ποινή τῆς καθαιρέσεως. Τάς τοιαύτας ὑποθέσεις παραπέμπουσι πρός τήν

163

Ί. Σύνοδον τῆς Ἀρχιεπισκοπῆς.

Αἱ ἀποφάσεις τῶν Ἐπισκοπικῶν Πνευματικῶν Δικαστηρίων, αἱ καταγινώσκουσαι ποινήν ἀργίας ἄνω τῶν δύο μηνῶν, ὑπόκεινται ἐντός προθεσμίας 31 ἡμερῶν ἀπό τῆς κοινοποιήσεώς των, εἰς ἔκκλησιν ἐνώπιον τῆς Ί. Συνόδου τῆς Ἀρχιεπισκοπῆς· εἰς ἔκκλησιν δέ ἐνώπιον τῆς Ί. Συνόδου τοῦ Οἰκουμενικοῦ Πατριαρχείου ἐντός προθεσμίας 91 ἡμερῶν ὑπόκεινται αἱ ἀποφάσεις τῆς Ί. Συνόδου τῆς Ἀρχιεπισκοπῆς, αἱ καταγινώσκουσαι ἤ καθαίρεσιν ἤ ἀργίαν ἄνω τοῦ ἑνός ἔτους.

ΑΡΘΡΟΝ ΙΒ΄

Τοπικαί Ἐκκλησιαστικαί Συνελεύσεις

Παρ᾽ ἑκάστῃ τῶν Τεσσάρων Ἐπισκοπῶν ὑπάρχει ἰδία Ἐκκλησιαστική Συνέλευσις, ἀποτελουμένη ἐκ πάντων τῶν κανονικῶν Κληρικῶν τῆς Ἐπισκοπῆς καί ἐξ ἑνός αἱρετοῦ λαϊκοῦ ἀντιπροσώπου ἑκάστου Ἱεροῦ Ναοῦ, ὠργανωμένου εἰς Σωματεῖον, ἐκλεγομένου ὑπό τοῦ ἐνοριακοῦ Συμβουλίου.

Ἑκάστη τῶν Ἐκκλησιαστικῶν τούτων Συνελεύσεων συνέρχεται τῇ προσκλήσει τοῦ οἰκείου Ἐπισκόπου, τοῦ καί Προέδρου αὐτῆς, ἤ τοῦ ἀντιπροσωπεύοντος καί ἀναπληροῦντος αὐτόν εἰς περίπτωσιν ἀπουσίας του, τακτικῶς μέν ἅπαξ τοῦ ἔτους, κατά μῆνα Μάϊον, ἐκτάκτως δέ ὁσάκις ἤθελε κρίνει τοῦτο εὔλογον ὁ Ἐπίσκοπος. Ἀπαρτία ὑπάρχει ὅταν, πρός τῇ παρουσίᾳ τοῦ Προέδρου Ἐπισκόπου ἤ τοῦ Προεδρεύοντος Ἀναπληρωτοῦ αὐτοῦ, ὦσι παρόντα αὐτοπροσώπως καί τά 12 ἐκ τῶν μελῶν, ἐξ ὧν τά 6 τουλάχιστον πρέπει νά εἶναι κληρικοί.

ΑΡΘΡΟΝ ΙΓ΄

Γενική Ἐκκλησιαστική Συνέλευσις

Ἡ Γενική Συνέλευσις τῆς ὅλης Ἑλληνικῆς Ἀρχιεπισκοπῆς Ἀμερικῆς Βορείου καί Νοτίου ἀποτελεῖται ἐκ τοῦ Ἀρχιεπισκόπου, τῶν Ἐπισκόπων καί 24 αἱρετῶν μελῶν, 12 κληρικῶν καί 12 λαϊκῶν. Καί οὗτοι καί ἐκεῖνοι ἐκλέγονται ἀνά ἕξ ὑφ᾽ ἑκάστης τῶν τεσσάρων Ἐκκλησιαστικῶν Συνελεύσεων.

Ἡ Γενική Συνέλευσις συνέρχεται τακτικῶς μέν ἀνά πᾶσαν διετίαν, κατά μῆνα Σεπτέμβριον, προσκλήσει τοῦ Ἀρχιεπισκόπου, ἐκτάκτως δέ μετ᾽ ἀπόφασιν περί τούτου τῆς Συνόδου τῆς Ἀρχιεπισκοπῆς.

Πρόεδρος τῆς Γενικῆς Συνελεύσεως εἶναι ὁ Ἀρχιεπίσκοπος, ἐν ἀπουσίᾳ δ᾽ αὐτοῦ ὁ ἐκ τῶν παρόντων Ἐπισκόπων ἔχων τά πρεσβεῖα τῆς χειροτονίας. Ἡ ἀπαρτία ὑπάρχει ὅταν, πρός τῇ παρουσίᾳ τοῦ Προέδρου Ἀρχιεπισκόπου ἤ τοῦ Προεδρεύοντος Ἐπισκόπου καί ἑνός τουλάχιστον ἄλλου Ἐπισκόπου, ὦσι παρόντα αὐτοπροσώπως καί 12 ἐκ τῶν μελῶν, ἐξ ὧν τά 6 κληρικοί.

ΑΡΘΡΟΝ ΙΔ΄

Ἀντιπροσώπευσις λόγῳ ἀποστάσεως

Καί εἰς τάς ἰδιαιτέρας ἑκάστης Ἐπισκοπῆς Συνελεύσεις καί εἰς

164

τήν Γενικήν Συνέλευσιν τῆς ὅλης Ἀρχιεπισκοπῆς, οἱ δικαιούμενοι εἰς συμμετοχήν Κληρικοί καί Λαϊκοί, δύνανται νά ἀντιπροσωπευθῶσιν οἱ Κληρικοί ὑπό Κληρικοῦ καί Λαϊκοί ὑπό Λαϊκοῦ, ἁρμοδίως πληρεξουσιοδοτημένου.

Κατά τάς ψηφοφορίας, ὁ ἀντιπρόσωπος ἀπόντων μελῶν διαθέτει καί τάς ψήφους των· ἐξαιρεῖται ἡ ψηφοφορία πρός ὑπόδειξιν Ἐπισκόπων, ὅτε ἕκαστος τῶν παρόντων διαθέτει μίαν μόνον ψῆφον.

Ἡ ἐντολή ἀντιπροσωπεύσεως ἅπαξ δοθεῖσα καί ὑπό τῆς Συνελεύσεως ἀναγνωρισθεῖσα δέν ἀνακαλεῖται καί ἰσχύει καθ᾽ ὅλην τήν διάρκειαν τῶν ἐργασιῶν τῆς Συνελεύσεως.

Ἐφ᾽ ὅσον ἡ ἁρμοδία Συνέλευσις δέν προνοήσῃ, πᾶσα λεπτομέρεια, ἀφορῶσα τήν συγκρότησιν τοῦ Σώματος καί τήν διεξαγωγήν τῶν ἐργασιῶν αὐτοῦ καθορίζεται ὑπό τοῦ οἰκείου Ἐπισκόπου ἤ τοῦ ἀναπληρωτοῦ αὐτοῦ.

ΑΡΘΡΟΝ ΙΕ΄

Τά ἔργα τῶν Συνελεύσεων

Δικαίωμα καί καθῆκον τῆς παρ᾽ ἑκάστῃ Ἐπιτροπῇ Ἐκκλησιαστικῆς Συνελεύσεως εἶναι νά ἐλέγχῃ τήν ὅλην Ἐκκλησιαστικήν διαχείρισιν, ψηφίζῃ δέ ἐκ συμφώνου μετά τοῦ οἰκείου Ἐπισκόπου διατάξεις, διά τήν διοίκησιν τῶν Ἐκκλησιαστικῶν ἱδρυμάτων, συμφώνως μέ τούς Ἱ. Ἀποστολικούς καί Συνοδικούς Κανόνας καί κατά τούς σχετικούς Νόμους τῶν Ἡνωμένων Πολιτειῶν. Ἵνα τεθῶσιν ἐν ἰσχύϊ αἱ διατάξεις αὗται προαπαιτεῖται ἡ ἔγκρισις αὐτῶν ὑπό τῆς περί τόν Ἀρχιεπίσκοπον Συνόδου.

Διατάξεις ἰδιαιτέρων ἐνοριακῶν Καταστατικῶν, ἀντιβαίνουσαι πρός τούς θείους καί Ἱ. Κανόνας καί τούς Νόμους τῶν Ἡνωμένων Πολιτειῶν εἶναι αὐτοδικαίως ἄκυροι.

Ἡ Γενική Συνέλευσις λαμβάνει ἀποφάσεις καί ἐγκρίνει μέτρα κοινῆς δράσεως, καθ᾽ ὅλην τήν Ἀρχιεπισκοπήν, πρός αὔξουσαν ἀείποτε πρόοδον ταύτης ἐν τῇ πραγματοποιήσει τοῦ ἐν τῷ Β΄ ἄρθρῳ καθοριζομένου θρησκευτικοῦ, ἠθικοῦ καί κοινωνικοῦ ἔργου της.

ΑΡΘΡΟΝ ΙΣΤ΄

Ἐκλογή Ἀρχιεπισκόπου καί Ἐπισκόπων

Ὑπάρχοντος σήμερον μόνον Ἀρχιεπισκόπου, ἐκλεγέντος κανονικῶς ὑπό τῆς ἐν Κωνσταντινουπόλει Ἱ. Συνόδου, ἡ πλήρωσις τῶν Πατριαρχικῇ καί Συνοδικῇ ἀποφάσει ἱδρυομένων, πρός τῇ Ἀρχιεπισκοπῇ Νέας Ὑόρκης, τριῶν Ἐπισκόπων, Σικάγου, Βοστώνης καί Ἁγίου Φραγκίσκου, γενήσεται διά πρώτην φοράν ὡς ἑξῆς:

Μετά τήν νόμιμον κύρωσιν τοῦ Καταστατικοῦ τούτου τοῦ καθορίζοντος καί τά ὅρια ἑκάστης τῶν Ἐπισκοπῶν, θέλει συνέλθει πρώτη ἡ ἰδιαιτέρα Ἐκκλησιαστική Συνέλευσις Σικάγου, τῇ προσκλήσει καί ὑπό τήν Προεδρίαν τοῦ Ἀρχιεπισκόπου, πρός πρότασιν τριῶν ὑποψηφίων διά τήν Ἐπισκοπήν ταύτην, ληφθησομένων ἐκ καταλόγου Κληρικῶν τῆς ὅλης Ἀρχιεπισκοπῆς, κεκτημένων δίπλωμα ἀνεγνωρισμένης Ὀρθοδόξου Θεολογικῆς Σχολῆς, ἄμεμπτον βίον καί ἐκκλησιαστικήν ἐμπειρίαν, προεγκεκριμένου δέ ὑπό τοῦ Οἰκουμενικοῦ Πατριάρχου καί τῆς περί αὐτόν Ἁγίας καί Ἱερᾶς Συνόδου. Ἐκ τῶν τριῶν δέ τούτων ἡ Σύνοδος αὕτη θά ἐκλέξῃ διά κανονικῆς ψήφου τόν Ἐπίσκοπον.

ΑΡΘΡΟΝ ΙΖ´

Μετά τήν οὕτω γενησομένην πλήρωσιν καί τῶν τριῶν Ἐπισκοπῶν, ἡ ἐκλογή ἐν τῷ μέλλοντι, εἰς περίπτωσιν κενώσεως τινός ἐξ αὐτῶν, ἤ καί τῆς Ἀρχιεπισκοπῆς, θά γίνηται ὡς ἑξῆς: Ἐντός τριῶν μηνῶν ἀνυπερθέτως, ἡ ἰδιαιτέρα Ἐκκλησιαστική Συνέλευσις, συνερχομένη ὑπό τήν Προεδρίαν τοῦ Τοποτηρητοῦ καί μετά πρόσκλησιν παρ᾽ αὐτοῦ, κατόπιν ἐντολῆς τῆς Συνόδου, θά προτείνῃ τρεῖς ὑποψηφίους ἐκ τοῦ καταλόγου τοῦ προεγκεκριμένου ὑπό τῆς ἐν Κωνσταντινουπόλει Ἁγίας καί Ἱερᾶς Συνόδου τοῦ Οἰκουμενικοῦ Πατριαρχείου. Ἐκ τούτων ἡ περί τόν Ἀρχιεπίσκοπον Σύνοδος θά ἐκλέγη τόν Ἐπίσκοπον, ἤ, προκειμένου περί τῆς Ἀρχιεπισκοπῆς, τόν Ἀρχιεπίσκοπον. Ἡ δέ τοιαύτη ἐκλογή θά γνωρίζηται ὑπό τῆς Συνόδου πρός τό Οἰκουμενικόν Πατριαρχεῖον διά τήν παροχήν τῆς ἐγκρίσεως.

ΑΡΘΡΟΝ ΙΗ´

Τόν οὕτως ἐκλεγέντα καί ἐγκριθέντα, ἄν μέν εἶναι ὁ Ἀρχιεπίσκοπος χειροτονεῖ ὁ Οἰκουμενικός Πατριάρχης· ἄν δέ Ἐπίσκοπος, δίδει ὁ Πατριάρχης μόνον τήν πρός χειροτονίαν ἄδειαν.

ΑΡΘΡΟΝ ΙΘ´

Ἐάν πρό τῆς πληρώσεως κενωθείσης Ἐπισκοπῆς, ἤθελε κενωθῇ καί ἑτέρα, οὕτως ὥστε οἱ ἀπομένοντες δύο Ἐπίσκοποι νά μή δύνανται πλέον μόνοι αὐτοί νά ἀπαρτίσωσι Σύνοδον, ἡ ἐκλογή μετά τῶν προταθέντων ὑπό τῆς οἰκείας ἰδιαιτέρας Συνελεύσεως, διά τήν πρώτην χηρεύσασαν Ἐπισκοπήν γενήσεται ἐν τῇ περί τόν Οἰκουμενικόν Πατριάρχην Συνόδῳ, συμψήφων γινομένων καί τῶν ἐν Ἀμερικῇ Ἐπισκόπων, μεθ᾽ ὅ ἡ ἐκλογή τοῦ ἑτέρου Ἐπισκόπου γενήσεται κατά τά ἐν τῷ Ἄρθρῳ ΙΖ´ ὑπό τῆς περί τόν Ἀρχιεπίσκοπον λειτουργούσης πάλιν Συνόδου.

ΑΡΘΡΟΝ Κ´

Διευθύνον Συμβούλιον

Καί παρά τῇ Ἀρχιεπισκοπῇ Νέας Ὑόρκης καί παρ᾽ ἑκάστῃ τῶν τριῶν Ἐπισκοπῶν λειτουργεῖ Διευθύνον Συμβούλιον. Τό Συμβούλιον τοῦτο ἀποτελεῖται ἐν τῇ Ἀρχιεπισκοπῇ Νέας Ὑόρκης, ἐκ τοῦ Ἀρχιεπισκόπου αὐτοῦ, ὡς Προέδρου καί τεσσάρων βαθμούχων Κληρικῶν τῆς Ἀρχιεπισκοπῆς, ὑπό τοῦ Ἀρχιεπισκόπου διοριζομένων, ὡς καί τεσσάρων ἐγκρίτων Λαϊκῶν μελῶν, προτεινομένων μέν ὑπό τοῦ Ἀρχιεπισκόπου, ἐγκρινομένων δέ ὑπό τῆς οἰκείας Ἐκκλησιαστικῆς Συνελεύσεως.

Τό αὐτό ἰσχύει καί δι' ἕκαστον Διευθύνον Συμβούλιον ἑκάστης τῶν Ἐπισκοπῶν· μέ μόνον τήν διαφοράν ὅτι, διά ταῦτα, τά μέλη, Κληρικά καί Λαϊκά, δύνανται νά εἶναι καί ἀνά τρία μόνον, ἐν ἐλλείψει πλείονος ἀριθμοῦ Ἱερέων ἐν τῇ ἕδρᾳ τῆς Ἐπισκοπῆς καί τοῖς πλησιοχώροις αὐτοῖς.

ΑΡΘΡΟΝ ΚΑ΄

Ἡ ἐντολή καί τῶν Κληρικῶν καί τῶν Λαϊκῶν μελῶν τοῦ Διευθύνοντος Συμβουλίου διαρκεῖ ἐπί διετίαν. Ἀντιπρόεδρος τοῦ Συμβουλίου εἶναι ὁ ἐκ τῶν μελῶν αὐτοῦ κατά βαθμόν ἀνώτερος Κληρικός, ὅστις καί διατηρεῖ τήν ψῆφόν του ὡς μέλος. Τόν Ταμίαν καί τόν Γραμματέα ἐκλέγει τό Διευθύνον Συμβούλιον ἐκ τῶν μελῶν αὐτοῦ.

Ἡ δικαιοδοσία τοῦ Διευθύνοντος Συμβουλίου ἐπεκτείνεται ἐπί πάσας τάς ὑποθέσεις τοῦ Σωματείου, ἐκτός ἐκείνων, αἵτινες, κατά τούς Κανόνας τῆς Ἐκκλησίας, θεωροῦνται ἀποκλειστικόν δικαίωμα τοῦ Ἐπισκόπου, μόνου, ἤ μετά τοῦ Πνευματικοῦ Δικαστηρίου ἐνεργοῦντος· δικαιοῦται δέ νά θέτη ἐν ἐνεργείᾳ Κανονισμούς ἐν τῷ κύκλῳ τῆς δικαιοδοσίας αὐτοῦ, μή ἀντιβαίνοντας εἰς τό παρόν Καταστατικόν.

ΑΡΘΡΟΝ ΚΒ΄

Τήν Ἀρχιεπισκοπήν χηρεύουσαν τοποτηρητεύει ὁ Ἐπίσκοπος, ὁ ἔχων τά πρεσβεῖα τῆς χειροτονίας. Τάς Ἐπισκοπάς Βοστώνης καί Σικάγου, χηρευούσας, τοποτηρητεύει, διά τό ἐγγύτερον, ὁ Ἀρχιεπίσκοπος· τήν δέ τοῦ Ἁγίου Φραγκίσκου, διά τόν αὐτόν λόγον, ὁ τοῦ Σικάγου.

ΑΡΘΡΟΝ ΚΓ΄

Ἱερατική Σχολή

Ἡ Ἑλληνική Ἀρχιεπισκοπή Ἀμερικῆς Βορείου καί Νοτίου ἔχει εἰδικήν διά τήν μόρφωσιν τοῦ Κλήρου Σχολήν. Τῆς Σχολῆς ταύτης προΐσταται, ὑπό τήν Προεδρίαν τοῦ Ἀρχιεπισκόπου Ἐφορία, περιλαμβάνουσα τέσσαρα Κληρικά μέλη καί τρία Λαϊκά, διοριζόμενα ὑπό τοῦ Ἀρχιεπισκόπου ἤ ὑπό τῆς Συνόδου, μετά τόν σχηματισμόν τοιαύτης, ἐκ τῶν μᾶλλον μεμορφωμένων Κληρικῶν καί Λαϊκῶν τῆς Ἀρχιεπισκοπῆς Νέας Ὑόρκης.

Τόν Διοικητικόν Ὀργανισμόν τῆς Σχολῆς καί τό πρόγραμμα τῶν μαθητῶν καταρτίζει ἡ Ἐφορία καί ἐγκρίνει ὁ Ἀρχιεπίσκοπος, μετά δέ τήν συγκρότησιν αὐτῆς, ἡ Σύνοδος τῆς Ἀρχιεπισκοπῆς.

Ἡ Σχολή ἔχει ἰδιαίτερον Ταμεῖον, οὗ αἱ πρόσοδοι χρησιμοποιοῦνται ἀποκλειστικῶς διά τόν σκοπόν ἐκπαιδεύσεως, πρός μόρφωσιν Κλήρου καί Διδασκάλων.

ΑΡΘΡΟΝ ΚΔ΄

Ἀκροτελεύτιοι Διατάξεις

Οἱ ἀπό χειροτονίας Κληρικοί τῆς Ἑλληνικῆς Ἀρχιεπισκοπῆς Ἀμερικῆς Βορείου καί Νοτίου ἀποκαθίστανται εἰς τάς διακονίας αὐτῶν, ἄνευ ἀναμίξεως οἱασδήποτε πολιτικῆς Ἀρχῆς. Νόμος δέ χειραγωγός εἰς ὅλην τήν διοίκησιν αὐτῆς θά εἶναι οἱ Ἱεροί Ἀποστολικοί καί Συνοδικοί Κανόνες καί οἱ ἁρμοδίως τιθέμενοι εἰς ἐνέργειαν κανονισμοί, συμφώνως πρός τούς Ἱ. Κανόνας καί τούς Νόμους ἑκάστης Πολιτείας, ἐν ᾗ ἐκτείνεται ἡ τῆς Ἐκκλησίας ταύτης διοίκησις.

ΑΡΘΡΟΝ ΚΕ΄

Πᾶσα διάταξις οἱουδήποτε Ἐκκλησιαστικοῦ Ὀργάνου, ἀντιβαίνουσα εἴτε πρός τούς Ἱερούς Κανόνας τῆς Ὀρθοδόξου Ἐκκλησίας εἶναι αὐτοδικαίως ἄκυρος.

ΑΡΘΡΟΝ ΚΣΤ΄

Μέχρι τῆς ἐκλογῆς τῶν Ἐπισκόπων καί τοῦ καταρτισμοῦ τῆς Ἱ. Συνόδου, διά τήν Γενικήν Συνέλευσιν, τό Πνευματικόν Δικαστήριον, τό Διευθῦνον Συμβούλιον, καί τήν Ἱερατικήν Σχολήν, θέλουσι ἰσχύει αἱ διατάξεις τοῦ προηγουμένου προσωρινοῦ Καταστατικοῦ.

ΑΡΘΡΟΝ ΚΖ΄

Τό παρόν Καταστατικόν ἐψηφίσθη ἐν Γενικῇ Συνελεύσει, συγκροτηθείσῃ, κατά τό Δ΄ Ἄρθρον τοῦ Προσωρινοῦ Καταστατικοῦ τῆς Ἑλληνικῆς Ἀρχιεπισκοπῆς Ἀμερικῆς Βορείου καί Νοτίου, ὑπόκειται δέ εἰς ἀναθεώρησιν, ἄν κριθῇ εὔλογον, μετά διετῆ ἐφαρμογήν ἀπό τῆς νομίμου κυρώσεως αὐτοῦ.

ΝΕΑ ΥΟΡΚΗ, 11 Αὐγούστου 1922.

ΣΗΜΕΙΩΣΕΙΣ

[1]Τζωρτζάτου Β., Μητροπολίτου Κίτρους, *Ἡ εἰς τήν Ἐκκλησίαν τῆς Ἑλλάδος, ὑπαγωγή τῶν ἐν Διασπορᾷ Ἑλληνικῶν Ἐκκλησιῶν καί ἀνάκλησις αὐτῆς.* Ἐν Ἀθήναις 1977, σ. 3.

[2] Ὁ κη΄ κανών διαλαμβάνων καί τά ἑξῆς: «... τούς τῆς Ποντικῆς καί τῆς Ἀσιανῆς καί τῆς Θρακικῆς διοικήσεως μητροπολίτας μόνους, ἔτι δέ καί τούς ἐν τοῖς βαρβαρικοῖς ἐπισκόπους τῶν προειρημένων διοικήσεων χειροτονεῖσθαι ὑπό τοῦ προειρημένου ἁγιωτάτου θρόνου τῆς κατά Κωνσταντινούπολιν ἁγιωτάτης Ἐκκλησίας ...» (Γ. Ράλλη — Μ. Ποτλῆ, *Σύνταγμα τῶν Θείων καί Ἱερῶν Κανόνων*, τόμ. Ι-VI, Ἀθῆναι 1852 - 1859, τόμ. ΙΙ, σελ. 280 - 286) ἀπονέμει εἰς τό Πατριαρχεῖον Κωνσταντινουπόλεως de jure ὅ,τι τοῦτο de facto ἤδη κατεῖχε.

[3]βλ. Τζωρτζάτου Β., ἔ.ἀ., σ. 4.

[4]πβ. Καλογήρου Ι., *Ἡ ἐγκαθίδρυσις Ἕλληνος Ἀρχιεπισκόπου εἰς Βενετίαν κατά τά τέλη τοῦ ις΄ αἰῶνος.* Padova 1973.

⁵Γερμανοῦ Ἱεραπόλεως, *Ἡ Ὀρθόδοξος Ἑλληνική Ἀρχιεπισκοπή Β. καί Ν. Ἀμερικῆς*, ΘΗΕ, τ. 2, σ. 330.

⁶Τζωρτζάτου Β., ἔ.ἀ. σ. 4.

⁷Ὁ τόμος οὗτος δημοσιεύεται ἐν Τζωρτζάτου Β., ἔ.ἀ. σ. 5 - 9.

⁸βλ. Κονιδάρη Γ. Ι., *Ἐκκλησιαστική Ἱστορία τῆς Ἑλλάδος*, Τόμος Β΄, Ἐν Ἀθήναις 1970, σ. 266.

⁹Κωνσταντινίδου, Ι. Χ., *Μελέτιος ὁ Μεταξάκης*, ΘΗΕ, τ. 8, στ. 967 - 968.

¹⁰βλ. ἐν Παραρτήματι. *Ἐπίσημα ἔγγραφα*, τό ὑπ᾽ ἀριθμ. 1.

¹¹βλ. ἔγγραφον ὑπ᾽ ἀριθμ. 2 εἰς τό Παράρτημα.

¹²βλ. ἔγγραφον ὑπ᾽ ἀριθμ. 3 εἰς τό Παράρτημα.

¹³βλ. ἔγγραφον ὑπ᾽ ἀριθμ. 4 ἐν Παραρτήματι.

¹⁴βλ. ἔγγραφον ὑπ᾽ ἀριθμ. 5 ἐν Παραρτήματι.

¹⁵Β.Φ.Α., *Φιλάρετος, Μητροπολίτης Σύρου*, ΘΗΕ τ. 11, στ. 1067.

¹⁶Πβ. Τζωρτζάτου Β., ἔ.ἀ., σ. 13.

¹⁷βλ. τό ὡς ἄνω γράμμα ἐν «*Ἐκκλησίᾳ*» Α (1924) 459 - 460.

¹⁸βλ. Στράγκα Θ., *Ἐκκλησίας τῆς Ἑλλάδος Ἱστορία ἐκ πηγῶν ἀψευδῶν* (1817 - 1967), τ. Β΄ Ἀθῆναι 1970, σσ. 1265 - 1266.

¹⁹βλ. ταύτην ἐν «*Συνοδικαί Ἐγκύκλιοι*», τ. Α΄ (1901 - 1933). Ἀθῆναι 1955, σσ. 422 - 423.

²⁰βλ. τοῦτο ἐν «*Ἐκκλησία*», Α (1924), σ. 460.

²¹βλ. τήν ὡς ἄνω ἐγκύκλιον ἐν τῷ Παραρτήματι ἐν τῷ ἀριθμῷ 6.

²²βλ. τοῦτο ἐν τῷ Παραρτήματι ὑπ᾽ ἀριθμ. 7.

²³βλ. τοῦτο ἐν τῷ Ἐκκλησιαστικῷ Κήρυκι, Α (1921), σσ. 75 - 77.

²⁴βλ. περί *Μελετίου Μεταξάκη* ἐν ΗΘΕ, ἔ.ἀ.

*Archbishop Athenagoras
and Ecumenical Patriarch
of Constantinople
served as Archbishop 1930 - 1949.*

*Archbishop Michael served
as Archbishop 1949 - 1958.*

*Archbishop Iakovos
has been serving the Archdioce
1958 to the present.*

Our Three Hierarchs

Evaluating the Ministry of
of Archbishops Athenagoras, Michael and Iakovos

By
Peter T. Kourides

I

he Greek Archdiocese was legally incorporated on September 19, 1921 and its existing Synodical Exarch, the Bishop of Rodostolon became its first primate as Archbishop Alexander. However, there was no functioning archdiocesan administration until nearly a decade later when Archbishop Athenagoras arrived.

Unfortunately, the nine year reign of Archbishop Alexander was literally an undisciplined nightmare that he was unable to control or even restrain. During this interval, our parishes and communities became so politicized and divided that Venizelists and Royalists used physical violence even within the sanctuary of the holy altar. Often, the police were called in to take strategic positions in the community center and even in some of our churches to avert bloodshed. And, of course, these disgraceful incidents were duly reported on the front pages of the American press to the humiliation and irreparable damage of the reputation and good name of our people throughout the nation. The reign of Archbishop Alexander was finally terminated on June 19, 1930 by his reassignment abroad at the direction of the Ecumenical Patriarchate. Therefore, this article will be limited to the three archbishops that followed.

With the ascendancy of Photios II to the Ecumenical Throne, the Metropolitan of Corinth Damaskinos was dispatched in May 1930 in an effort to bring peace amongst the two warring ghettocized factions that had devastatingly disarrayed and virtually destroyed our Church, and its image, in this country. Fortunately, on August 13, 1930 the then Metropolitan of Kerkyra Athenagoras was elected as our Archbishop.

Into this depressingly devastating and hopelessly chaotic state of our Church there came, on February 24, 1931, this dedicated, unsophisticated and ascetic-looking churchman. His only prayerful message was that the love of God for man required the prompt conciliation of the existing differences between parishioners so that the Church might again be united.

The Archdiocese was then situated in an antiquated frame house in Astoria which could be neither water-proofed from heavy rains nor heated in the winter months. The entire archdiocesan budget in 1932 was $22,000. The Archbishop's compensation was $200 per month which was subsequently increased until it reached $400 at the time that he left for Constantinople, on January 23, 1949, for his enthronement as Ecumenical Patriarch.

The first three-year period of the Athenagoras reign was truly martyrdom. It is doubtful whether any other prelate would have put up with the insurmountable difficulties and bitter opposition that he faced with such Christ-like equanimity, tact and patience.

Incredibly, Archbishop Athenagoras discovered at the very outset that the biggest obstacle in the path of a disciplined operative church organization was the then existing clergy. Some of them, unfortunately, looked to the priesthood as an occupation to be conducted at all costs for the largest possible personal monetary intake. A number of these priests had little education and several of them had been ordained for a few hundred dollars by various hierarchs of questionable legitimacy. Many of these clerics looked at Archbishop Athenagoras with overt distrust and suspicion, a few would not even permit him to officiate in their churches, which they considered as their own personal properties.

Slowly but surely, Archbishop Athenagoras began to win the confidence of our people. His tactful patience and ascetic self denial impressed our communicants to such an extent that the dissident clergy had no choice but to reluctantly go along. After establishing some semblance of discipline in our Church and clergy, he started the equally difficult task of uniting our battling communities by personally visiting churches in many cities. By 1940, aided by a quieting down in Greek politics, Athenagoras succeeded so thoroughly in this objective that strife and factionalism were completely eliminated.

It became apparent to Archbishop Athenagoras at the very outset that little would be accomplished for Greek Orthodoxy in this hemisphere unless it became possible to select our own clergy from young men who were born and educated in the Americas. He was convinced that only such clergy would be able to communicate with and serve our younger generation. In 1937, he succeeded in raising $35,000 with which he bought the Clark Estate in Pomfret Center, Connecticut. On October 3, 1937 under the deanship of the then Bishop of Boston (Athenagoras Cavadas) and later the Metropolitan of Thyateiron in Great Britain our theological school began functioning. It had a faculty of three and fourteen students. In 1947, the Holy Cross Theological Seminary relocated to its present multimillion dollar campus in Brookline, Massachusetts and subsequently became a part of Hellenic College.

Archbishop Athenagoras also felt a great need for a sanctuary for abandoned, ill-treated and orphaned children. In December 1943 there was made available the 400 acre estate of Jacob Ruppert, the deceased brewer and owner of the New York Yankees. This beautifully landscaped property was on the bank of the Hudson River at Garrison, New York, exactly opposite the United States Military Academy at West Point. As no purchaser could be found because of the heavy real property tax assessment on it, the property was bought for only $45,000. In September 1944, St. Basil Academy began operating after obtaining initially a provisional charter from the Board of Regents of the University of the State of New York. Subsequently, the school's accreditation became absolute and permanent.

Archbishop Athenagoras believed that women could make a strong, unifying, contribution to our parishes and was determined to find a way to bring them officially and permanently into the philanthropic affairs of the Church. The Philoptochos Society was the answer not only because it undertook the philanthropic mission of the Church but because in some instances the communities could not survive without the help and support of the women. The national Philoptochos organization to which all local units are affiliated was legally chartered by the State of New York on January 7, 1944.

On the tenth of anniversary of the arrival of Archbishop Athenagoras, the gross annual income of the Archdiocese in 1941 amounted to only $41,000. In the decade that intervened total annual funding had increased by only $19,000 from the $22,000 of 1931. In sheer desperation, at the Eighth Biennial Congress in Philadelphia on June 22, 1942 Archbishop Athe-

173

nagoras proposed the revolutionary *monodollarion* concept, a system calling for each Greek Orthodox communicant to contribute $1.00 per year for the support of the Archdiocese. Incredibly enough, there were several delegates who bitterly and vociferously opposed even this nominal contribution upon the purported in-principle objection that this was an unacceptable compulsory tax on our people. It took six years for this monodollarion fund to finally reach $100,320.25 for the calendar year that closed December 31, 1948, which was the last year prior to the Archbishop's departure to take over the Ecumenical Throne.

The outstanding achievement of Archbishop Athenagoras was his ability to unite the politicized warring factions in our communities. With this accomplishment, he truly came forth as a prophet of reconciliation. No other hierarch could have done it. He was God sent for he turned out to be the right man for the right time. He was basically a monk at heart even though he loved people and delighted to be with them. Nonetheless, he could have been happy and fully satisfied in one of the smaller monasteries at Mount Athos. He was typically Byzantine in his idiosyncracies. He disliked making particularly-unpleasant decisions. He would postpone them indefinitely in the hope that passing time would inevitably somehow resolve them. Incredibly enough, even this indecisiveness helped to unite the Church for it prevented him from making the kind of decisions that would ultimately erode his personal capacity to persuade and unify dissident communicants. Those trying and difficult times could not have used a strong, decisive, authoritarian primate for it was not the period to exercise ecclesiastical power. What was needed then was a soft, humble and entreating tone begging, and at the same time convincing, opponents who hated each other to unite for the sake of the survival of Greek Orthodoxy. He succeeded as no other churchman could have. Athenagoras towered gauntly over men like an El Greco saint, with luminous, searching and penetrating eyes. His loving arms engulfed you paternally to him in such an overwhelming manner that you felt the very presence of God within yourself. He somehow managed, without intent or artifice, to look god-like in his sheer etherealness and humility.

Another of his gifts was his uncanny memory for names and faces. He was able to remember literally thousands of people. He never forgot their particular personal problems, their family relations or their specific contributions to the Church.

He had the unbelievable capacity of making everyone a part of his own extended church family. To the day of his death, thousands of Greek-Americans made the pilgrimage to Fanar to visit with him as if they were intimate personal friends or relatives. And they were received as warmly and informally as if they were calling on one of their own.

The final estimate of the contribution of Archbishop Athenagoras to our Church can be judged from the multitude of openly and shamelessly weeping young and old Greek Orthodox communicants who crowded the airport when he left for his Patriarchal enthronement. They felt and sincerely believed down deep in their hearts that in President Truman's departing private plane (The Sacred Cow) was the personal and intimate friend of each and every one of them. Though he was thoroughly unknown when he came to New York harbor on February 24, 1931 and not a single American newspaper carried a line about his arrival, 18 years later when he left his photograph was on the cover of *Life* magazine.

II

The second modern era of our Church began with the arrival of Archbishop Michael on December 15, 1949. For a decade, he had been the Metropolitan of Corinth and for 12 years prior he had been the Dean of the Cathedral of St. Sophia in London.

Archbishop Michael's personality was totally different from that of his predecessor. He was an intellectual, a deeply-learned man, a theologian and an author. He was reserved in demeanor, a very quiet person and literally avoided publicity and personal popularity. His long tenure in Great Britain did not anglicize his approach to his religious assignment in the Americas. His attitude to his new responsibilities was no different than if he had been transferred from Corinth to Thessaloniki. He somehow felt that the United States was just another Greek *eparchia* assignment. His English was not very fluent and his mentality was entirely Greek. It was difficult to believe that this hierarch had lived in Great Britain for so long and yet no part of that experience had been imprinted on him.

He was a very pious, spiritual, man concerned intensely with the sacramental life of the Church. In succeeding a man of Archbishop Athenagoras' stature and charismatic gifts, Archbishop Michael was faced with a great challenge. He made

175

no effort to compete. He was not of competitive nature and felt no need to win the acclaim of the populace. As a result, his reign was particularly low key and totally different from the exciting and vibrant Athenagoras tenure. It is reasonable to conclude that after 18 years of Archbishop Athenagoras, the time had come for a period of consolidating the unity that was gained with so much struggle. Michael's era was the 1950 decade of consolidating and extending the unity and reconciliation that Archbishop Athenagoras attained with so much blood, sweat and tears during the 1930's and the 1940's.

Notwithstanding his lack of familiarity with the unique problems that confronted the administration and direction of our Church in the Americas, Archbishop Michael within a few months concluded that it was not possible to fund the archdiocesan responsibilities with the monodollarion. As a result, at the 10th Biennial Congress in November 1950 at St. Louis, Missouri he recommended that the $1.00 annual contribution to the Archdiocese be increased to $10.00, the dekadollarion. To the surprise of many, there was practically no opposition either in committee or at plenary sessions. By the time of his death eight years later, the annual income of the Archdiocese had increased from $100,320.25 during the last year of Archbishop Athenagoras' monodollarion to $585,698.99 under the dekadollarion for the 1958 calendar year. Because boom times had come to the U.S. and the Greek-American population had "matured" as citizens and economic units, Archbishop Michael within his short reign was able to increase the revenue of the Archdiocese six-fold.

Another significant contribution of Archbishop Michael was the organization in July 1951 of the Greek Orthodox Youth of America, popularly known as GOYA. During his reign, it mushroomed into nearly 250 communities in every part of the country. It was the first time that a serious centralized effort had been made to organize our young people under the aegis of our Church.

Archbishop Michael continued the efforts that were begun by Archbishop Athenagoras to obtain formal recognition of our Church through uniform incorporation statutes or through resolutions adopted by the state legislatures directing that all references to major faiths, which were heretofore limited to Protestant, Catholic and Jewish, would hereinafter include the Greek Orthodox as well. As a result of this concerted action, 27 states adopted either the uniform incorporation act or the major faith resolution. This activity was begun in 1943 in New

York in an effort to legitimatize Greek Orthodoxy, to blot out its ghettocized disarray during the pre-Athenagoras period and to give it sufficient respectability so as to make it a part of the religious and social mainstream of the United States. The crowning event of this effort occurred in 1956 during the 13th Biennial Congress. President and Mrs. Eisenhower not only participated in the dedicatory ceremonies of the Cathedral of St. Sophia in the nation's capital but joined us for nearly two hours in worshipping there. Then Vice-President Nixon addressed one of its plenary sessions and Secretary of State John Foster Dulles was the principal speaker at the banquet. There is no record of any other religious conclave in the history of the United States that had been so signally honored by the attendance at its functions, of the three highest ranking officers of the nation: President, Vice-President and Secretary of State.

III

The third and still ongoing era began with the arrival of Archbishop Iakovos on April 1, 1959. The great difference between him and his two predecessors was that he had been in this country for 15 years. He knew this country intimately. He knew our people and our problems having served amongst us. He knew the language fluently having a master's degree in sacred theology from Harvard. He taught at our Holy Cross Theological School in both Pomfret Center, Connecticut and at Brookline, Massachusetts. He was for 12 years the dean of our Boston Cathedral. Thereafter, for four years he was the liaison officer of the Patriarchate to the World Council of Churches in Geneva. He had become Americanized on his own time prior to his ascendancy to the archdiocesan throne. He was truly equipped by training, education and temperament to exercise his office on the very day of his assignment.

It was fortunate that he came at that time for the problems of our Church had become much more complex and manifold. In the Michael era, the major task was of consolidating a unified Church and expanding its facilities so that it could reach more of our communicants throughout the land. It was basically a matter of organizational activity and enthusiasm and hard work were quite sufficient to carry us through.

With the arrival of Archbishop Iakovos, we had reached the time when the adequacy of our Church and its programs

177

were being actively challenged by native born, well-educated young men and women of Greek extraction. It was not enough to point out to them the sacrifices and expenditures that went into the formation of our communities and the churches, schools and community centers that their immigrant parents built. They were demanding a lot more from their Church. They wanted a thorough and agonizing reappraisal of its ecclesiastical structure and policy to be sure it could safely encompass within its fold the generations that were yet to come.

Fortunately, Archbishop Iakovos recognized at the outset the magnitude of the challenge. He realized that Greek Orthodoxy in this country was still inhibited by the confining parameters of its immigrant origins. He promptly set as his top objective the liberation of our Church and our people from their own self-induced inferiority limitations.

With intelligence and innovative imagination Archbishop Iakovos commenced a public relations and communications program to create an image of Greek Orthodoxy that could make it an integral part of the religious life of this nation. He achieved his goal by publicizing his prior ecumenical assignment to the World Council of Churches in Geneva, by serving as one of its presidents for a nine year period, by joining Martin Luther King on his dramatic march to Selma (making the cover page of *Life* in an historical photograph showing their joint march for human dignity, and by participating at all levels of religious cooperative endeavors. He solidified friendships with religious editors of the press associations and of influential newspapers, such as The New York Times, Washington Post, Los Angeles Times and Chicago Tribune. He cultivated close personal relationships with the executives controlling religious programming on our national television and radio outlets. He retained the George Gallup organization to poll our people's requirements, demands and aspirations for our Church. He inaugurated The Orthodox Observer in a newspaper format which now has a circulation of 120,000 subscribers and reaches more than 500,000 readers in each issue. He created within the Archdiocese a public relations office that was sufficiently effective to obtain on the NBC and CBS national networks from coast to coast telecasts of our religious services and special events. He succeeded through maximized utilization of available communications media to project so proud and a secure portrayal of Greek Orthodoxy that our Church became universally accepted as the fourth major faith in this country.

His next priority was to marshal into service for Greek Orthodoxy the very best men and women of Greek background that were available throughout the nation. This element was, unfortunately, inactive in the Church and the Greek-American community. In some instances the exceptional citizens were unknown to their co-religionists and other Greek-Americans. Notwithstanding, he searched for them, found them and convinced them that they had a sacred obligation to lead their people in determined effort to become an effective force in American life. On a national scale, he brought to the Archdiocesan Council the most successful Hellenic-American professionals, industrialists, academicians and politicians. He made each of them proud to be a part of our ethnic community. He encouraged and inspired them to undertake positions of leadership and render to our Church and to our people the unique services of which each was capable. On a local scale, he similarly inspired the American-born university graduate, the affluent and successful, to assume the leadership roles from his or her immigrant forebearers. Doctors, lawyers, university professors and business men and women now direct the destinies of our numerous parishes. Nearly two-thirds of our lay delegates at the recent San Francisco biennial congress were college graduates.

Archbishop Iakovos also focused his attention on our young people. He realized Greek Orthodoxy could not survive unless they were convinced their Church was a viable part of the very blood stream of this country and until it was conclusively demonstrated to them that their ethnic and cultural background was something they could be truly proud of. To enrich them and take them back to their roots, he built the impressive Ionian Village at Bartholomeo in the Peloponnese. Every summer, nearly a thousand third- and fourth- generation youngsters of Greek extraction spend their vacations in this beautiful and comfortable environment and learn for themselves the source of their ancestry, their faith and their culture. This program has been expanded and raised to a sophisticated and intellectual level so as to include college and university students who travel and study the antiquities and the grandeur of our Hellenic civilization. In this manner, he has propagated amongst our youth a profound love and affection for the land and the creed of our forefathers and a strong attachment to them that is reflected in our youth's continued interest in and devotion to our Church.

Archbishop Iakovos perceived at the very outset that he could not succeed in his determination to make Greek Orthodoxy a major faith unless the financial means were available so that Archdiocese could adequately fund the requisite supporting programs. He realized that this could not be achieved on a parish level. This had to be done nationally. To accomplish this objective it became necessary to radically and fundamentally change the parochial outlook of our people who viewed the needs of our Church on a local parish perspective only. Our Greek Orthodox communicants had to be educated and instructed to accept the broader concept that their parish was only a small part of the Church and that their Church allegiance was not only to their local parish but also to its national hierarchy. That this revolutionary view has been finally accepted in the Americas is evidenced by the fact that the 1983 annual budget of the Archdiocese was $10,000,000 and that most of it was voluntarily subscribed to by parishes on a fair share basis. When one recalls the internecine warfare that was waged against the monodollarion sought by Archbishop Athenagoras, only then can be truly conceived the extent of the progress that has been made. To administer this substantial financial program Archbishop Iakovos has created a sophisticated economic development department with computerized support and Price Waterhouse & Co. auditing that can compete with the very best available in any other religious, educational and philanthropic institutions.

Having practically doubled the number of parishes during his primacy, Archbishop Iakovos realized the necessity of strengthening Hellenic College and the Holy Cross Theological Seminary so that it could supply the additional priests that were required. Both the college and the theological seminary were fully accredited by the necessary educational and governmental agencies. The physical plant was enlarged to increase both its residential and recreational facilities. The beautiful Pentelic marble-fronted Maliotis Cultural Center was built to supply a modernized spacious exhibit area, a much needed theatre and sufficient conference space for symposiums, meetings and similar activities. An ambitious athletic program to attract students was instituted with such success that the school's basketball team achieved championship status and received national attention in the sports pages of metropolitan newspapers. Through the S. Gregory Taylor Foundation, post-graduate education was encouraged for our seminarians so that nearly 50 of our parish priests have obtained Ph.D.'s from the divinity schools at Harvard, Yale, Princeton and other major universities throughout the nation. This institution has now 200 students and an annual budget for next year of $3,160,000.

His most recent legacy to our Church has been administrative restructuring. He recognized the Church had become top heavy on the archdiocesan level and that selective administrative decentralization was required. This has been imposed gradually so as not to weaken or diminish precious unity won through decades of struggle.

Entirely on his own initiative, Archbishop Iakovos recommended structural changes in our Patriarchal charter so that our relationship with the Mother Church of Constantinople could be even more buttressed. Also our diocesan bishops, freed from their auxiliary limitations, could independently administer their dioceses, as canonically provided for, without however threatening or endangering indispensible unity required for survival in the Americas. He asked for the establishment of a Synod of Bishops, which together with the Clergy-Laity Congress and the Archdiocesan Council, could collegially and democratically govern the Greek Orthodox Church on this continent. And to its everlasting credit, the Patriarchate, exercising the broadest discretion possible, acknowledged the singular and particular uniqueness of our Church in the Americas and assured its survival and growth in the New World through a new charter.

IV

In reviewing the reigns of the three hierarchs that administered our Church during the past half century, it is significant to note the manner in which each was the right person for his time. Only the equanimity, tact and patience exercised by Athenagoras could have enabled our Church to survive during the political chaos of the Venizelist/Royalist era. Only he could convince warring partisans to unite by literally engulfing them in his bosom and beseeching them, god-like, with his entreating and searching eyes. After the exciting and stimulating 18 years of his reign, which catapulted him to the Ecumenical Throne, it was impossible for any successor to compete with him. As a result, it required for a short interim, a lower key level of activity during which the Athenagoras era could be solidified, consolidated and become the base for the Church's next advancement. Archbishop Michael— imbued with simple, spiritual, goodness and scholarship— fulfilled the need of that period. When it came time for Greek Orthodoxy to reach its fruition in the Americas, Iakovos appeared. He offered professionalism, sophistication, imagination and the quality leadership that our modern times required. If the timetable was changed, or the order of succession inverted, it is my view that the extraordinary triumph of Greek Orthodoxy in the Americas during the past 60 years would not have occurred. The three hierarchs involved met the needs and demands of the particular requirements of the times each served.

God has been good to our Church for He made it possible for us to have the kind and the quality of leadership that it had to have during each of its three different critical junctures in its establishment and development in the western world.

A Reflective Portrait of His Eminence

By

Julie Charles

eligion was an all-pervading force on the Aegean island of Imbros where in 1911, His Eminence Archbishop Iakovos was born and baptized as Demetrios Coucouzis. He was lucky enough to be born there when, from 1912 to 1923, Imbros was back in the fold of her motherland and the majestic white-cross-on-blue-field flag of Hellas proudly fluttered in the wind atop staffs in seven villages. The devout islanders worshipped in 25 churches and in more than 200 *exocclesia* scattered around the island's 108 square miles. In keeping with Greek Orthodox traditions, no house on Imbros was raised, nor any field tilled for a new crop, without the blessings of the parish priest. The people truly believed in a personal, immanent God who loved them and the whole human race.

In the village of Haghioi Theodoroi, a lively community noted for its production of olive oil, there were two dominant churches, one devoted to St. George and the other to the Virgin Mary. St. George—a modest, white structure typical of the Greek islands, without the faintest notion of its future claim to prominence—occupied a commanding position on a grass-covered knoll. The laughter of boys and girls playing *krifto, tselekia, omathes* and other games filled the air during recess periods of an elementary school nearby. Children, of all ages, were taught language and religion thoroughly as well as Greek mythology and the full range of classical masterpieces.

Among the youngsters who attended St. George, and the adjacent school, were the four children of Maria and Athanasios Coucouzis—Panayiotis, Virginia, Chrysanthi and Demetrios. Above all else, this devoted couple was concerned that their children learn, practice and defend the true meaning of love, the essential element of Christianity.

Consistently a student of merit, Demetrios assisted as an altar boy from the time he was old enough to hold a *xefteri*.

Carrying out the duties assigned to him before the altar was not a chore to the sensitive pre-teenager nor mere obedience to his devout parents. Demetrios was a child drawn naturally to divine worship. He had intuited from a very early age that the Sacraments disclose and reveal God to us and also make us receptive to Him.

In awe of the *ieron* and its unique purpose, the child often gazed with wonder at the large cross upon which hung the iconographic figure of the crucified Messiah. Also, the youth often felt the sensation of being touched by the power and presence of his namesake, St. Demetrios. A 12th-Century icon of that saint was displayed in the church where the adolescent served and, gradually, he developed a spiritual bond with the martyred saint represented in the icon.

As Demetrios grew in years so did his persuasion that the power of the Cross is the word of God. Even so, he had no premonition that he was destined to follow a plan which would bear him far away from his beloved Imbros and that, eventually as Archbishop Iakovos, he would become the spiritual leader of more than three million Orthodox Christians in North and South America.

For the most part we know our Primate for his national and international statements and the innumerable honors bestowed upon him by universities, Congress and others—a large topic in itself. We know His Eminence for his leadership in ecumenical efforts and his endeavors to achieve lasting world peace. Tending to the principles of our unique heritage, which determines the moral quality of our actions and our traditional involvement in the rights of mankind, we know that His Eminence has continually used his high office to defend the cause of human rights. How proud we were when, in manifesting our own thinking, our Primate nobly led us forward with a proclamation that will forever echo around the world in the annals of the brotherhood of man, "The Greek Orthodox Church is against segregation!"

There is about His Eminence an individual quality of greatness that spontaneously evokes respect and the natural desire to know more about him. To those outside his realm, the tall, imposing, white-bearded figure is principally identified by his traditional Orthodox vestments, either the brilliant robes, crown and silver episcopal staff of the celebrant bishop or his more usual mode of dress—a plain black cassock, tall hat and large pectoral cross. The world also knows him as a news-

As Student.

As Deacon

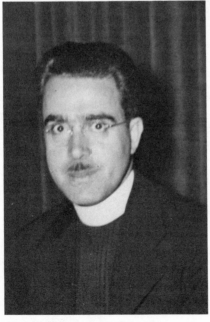

As Priest.

185

maker and strong moral voice in America.

And especially we, who are Greek Orthodox, cannot but wonder what his personal thoughts are on many complex matters— our heritage, language, religion— and on quotidian affairs as well—art, music, food, leisure, media, sports, hobbies, etc. These are the core of this article, based upon several audiences with His Eminence and subsequent research.

In 1984 two eventful anniversaries mark significant moments in His Eminence's journey of life: Fifty years ago he entered the priesthood and twenty-five years have passed since his enthronement, on April 1, 1959. Therefore, this is a good time to record the Archbishop's insights and reflections.

In his office at the Archdiocese in New York we began at the beginning—his formative years. His Eminence lovingly alluded to his birthplace, his family and childhood; his recollections were so clear he virtually painted a self-portrait:

"Here is my most cherished possession," His Eminence announced early in the discussion, directing my attention to the icon of St. Demetrios, described earlier. "The icon was presented to me in 1957 when I was offered my choice of any religious artifact from St. George, the church of my childhood." Obviously, he had developed with that sacred image and indissoluble affinity that has provided him spiritual companionship and security. His Eminence elaborated by stating the icon was believed to have been made during the era of Anna Comnina, daughter of Byzantine Emperor Alexius I. . . " but more importantly Comnina was a historian of works of great value."

On a nearby wall, an icon of the Holy Virgin, ever vigilant, looks on. With his thoughts transported back to his early years, His Eminence reminisced: "In those days intellectual and spiritual freedom were the roots of our peaceful society. We Imbriotes were a sturdy folk. In spite of a lack of industry we were an amazingly industrious people. Following the precepts of our worthy ancestors we viewed life philosophically. This expedient attitude enabled us to accept the turbulent as well as the calm aspects of life.

"We were Greek and we lived on Greek soil. The past and present fused completely. Our daily life was a discipline that had evolved from many years of history. And our language, the most perfectly elaborated language ever spoken by human lips, was valued as a light from heaven."

With unconcealed nostalgia His Eminence described the

years when Imbros was totally free from any external control. "During the time of my boyhood, a succession of historical events took place. Imbros was a fast-growing island because of the great influx of Greek men, women and children who sought refuge from Turkey and Bolshevik-dominated Russia. The saving of a life was, of course, our prime concern and it truly mattered because every life is sacred. It was, and is, the moral obligation of every man who stems from our civilization to be philanthropic and to continue the ideals entrusted to us."

He expounded on the effect of his island's location off the Gallipoli Peninsula. During World War I, Imbros was host to the British Navy and Air Force, which used the Greek island as an advance base during the Dardanelles campaign.

"In addition to being historically meaningful," His Eminence explained, "that particular period was an exciting interlude as well. The refugees from Russia who spoke Russian in addition to Greek, their mother tongue, brought with them some delightful customs, chiefly related to festive days. Having been forced to flee from a land steeped in Orthodoxy, since the 10th Century, they told harrowing stories of the Bolsheviks' attempt to eliminate Christianity in Russia.

"Also, the Greeks who came to us from Turkey brought with them the Hellenic customs and culture only they had known for centuries. They were struck with horror by the events of that time but not at all surprised by the instinctive savagery of the Turks. As for our guests, the British, we thoroughly enjoyed having them. All in all it was a stimulating time. United by a common faith, Christianity, everybody was caught up in the friendly, warm-hearted enthusiasm, mutual trust and goodwill that prevailed."

With dramatic emphasis His Eminence underscored that love for one's fellow man is the inescapable requisite for the salvation of mankind: "There is nothing on this earth that transcends the combined power of faith, freedom and goodwill towards each other. And freedom—which is man's most treasured estate—is beyond assessment!"

It's a blight on the conscience of erring political strategists that Imbros is no longer free, no longer Greek soil. In 1922, this idyllic scene of rustic life and contentment was disrupted and destroyed by Turkey which claimed the island. With characteristic fanaticism Turkey attempted to bridle the Imbriotes by curbing their religion and trying to root out their allegiance to Greece and Hellenism.

His Eminence was clearly affected by a deep sense of emotion and disquiet when reminded of the turn in the history of modern Greece that divested Imbros of its distinctive Hellenic character. At this momentous stage of his life our Primate wishes desperately to visit Imbros. But Turkey has issued a mandate that precludes a pilgrimage to the final resting place of his parents and other loved ones, including his late brother, Panayiotis, and sister Chrysanthi, who died in the early weeks of 1984. The former Virginia Coucouzis (Mrs. Hallas) resides in Montreal, Canada, and frequently visits her brother.

"I look back now with amazement on that time," His Eminence said, "Our lives were so simple and uninvolved and as children our needs were so basic. How well I remember the exciting excursions with our parents to Mount Arasia where we played among the trees, wild flowers and springs. The bracing odor of the herbal vegetation of Arasia that permeated the air of Imbros is by no means lost in the past. How much I would like to smell that mountain air again and how grateful to God I am for the incomparable security of a happy childhood."

We then turned to the adolescent years of his life. With touching informality, His Eminence shared the memory of his decision to enter the church.

"While still in my teens, like any other teen-ager, I entertained a number of strong-willed ideas about my future. At various intervals I mulled over the prospects of a teaching career, of becoming a professor, a chemist, a physician. I was fully convinced that any one of these precise occupations would lead to vocational fulfillment. To this day I hold these endeavors in high regard.

"The point at which a decisive change took place and stemmed the course of assorted aspirations occurred a year after I graduated from high school. Since my plans no longer adhered to my earlier aims, I suddenly began examining the details of reality. By now I was wholly conscious of the physical power of God, which is far removed from merely being mindful of His power. I realized that it was the beginning of a spiritual crisis. In an attempt to subdue an overpowering persuasion I voiced a desire to become a professor of theology. All the same, an inner summons was not to be driven out.

"There was a dreadful period of nightmares. Terrifying *ephialtes* of a secular world crossed with visions of sacred

188

Archbishop Iakovos on Pastoral visit to Tokyo, Japan 1966.

At Mt. Sinai in 1966.

189

beauty. I withstood this severe contest of emotions for an entire year. Ultimately, although I had not grown to full maturity, it was with the strength and conviction of a man that I delivered myself from fluctuating options. To a considerable degree the advocacy and wisdom of Father Anesti, one of our best loved priests, were valued mediatorial elements in my definitive choice."

Demetrios, the youngest child of the Coucouzis family, was 15 years old when his name was inscribed for enrollment at the Ecumenical Patriarchal Theological Seminary at Halki, a Turkish-occupied island in the Sea of Marmara.

"My parents, of course, were overjoyed," His Eminence related. "The fact that my spiritual commitment was celebrated at Easter, the most precious event in Christianity, indulged all the members of my family. Friends gathered with us to celebrate and we felt a greater measure of happiness that year.

"Time has not at all dimmed the image of my mother preparing for the holiday, baking huge fancifully-shaped *kouloures*, dyeing red eggs, and my father setting a lamb on a skewer. Nor have time and distance diminished the sound of everyone wishing each other good health and a full life."

I asked, What element truly dominated the spiritual-mindedness of Demetrios Coucouzis—the child, the boy, the young adult? Was it the mysticism and Sacraments of Orthodoxy?

In reply, His Eminence said he had an absolute belief in one truth regarding his life: "In spite of my intentions and calculated objectives, and even though I proceeded to map out a variety of professional pursuits, I believe it is beyond dispute that there is an intervening source of power determining and channeling the terms of our existence."

He then referred to his feelings upon first coming to the U.S. and what it means to put yourself in the hands of Providence. The year is 1939. The German ocean liner *Bremen* is guardedly rolling her way across a raging North Atlantic. Newly ordained a deacon, Demetrios Coucouzis is a passenger aboard ship. Having taken the name Iakovos (which is James in Greek) he is bound for America to assume the duties of his new appointment, Archdeacon to Archbishop Athenagoras, our late Patriarch.

Caught up in the memory of that time His Eminence recalled that it was an extremely stormy voyage. "Throughout the turbulent crossing I could not but wonder if the boiling

190

sea and violent wind were predictive of the future. Up until that time my own world was a small, hedged-in area and my horizon did not extend much beyond the mountains of my native environment."

"What was your first impression of the United States, Your Eminence?" I asked.

"To begin with," His Eminence explained, "I was keen to learn what was *new* about the New World. Actually, I found nothing new, that is (in the exact sense) *original*. Geographically, yes, America was newborn. But I did not search for or find any great dissimilarities between this country and the world I left behind. Since my mission was a religious one, I conscientiously assessed what I perceived in the area of religion.

"As a boy, having personally witnessed and endured the agony of Turkey's unjust exercise of power in Imbros, I envisioned a land where Greek Orthodoxy—ethnically rooted in Greece and religiously in the Ecumenical Patriarchate—could grow untrammeled. I had always borne in mind the history of the time of struggle when the City of Constantinople and the vast lands of Asia Minor fell to the invading Muslims and the brutality that followed for centuries. I believed with full assurance that America was the land I dreamed of, where God intended Greek Orthodoxy to grow. This was a nation which from the beginning of its history, with indestructible authority, safeguarded the most fundamental of human rights—one's own rule of divine worship.

"As I looked back on Turkey's evil conspiracy to expunge our religion. I realized a positive sense of victory, a personal victory over a diabolic opponent. I contemplated strength, distinction and influence for Greek Orthodoxy in the United States, parallel to the inevitable advancement of this remarkable state. I remain eternally grateful for the opportunity to serve God by serving my faithful countrymen whose fortune it was to settle here.

"In addition to learning English, of course, and comprehending a mentality and attitude, both of which were in some respects at variance with mine, an urgent step was my acceptance of certain adjunctive conditions in our Church. A Greek Orthodox Church with organ music, social halls with secular activities, its administration—its everything—were modern supplements hitherto unknown to me. Futhermore, approximately five decades ago our Church was unreservedly parochial with little or no thought of ecumenism. Did I perceive a challenge? Not one but many."

Relative to duty and destiny in the United States, His Eminence expressed his abiding concern for young Americans of Greek descent. Constantly subject to his inquiry are the nature and course of their ideals, their concept of perfection, their expectations, their love for and the extent of their attachment to Greek Orthodoxy.

About the rising generation His Eminence is wont to quote from Aristotle: *All who have meditated on the art of governing men have been convinced that the fate of empires depends on the education of youth.*

"However," His Eminence said further, "we cann'ot hope to rely on our youth if their trust in the adult world is not absolute; if they are not made to feel secure within themselves. All human decency, ethics and principles stem from a happy family life. A youth will be lost, his or her life wasted, unless taught at home (by a caring household) the code of right and wrong."

His Eminence is merciless with regard to abortion and damns this murderous practice as "a crime against life, love and Christian motherhood." As for the issue of women becoming priests in the Greek Orthodox Church he feels it is incontrovertibly a closed topic. "Women have a God-ascribed mission," His Eminence has delivered formally, declaring that he will have none of this downgrading of women by offering them the priesthood or ministry as another "equal opportunity."

For half hour we moved to lighter and personal considerations to gain a bit more understanding of Archbishop Iakovos in his everyday activities. His Eminence begins his day at 5 a.m. with prayers at 5:30 and a simple breakfast at 7 o'clock. Plain, wholesome food is a meticulous rule of life. Taking meals that consist mainly of white meat, vegetables and fruit, His Eminence frequently favors a dinner dish of lamb prepared with green beans and celery. Inasmuch as it is a prevailing custom to present a wide range of native fare from around the world as "gourmet" concoctions it would be remiss not to introduce an Imbriote specialty *tzitziria*, a savory entree described as a "cheesefilled pizza."

In keeping with a vigorous workday schedule, His Eminence is an energetic traveler visiting local, out-of-town and foreign parishes. Asked how he copes with the pressure and demands of office our Primate's response though gentle in volume is mighty in persuasion: "Divine influence for strength and guidance is available to everyone through the Bible, Liturgy and prayer." In the Book of Psalms, His Eminence said he invariably will turn to the 50th Psalm which, he explained, "has always been an exalting influence and especially restorative."

The everyday business of running the Archdiocese, attending to pastoral and administrative responsibilities of approximately 550 parishes (extending from one end of the U.S. to the other and from Alaska in the north to Buenos Aires in the south), involvement in ecumenism and strenuous endeavors for world peace hardly allows His Eminence time for relaxation. But in this regard His Eminence said, "It's a serious misjudgment to believe that time out from the weight of daily pressure is not absolutely essential. It is, in fact, imperative for one's well-being to withdraw from the ups and downs, the anxieties and apprehensions of routine. Even if it is not an extended holiday, a weekend, a day or two, a few hours of spectator sport, participating sport—whatever—a brief respite is vitally important."

"Your Eminence, what is your favorite sport?" I asked.

"Actually I have two favorite sports and a hobby as well," His Eminence answered. With a smile, he changed the scene to an integral aspect of American life. "I'm a baseball fan," he added, "a loyal supporter of the Red Sox."

His Eminence was introduced to baseball soon after coming to the United States. His ordination to the priesthood took place a year later in Lowell, MA, a fact which explains his affection for the Boston team. One cannot but conjure up a picture of the newly arrived young Imbriote in clerical dress in Fenway Park briskly rooting: "Let's go, Sox! Home run! Strike him out!" He continued to follow that team for many years, while serving in various capacities in Boston.

"And then I enjoy fishing," His Eminence said. "Personally I prefer deep water fishing, and I've engaged in some first rate fishing in the waters of Martha's Vineyard and off the coast of Cape Cod."

Additionally, His Eminence takes pleasure in gardening which he pursues with considerable know-how. Picturesque flower beds and a profusion of flourishing shrubs at his home in upstate New York attest to his experience and skill. "The adage always applies," His Eminence said, "that a garden grows best with personal care and attention.

"But aside from sports and other outdoor pastimes there's a plethora of fine entertainment in the performing arts. And there is television and radio, both excellent leisure time diversions."

"Your Eminence, do you watch television?" I asked sceptically.

"I most certainly do," His Eminence responded with lively interest. "Such is the arrangement of this remarkable instrument that it is possible for almost everyone the world over to know what is going on at the same time. Admittedly, while there are certain condemnatory elements, particularly with regard to children, television has uses and advantages of unlimited potential.

"I whole-heartedly commend Public Television for its marvellous work in projecting programs of a social and religious significance. Their educational presentations are also first-rate. While these specific productions, whether on Channel 13 or another channel have priority on my television agenda, I am not at all averse to switching to another channel to watch a baseball game, moreso when the Red Sox are playing. I also enjoy watching Western films."

A citizen of the United States since 1950, His Eminence retains a scholarly interest in the early history of his adoptive country. "We must always keep fresh the memory of the men and women who pioneered this great land," His Eminence said. "How indebted we are to those trail-blazing frontiersmen who gave their lives to establish law and order for statehood and nationhood. In the company of millions of other Americans, if it is convenient to my schedule, I will eagerly turn on the television to watch a film that recreates the spirit and adventure of the old West. Of course, I know that the story of the West is quintessentially the story of America. Quite frankly, I find a good Western a particularly pleasant source of entertainment."

His Eminence casts a longing eye at more free time to listen to music. Among others noted for their glorious instrumentation, Passion Week compositions and requiems, he named Handel, Haydn, Bach, Beethoven, Mozart and Verdi. His Eminence commented: "A room, a vestibule, a library, an auditorium—whatever the size—becomes a church when sacred works are performed and listened to with reverence.

"I'm addicted to good music. Whoever the composer— Brahms, Schubert, Tchaikovsky—I'm a most attentive listener." His Eminence is also an opera buff. He has no favorite opera and admits to "loving them all." The late Maria Callas was

194

much admired by His Eminence for her natural talent and remarkable artistic achievements. Ultra modern in the sense that he is fully attuned to the movement and mood of today, His Eminence is well aware of an extraordinary interest in music beyond the category of sacred and classical.

"Your Eminence, do you ever listen to country Western music?" I asked.

"Yes. . . why not?" His Eminence responded. "Country Western songs have secured a very respectable niche in American entertainment. I can identify with this bright feature of the present day. I find the music harmoniously effective and the lyrics frequently convey a message that merits attention."

His Eminence advocates freedom for artistic self-expression and self-fulfillment. He is in sympathy with those who accord recognition to interpretations of long validated teachings, standards and values. But he also sympathizes with innovation.

"However," His Eminence explained, "this stance is not a judgment, nor easy acceptance, of certain cacophonous clangor that is presented under the guise of music, and which, quite obviously, is preferred by a large number of modern youth. Each generation obtains its own license for prerogatives and within limits, of course, is entitled to its original style of expression. Such inroads into society are a fad and this so-called 'last word' in entertainment generally eliminates itself as the succeeding generation comes in. But my position is one of total aggression against related insidious influences that tempt our youth and lure them into irreversible moral and physical decay."

His Eminence finds it infinitely interesting and enlightening to go from one form of art to another. An avid reader, he believes that the gulf that separates ancient literature from modern writing may be bridged if we read more.

"Literature is a social product," he said. "We must read if we desire to help ourselves from the sum total of finished thoughts in men's writings. The power of literature must not be minimized. Whatever the subject, all parts of it are food for the mind and this is how we acquire knowledge. While I personally routinely read the writings of St. Paul, St. Basil, St. Chrysostom, and St. Gregory, among others, I cannot but plead for their general study. There is consistently new light and fresh ideas to be found in these consecrated volumes. One need not be an ecclesiastic to appreciate the depth of thought of these holy men."

Stocked in His Eminence's voluminous library are the works of many nations. In addition to keeping abreast with readings of a social and religious nature, he enjoys biographies, autobiographies and historical narratives. "Above all," His Eminence said, "I admire clear thinking in prose and excellence of style."

Since the days of Pindar, the Greek—whose temperament acknowledges the distinctive advantage of imagination—is pressed to poeticize. His Eminence admits that he is no exception and, from time to time, he too composes metrical verse. "But in writing poetry," His Eminence commented, "I would not think of departing from the ancient view that the function of the poet is to teach, to make men better." His theory is elemental: *Tell the truth, which is the scope of poetry, and select a piece of life which bears an interesting relationship to life as a whole.*

Among others, His Eminence praises the versecraft of Robert Frost, T.S. Eliot and Henry Wadsworth Longfellow. He calls *The Village Blacksmith* by Longfellow "a good poem because it fills the conditions of what I have defined as a kind of beauty."

Although the Christian religion is the chief fact of his life, the heart of Archbishop Iakovos has plenty of room for classical Greek culture. From the moment of his enthronement our Primate naturally assumed proctorship in North and South America of ancient Greek culture. In these lands, far removed from the home of the first fine arts, while many other ideals are scarred by disaster or simply fade away His Eminence will not tolerate any underevaluation of our inherited estate.

"Let us not forget," His Eminence reminds us, "that the roots of permanence and power for resisting outside influence has been in our religion and then in our artistic legacy. The achievements of those men of long ago were no ordinary fragments. They are our keepsakes. They cry for life. The chief element, after all, in that wonderful product which is known to the world as Greek civilization has been the Greek genius itself. It was the mental aptitude of the Hellenes—that original, versatile, imaginative genius, that love of the beautiful and sense of proportion—that affords the only satisfactory explanation of their achievements in art, in literature and in philosophy. Undoubtedly, without the quickening power of the Greek genius, the germs of culture, transmitted to the West from the East would have remained dormant or developed into less perfect and less admirable forms."

Having couched his sentiments with unmistakable distinctness His Eminence cited a passage by the noted American journalist, Alexander Eliot, from his interpretation of Greece in the Life World library: "The people of the Icon have given the world a torchlight procession of great artists stretching back almost 4,000 years illuminating and reflecting that which endures."

Since the Greek way of life from that brilliant bygone era has always been imbued with art, His Eminence added: "I embrace with unfathomable pride Americans of Greek ancestry who have achieved universal acclaim in the arts, medicine, education, philosophy—whatever the sphere of endeavor—their accomplishments will endure as worthy achievements partially imbedded in the Hellenic past."

If philosophy seemed an extremely valuable pursuit to the ancients so it is with His Eminence. Among non-Hellenic theorists with whose pragmatic logic he is familiar, His Eminence named Francis Bacon, Roger Williams, Ralph Waldo Emerson and Alfred North Whitehead. "Of course, there is a connecting link with men of wisdom of other nations," His Eminence conceded, "but, since Greece was the fountainhead, the authority of the ancient Greeks will never cease to attract and influence the best minds in the world."

His Eminence invests Plato with the title "First Philosopher Activist" for endlessly probing for a thorough understanding of values and ideals with which to discipline the intelligence of man and in doing so cast a paradigm for the young. In all the generations that have passed His Eminence will transfuse the sense of old Greece by honoring the legendary figures of mythical lore for their unique treasury and effect on the world. "Without doubt," His Eminence said, "the most precious invention of the old Greeks, and one which distinguished them profoundly, is the myth. The essence of Greek mythology is the recognition of the divineness of nature. Clearly, mythology still has a voice."

From the ancient crowd of Hellenic divinities Athena and Apollo have long remained His Eminence's choice. "I have always admired Athena, the goddess of wisdom, for paving the way for patrons of the arts," His Eminence said. "To this day this gesture is of immense inspirational value. As for Apollo—in addition to being the god of the sun, medicine, music, poetry and the arts—his contribution was of immeasurable benefit to us. Aside from originating the founding of colonies, Apollo was the first to encourage the practice of mi-

197

gration. But above all, Apollo was the moral teacher of man. His slaying of the monster, Python, expresses in allegory, the triumph of good over evil."

His Eminence refers to the dauntless warrior, Achilles of Thessaly, as "a praiseworthy hero for his lion-hearted leadership, his concern for his compatriots away from home and his enduring efforts, though in vain, to make peace between nations at war." And he cites, *The Trojan Women* by Euripides as the greatest anti-war play ever written. In supporting the plea of the deathless playwright, whose tortuous cry contested the barbarism of war, His Eminence stressed: "The imploration of Euripides, in which we all share, can only be lulled into harmony when the spirit of compassionate patience and understanding brotherhood penetrates the hearts of all men, for all time, and becomes a reality. The whole vast picture of our contemporary chaotic age—and of all it means in terms of suffering and death—must end! But a lasting peace cannot be realized without God's presence. It we fail to grasp the meaning of war and possible annihilation we shall descend into total darkness and, as in *The Trojan Women*, there will not be a gleam of light anywhere."

In an attempt to further clarify his viewpoint, I asked, "Your Eminence, is today's youth closer to God by his rampaging demonstrations that he is opposed to war?"

"Absolutely not!" the Archbishop replied. "Closeness to God can never be identified with anything as negative as violent disorder. Opposition to war might well be identified as an attempt to draw closer to God *if, and when, implemented with genuine care and concern!* The increase of disrespect for law and order, and the violence and crime associated with some mass meetings, in no way reflects a testimony of reaching out for peace through God."

"Are we to despair of the world when there are revolutionary cries everywhere for equality and independence?" I asked.

"Nihilistic protests that fail to lead to conclusive action for equality and independence constitute the kind of confusion and chaotic conditions with which the First, Second and Third Worlds are confronted," the Archbishop replied and added: "Certain so-called leaders with their convulsive oratory for equality and independence which, after all, are mere concepts, would do themselves and the entire world an incalculable service by directing the course of a parade of hungry, weary men—their confidence in humanity utterly shattered—towards

Playing baseball with students at St. Basil's Academy 1966.

A Reflective Portrait of His Eminence.

the coping and solving of their problems by peaceful and ethical principles."

"Your Eminence how can the world balance the gross materialism and sensualism of a permissive society and Christianity?" I asked.

"What is called a permissive society is a *malaise*," His Eminence explained. "It is not unusual in a society of little faith and great wealth. The Church has always addressed itself to the spiritual self. The antidote she is using today is intensified evangelism, a call for spiritual renewal and appeal to one's conscience."

"Your Eminence, we cannot but be aware that the seeds of skepticism have germinated," I said. "Therefore, as we move into the computer age, is it possible that this generation—which has shown itself so great by its achievements in discovery and invention—could fall so low spiritually as to give up thinking?"

"Unfortunately, there are countless symptoms of such a calamity. For example, the average learning level is already far below of what is acceptable. Very definitely, there is a fear that humans may well entrust computers to do their thinking."

"How would our religion be affected? How could we use our religion to preserve what we have learned?" I asked

"This question is improperly phrased," he replied. "What we need to preserve is not our technological learning but the high moral and spiritual values that compose the essence and *raison d'etre* of all religions. Christianity's center of gravity is the soul of man. Accordingly, we should utilize its spiritual dynamics to discover what ennobles, not what perverts, human nature.

"It is said that the target towards which learning is aimed is truth. Not in these times. The target of science and techology is a trap which ensnares humans and puts them in an environment of inhumanity. And this inhumanity will eventually be consumed by its own greed."

"Your Eminence, how can we bind ourselves to truth when we live in a civilization which has no ethical principles behind it?" I asked.

"But it is not true that civilization has no ethical principles behind it!" Archbishop Iakovos retorted. "The culture cult may not be bound by ethical principles but civilization is more than culture. Civilization is the expression of refinement of the human soul, in art, in everything pertaining to man. While culture may increase the appetite for sensual pursuits, civilization wants nothing more than the continued humanization of man."

"Must we study the historical Jesus in order to retain the spirit of Jesus in our hearts?" I asked.

"Yes, we must." he asserted. "The more we study the historical Jesus the more we understand, appreciate and love Him. Other than Jesus, there is no other human who alleviated suffering and elevated man and woman to the exalted status of sonship and daughtership."

'Your Eminence, what good is the use of religious history to us? How can this knowledge influence, for example the doctrine of forgiveness?" I asked.

"The study of religious history provides us with an opportunity to delve into the practice of religionists and draw our own conclusions as to which one will help us form and convert our ethos. As for forgiving. . . .it is hardly a doctrine for it can neither be taught nor indoctrinated. Forgiveness is a Christian gesture, the truest act of humanity, of self knowledge."

"If we do not agree with some historical religious fact or decision, are we less Christian?" I asked.

"This is a thought that should be banished forever. In religion, and especially in the Orthodox religion, there is no room for personal preferences, the acceptance of one event and not another, or any compromising attitudes. A true Christian is one totally committed to Christ's love and concern for man's perfection and salvation."

"Why is it imperative for us to attend church services?" I asked.

"So that you may detach yourselves from worldly excesses and look for God," His Eminence elucidated. "A church is more than a building. A church where we gather together is an element of utmost significance in the experience of the soul. Living away from God, and from one another, leads us to complete estrangement from the reality of life. St. Luke in the *Book of Acts*, 17:28, presents St. Paul citing Aratus, a poet and philosopher from Sicily (c. 280 B.C.), in one of the most beautiful definitions of man's relationship to God: 'Regardless if no man want it or not it is in God that we live, love and have our being.' This will always hold true."

"Your Eminence, as Greek Orthodox, what is the most significant reason why we must cling to our faith?" I asked.

"Our religion is our most powerful weapon for self preservation and self-respect," the hierarch replied. "A man's spiritual being consists of his mind and soul which, together have the might to counteract falsehood and illusion in man's concept of God and his fellowmen."

"In today's world which is advancing with annihilative determination what other commandments might be appended to the original ten as we know them?" I asked.

"The two commandments articulated by Jesus, the highest voice ever heard on this earth and called the epitome of Christianity, 'Thou shall love thy Lord with all thy heart with all thy mind and with all thy might and thy neighbor as thyself.' We must always remember that God is the soul of the world and its history; that man's actions reach high as heaven or downward, very low. We shall not see into the true heart of anything if we merely look at the delusions, confusions and falsehoods of life."

The interview was over. I summarized my impressions, concluding that the questions and answers had given me the insights of an exceptional world leader—who in addition, down deep, was *the highest exponent of the Christian priesthood.* To those of us who are Greek Orthodox, a priest naturally commands our deepest love and respect. As Greeks we are mindful of the efforts of our priests in preserving our language, national ideal and religion through four centuries of barbaric rule. Therefore, for us a priest symbolizes freedom. We understand a priest to be a kind of prophet. He presides over our worship and celebrates in the Mystery that unifies us with God. We do indeed look to our priest to guide us through our life. These were my thoughts and they continue to express my opinion that (in his 50th year of priesthood) His Eminence remains the ideal priest. As Archbishop, his enormous pastorate is the New World and—through example and involvement in so many areas of religious and political affairs—he influences the world.

"Have we overlooked any major area of concern; is there some other area of interest you have not yet commented upon?" I asked as a formal close. The white-bearded figure paused, obviously weighing all the questions and answers. Finally, he said:

"We must stress our concern for the children. I have a special fatherly interest and love for them and I'm not sure we have covered the topic fully. We must reach our children, touch the hearts of our children now. If, as Aristotle reasoned, the fate of empires depends on the education of youth, so does

Christianity. For us, Greek Orthodoxy.

"These are rapidly changing times. The forecast for tomorrow depends upon the shifting and reshaping of ideals. For our religion, our children are our only trusted security. Our only hope."

Sensitized to His Eminence's deep love and pastoral concern for our children, I inquired further at the Archdiocese and found one unusual, recent, dramatic letter from one of the many children who write to him every year. The writer is now an adult women living in California, but was once a child in his Boston parish. Expressing the feeling of many of us, she wrote in part, the following to our Archbishop:

> Since you preached about love when you were here (Los Angeles), I've decided to write you this letter in appreciation of the man who was my parish priest as a child, of the man who was my confessor, whose words in the confessional I still remember, whose pinch on the check I can still feel and whose faith and intensity I felt so poignantly.
>
> All of eleven or twelve years old, I once went into your office to ask if we could eat peanut butter during Lent and I heard, "Lemonia, do you really like peanut butter?" "Yes," was my response. "Well, then, you shouldn't eat it during Lent." A lesson succinctly taught and well remembered.
>
> There are so many little stories like this about the much beloved priest of my childhood but more than what was said was the precious example and heartfelt passion with which everything was done. I loved you so much and, indeed, I'm certain that all the Cathedral kids did. In a very special way, you were our father. It was a special blessing for me to have been one of your Cathedral kids.

President Carter's Homily, April 1, 1984.

President Carter with His Eminence.

President Carter's Homily

At the 25th Anniversary Ecumenical Doxology
Greek Orthodox Archdiocesan Cathedral
of the Holy Trinity, New York City, April 1, 1984

irst of all let me say how delighted I am to be here to honor a distinguished servant of Christ and my personal friend, Archbishop Iakovos. I've had a great admiration for him since I first learned about his distinguished service and his leadership, and this morning I want to talk not about his pride in his great achievements, not even his great ambitions for the future, not his gratification at being able to serve but of some of the frustrations and disappointments which he has expressed to me personally, and to some of the news reporters who have relayed his disappointments to an interested world. I tried to think of something that might make him feel better at being frustrated about not being able to reach all his noble goals rapidly.

I looked back and read some of the remarks my predecessors in the White House have made about the presidency itself: I'll read just a few brief ones—Woodrow Wilson: "The Office of the President requires the constitution of an athlete, the patience of a mother and the endurance of an early Christian." Abraham Lincoln: "In God's Name, if anyone can do better in my place let him try his hand at it. I wish I had never been born." Thomas Jefferson: "I'm tired of an office that brings nothing but unceasing drudgery and daily loss of friends." John Quincy Adams: "The four most miserable years of my life," and John Adams: "Had I been chosen President again, I am certain I could not have lived another year." I might point out that I think all these men ran for a second term. But it is important for those who serve in public life, in Christ's Kingdom, or in other service, not to be proud and self-satisfied with achievements, but to constantly struggle to overcome difficulties, to recognize the inherent uncertainties of life, but still not to be so disappointed that one yields in the achievement of great goals.

I've traveled extensively throughout the world and in whatever country I have visited I tried to go to church and talk with Christian leaders about the vitality of Christ, of the community of believers in that particular region. It's sad to report that in many places, the Church of Jesus Christ is not much more than a hollow shell. People believe but there is no activity, no drive, no enthusiasm, no evangelistic zeal and no dedication. In other places, I think particularly in this hemisphere, there is a much greater determination to put into practical effect in our own individual lives, and within the community around us, the teachings of Christ.

This morning for example, I went to the lower east side to visit an old dilapidated building, which was constructed about the turn-of-the-century. This building is being acquired by a group of Christians from a large number of churches and it will be rehabilitated completely without any government funds at all, but with volunteers coming into New York from throughout the nation, including Christians of many denominations from Georgia. Over the next two-years it will provide 16 nice apartments for some of those who live in the lower east side, not too far from very wealthy Wall Street. Government in many ways, in spite of individual projects of this kind, has perhaps in my own political life-time done more to alleviate suffering, and to carry out the teachings of Christ, than have our Churches.

Quite often, those who seek public office undergo intense competition to evolve ways to address the needs of a society which they hope to represent by exploring new ways to alleviate suffering, to provide better health care, better education, better housing, more freedom and the alleviation of unwarranted incarceration, imprisonment without trial, torture or even murder by one's own government. These goals and ambitions of government obviously are not always positive because people are abused by governments as well. But there need be no sharp, distinctive, division between the organized Church of Christ and individual citizens or governments comprised of citizens. But there should still be a competition, an enthusiasm, a dynamism of dedication, in exploring better ways to carry out the commandments of Christ. The marks of greatness of a government and of a Church are similar, in many ways, I think each one of those Presidents who served before me, and including myself, tried to say when going into the White House: "What can I do to make my nation greater, for its own people and in the eyes of the world?" What is the measurement of greatness? Is it just military strength? Is it just the imposition of our will on those around the world who are weaker than we? Obviously, no! A government is great if it searches for peace, not just peace between itself and potential adversaries but peace for others on earth who suffer from hatred and also from bloodshed.

Our nation as a superpower is great to the extent that it seeks to remove the threat of increased armaments—in particular nuclear destruction, which is an impending danger for us all. A nation is great to the extent that it enhances civil rights at home, human rights abroad, lets a banner of freedom and democracy be raised high. We know from experience that when (our nation) is silent it's quite unlikely that others will speak or act for the benefit of those who suffer. A nation or government is also great when it increases or improves the quality of life of those who do not share the great material blessings which everyone in this room has to enjoy.

The goals of a Church are not dissimilar from such national goals. I listened this morning to the readings of the Doxology and I especially noted a few things that were said: "They shall beat their swords into plow-shares and their spears into pruning hooks. Nation shall not lift up sword against nation, neither shall they learn war anymore." And again, "This is my commandment," Christ said, "That you love one another as I have loved you," and again Paul said to the Romans: "Love one another with brotherly affection;" and again, "Live in harmony with one another; do not be haughty but associate with the lowly; never be conceited. . . live peaceably with all. Do not be overcome by evil but overcome evil with good."

I think that the frustrations of Archbishop Iakovos are a great credit to him because it shows that this dedicated, humble and courageous man is not satisfied with his prestige, with the power and authority of his office or even with the esteem of other Christians who admire and appreciate what he has done. It's not good for any Christian to be satisfied. We know that even Christ was a man of sorrows. They refer to Christ as one who was scorned and despised, even unattractive in appearance—a man of sorrows. We all learned as children that Jesus wept at Gethsemane. Christ, himself of God, was tortured and he said: "My soul is overwhelmed with sorrow to the point of death." And on the cross Christ was also tortured in spirit as a human being. He even felt the absence of

207

God as he bore the stigma of our sins and he said, "Why hast Thou foresaken me."

In the Doxology this morning, also, Paul himself said: "For three years I did not cease night or day to admonish everyone with tears." God does not expect all Christians to be happy, satisfied, proud. He expects us to share the sorrows of those who grieve, and He blesses those who weep because of unrighteousness. One of the most grievous things that we Christians have to face is that we have a divided Church of Christ—indicated by the signals that we send out to unbelievers, whom we would like to bring to our Savior; when we are filled with hatred for one another, incapable of close. friendship or cooperation, much less love; when we (as different denominations) hold each other at arms length, and even kill, ostensibly in the name of Christ.

We also heard in the Doxology this morning: "But you are a chosen race, a royal priesthood, a holy nation, God's own people, that you may declare the wonderful deeds of him who called you out of darkness into his marvelous light." And again in the words of Paul: "For as in one body we have many members, and all the members do not have the same function, so we, though many, are one body in Christ, and individually members of one another." So the admonition of God through Christ himself, and through his disciples, followers like Paul have told us to avoid divisions among ourselves. But we as leaders have not yet been successful in doing so, because there is yet an element of pride (which is a sin), which prevents the unification of brothers and sisters with Christ under God the Father. What can we do to close matters more?

Again in the Doxology, Paul says: "I have suffered the loss of all things and count them as refuse, in order that I might gain Christ." And again we are admonished to press on: "I press on toward the goal for the prize of the upward call of God in Christ Jesus." We, as Christians, have no cause to be satisfied. We should share with Archbishop Iakovos his frustrations and disappointments, similar to that of Paul, Peter, John and Christ himself, at the goals still not reached.

But we also know with Christ nothing is impossible. Our honoree this morning, having served for 25 years in this important position in Christ's Church in this hemisphere, leads two million Christian believers. In this country alone, we have 14 million Southern Baptists and tens of millions of other Christians who profess a dedicated, unselfish commitment to the teachings of our Savior and our Lord. Building upon unselfishness, dedication, courage, confidence or faith, brotherhood and sisterhood, compassion and love—with the spirit of Christ in our hearts—we can change the world.

With President Eisenhower, Spyros Skouras and Tom Pappas.

With former President, Harry S. Truman.

210

With President Kennedy and Patriarch Benedictos of Jerusalem.

At dedication of AHEPA Truman Memorial, Athens, Greece, May 29, 1963.

With Martin Luther King, Jr. in Selma, Alabama.

At Inauguration of President Johnson, January 20, 1964.

With Senator Henry (Scoop) Jackson.

With President Karamanlis of Greece.

215

Receiving the "Medal of Freedom Award" from President Carter.

His Eminence looks on as President Reagan receives cross from Patriarch Diodoros of Jerusalem, September 1983.

Secretary of Education, Terrel Bell, at Archbishop Iakovos dinner.

With Vice-Presidential Candidate of the Democratic Party, Geraldine Ferraro 1984.

San Francisco Mayor Diane Feinstein presents His Eminence with a Key to the City at the Grand Banquet of the 26th Biennial Clergy-Laity Congress, 1982.

217

With Archbishop of Canterbury,
Geoffrey Fisher.

With the late Terrence Cardinal Cooke
at St. Patrick's Cathedral.

Receiving Templeton Award. Seen in photo with Archbishop Iakovos is Rabbi
Marc Tannenbaum, President American Jewish Committee; Reverend Billy
Graham; Bishop John Allin, presiding Bishop of the Episcopal Church; Jean
Kirkpatrick, U.S. Representative to the United Nations.

218

Welcoming Armenian Orthodox Patriarch Karekian II in St. Paul's Chapel at the Greek Archdiocese of North and South America in New York.

With Rumanian Patriarch Justin on his visit to the United States 1979.

219

Praying together during the Papal visit to the United States in 1979.

His Eminence walking with Pope John Paul II, 1979.

220

The Administration of the Archdiocese

By

Basil Foussianes

he Archdiocese today is a reflection of the understanding and vision of His Eminence Archbishop Iakovos and is characterized by a transformation brought about by his stewardship of the Church in the Americas—a transformation that has affected all aspects of the Church including the administrative and financial aspects. It is the purpose of this writing to review briefly how the stewardship of His Eminence Archbishop Iakovos has affected the administration of the Church.

In order to place his ministry in perspective, we need only look at the status of the Church at the beginning of his ministry in 1959. As a Church built through sacrifices and dedication of immigrants, His Eminence found the Church in a position where the Church was parochial, had not grown administratively, unable to meet its own needs and financially weak.

The educated first generation which was inheriting the helm of lay leadership was sincere and able but with an uncertain spiritual basis and foundation. As a young Church, she was still quite self-centered and generally out of the mainstream of American society.

To this Church, His Eminence brought a vision of a growing, vibrant, living Church.

His tolerant and inspiring leadership began to make the communicants more conscious that Church is one, neither parish nor Archdiocese, clergy nor laity, but one—that only if we are united can the Church survive and grow—that we must find and recognize our problems and work together toward a solution—that we must strive to become dedicated churchmen and women who recognize our responsibilities to our Church, our fellow man, our country, and that this was a title to be earned and not bestowed, and that above all we must seek spiritual renewal.

Throughout all this, His Eminence has kept us together even

though forces may have sometime sought to divide us.

His charismatic personality has not only brought unity and understanding but has brought a better understanding between our Church and others.

In his active role in the World Council of Churches (WCC) as President, he articulated the position and role of Orthodoxy in the Ecumenical movement and at the same time made us aware of our obligations to those around us—in the National Council of Churches (NCC) by speaking as a voice of conscience. In his founding of the Standing Conference of Orthodox Bishops (SCOBA) he has served as a catalyst to unite efforts of all Orthodox. SCOBA brought together the prelates of twelve ethnic jurisdictions — many of whom had hardly faced each other previously—around the same table, to consider the need of being together, speaking with a common voice to the society in which we live and looking together at the problems and issues which each too often had been unable to face alone.

He has brought our Church to a status undreamed of in the post World War II period. One need only read addresses of His Eminence at the Clergy-Laity Congresses since 1960 and to read the decisions to see how he has brought the Church from the seeming total preoccupation with the necessary administrative and financial problems to a point wherein, at the Congress in Chicago in July 1974 and thereafter, the entire proceedings, every deliberation and decision were conducted with the basic and underlying perspective of spiritual growth and renewal, wherein each item of discussion was focused on the question—*How does our dedication to our spiritual growth require us to view this question before us?*

Fundamental to all that His Eminence has said and done is his conviction that because of her religious and cultural heritage, Orthodoxy has and must make a contribution to America and the world. All of this he has done for Christianity and mankind at great personal sacrifice, physically and emotionally.

His Eminence through his own courageous actions and words proceeded to mold a new image of Orthodoxy in the Americas, which raised the consciousness of all to a new awareness of the Church. His Eminence led most Greek Orthodox communicants to the realization that they were indeed a part of an indigenous Church which was an integral part of the American scene. Understanding the fact that the Church is one is no more aptly stated than by His Eminence is his keynote address to the 1976 Clergy-Laity Congress:

> *"A fourth aspect of the current state of the Church I consider the growing awareness of the notion and the conviction that our Church, the Holy Archdiocese, all of us are one—a unified indivisible organism."*

ARCHDIOCESAN ADMINISTRATION

When His Eminence undertook his ministry in 1959, he found that the Archdiocesan staff comprised a small group of loyal and dedicated persons, many of whom had served the Church for many years, were underpaid and had minimal office equipment and poor facilities. The staff members were attempting to perform their duties as best they could but obviously could do only the minimum. It was therefore necessary to restructure the Archdiocese internally not only to administer the Archdiocese properly but also to convince the parishes of the viability of the Archdiocesan functions for the parishes. The parochial attitudes of the parishes and communicants affected their views regarding their obligations to the Church and a factor in any change on this attitude would be the credibility and viability of the administrative structure of the Archdiocese.

Thus, early in his ministry, Archbishop Iakovos undertook to analyze the organization of the Archdiocese staff. To do this, he called on all the resources which he could muster. Top administrative and financial officers of other faiths, responding to the friendships developed by His Eminence in his work in the Ecumenical movement, came to the Archdiocese and made detailed reports to the Archbishop and the Archdiocesan Council. An internal structure committee of the Archdiocesan Council began its work that carried on for several years - interviewing all employees, analyzing the requirements and making recommendations. His Eminence carefully reviewed all these studies and systematically implemented them on the basis of priorities, resources and personnel.

The restructuring involved the adoption of new additional facilities, accounting procedures, expanded staff culminating in obtaining a qualified Director of Economic Development and the retention of internationally known auditors. This did not occur overnight but continued into the second decade of the Archbishop's ministry. Indeed it continues to this day with a continued monitoring of the organization and making changes and refinements as the need is identified.

The Archdiocesan staff today comprises a professional and dedicated group. The facilities compare favorably with those of other churches and make it possible to meet the diverse demands on the Church.

The present structure of the Department of Economic Development is able to manage and implement the financial programs of the Church and has established its credibility to the parishes and communicants. Equally as important, it has shown an empathy to the parishes in financial matters that has helped establish the program of Total Commitment. This permits funding and sustaining the National Ministries budget by apportionment of the National Budget to all parishes on the basis of a fair share and stewardship inspired system.

TOTAL COMMITMENT PROGRAM

When His Eminence undertook his ministry, the financial structure of the Church in the Americas reflected the parochial history. The financial obligations of the communicants comprised a per capita payment—known as the Dekadolarion—which was intended to take care of the needs of the Archdiocese and all its institutions.

His Eminence realized that such a financial system not only did not provide the necessary funds but, more importantly, such a system of per capita payment was not in accordance with Christian giving and did not promote unity. It tended to create an atmosphere of "we and them." The path of change from the per capita to the present Total Commitment was not a direct one but a progressive series of changes brought about by the patient and fatherly guidance of His Eminence and the ever increasing number of communicants who gradually began to realize their responsibilities under his leadership.

The implementation of the Total Commitment program coupled with the greater awareness of the needs of the Church have made it possible to progress from a budget of less than $1,000,000 in 1959 to over $10,000,000 in 1984.

Today, the National Ministries Budget serves the worldwide ministry of the Church, including its institutions, dioceses, departments, community services, benefits programs and missions.

224

His Eminence surrounded by his co-workers and department directors of the Archdiocese of North and South America during the Christmas holidays.

Initially, His Eminence recognized that the parochial nature of the parishes also posed serious problems for the parishes themselves. Almost without exception each parish of the Archdiocese in 1959 had a membership dues system of meeting its financial needs coupled with numerous special affairs for fund raising. Thus, among the first priorities of His Eminence was to encourage the parishes through his writings and visitations and through legislation at the Clergy-Laity Congresses to educate their parishioners to meet the financial needs of the parish by having the parishioners assume more directly and personally their financial obligations to the parish.

As stated by His Eminence in the workbook of the 1972 Clergy-Laity Congress:

"The Parish is the concern of the Archdiocese not only as an organism, but as the soul and the spirit of our total ecclesiastical existence—It is the Church, the universal Church in a given and specific place and time."

* * *

"Just as the Church cannot be understood without Christ, in like manner the Parish cannot be understood without the Bishop, whom the assigned priest or pastor represents."

His Eminence's continual concern was voiced many times including his keynote address to the 1974 Clergy-Laity Congress:

"He who is a member of a Greek Orthodox Parish, and has this belief deep in his conscience, no longer is a member merely of an organization. He becomes a member of the Body of Christ, around which the Parish is formed."

Thus, while the legislation and environment were being created, on the one hand, to make possible the change of the parish financial programs from a membership dues system to what eventually became a Fair Share System, on the other hand, the leadership of the parishes was also becoming cognizant that a comparable change from a capita system to a Total Commitment program was needed on the Archdiocesan level. The two changes, on the parish and Archdiocese levels, were inextricably associated as the communicants gradually came

226

to the full realization that the Church was indeed one and their commitment including their financial obligation was to the totality of the Church.

THE CLERGY BENEFITS PLAN

Other occurrences also contributed to the transformation in the thinking of the communicants - all brought about due to the visionary and persevering leadership of His Eminence.

One of these was the establishment of the Clergy Benefits Plan. Early in his ministry, His Eminence expressed his concern that the Church had not met its responsibility toward its clergy. Therefore, His Eminence directed studies to be undertaken and presented the results to Clergy-Laity Congresses. Repeatedly, the Congresses legislated the establishment of programs, without legislating the manner of funding. At the same time, many clergy and laity doubted the willingness of the communicants to support such programs and raised obstacles to their implementations. His Eminence persisted that the program should be fair and viable and adequately provide for changes such as inflation. He continued to direct studies and patiently pursued the educational process, assisted greatly by the unselfish efforts of a dedicated layman, until finally the clergy and laity authorized in 1973 both the establishment and financing of the outstanding Clergy Benefits Plan which we have today. This not only provides for the future but gives past service credits to the deserving clergy who had served when there was no program.

The implementation of the Clergy Benefits Plan not only fulfilled the long overdue obligation to the Clergy but also served to convince the communicants that it was absolutely essential that all join together in their obligation to the Church. This served to speed the establishment of the Fair Share programs on the parish level and the Total Commitment financial program on the Archdiocesan level.

Since the establishment of the Clergy Benefits Plan in 1973, the Archdiocesan budget has reflected the continued and increasing benefits to a point where the budget exceeds $2,500,000 for the pension plan, life insurance, disability insurance, and medical and dental insurance.

The concern of His Eminence for the clergy is also evidenced by the establishment of a National Presbyters Council, his systematic meeting with the clergy to discuss their concerns.

227

Early in his ministry, the parish regulations were changed to delineate the financial obligations of each parish toward its clergy and changes have been made in these regulations as warranted by further studies.

LAITY

As he made his pastoral visits throughout the Archdiocese, His Eminence has sought to find the best and most talented persons and encourage them to serve in the Archdiocesan Council, the boards of our institutions, the various commissions of the Archdiocese and the dioceses.

As a result, today we find a greater involvement of the laity in the life of the Church and the quality of lay leadership is continually improving. More and more Orthodox lay persons are offering their talents to the service of the Church at all levels and in all the functions of the Church.

THE ARCHONS

Through the efforts of His Eminence, the ancient practice of the Church in recognizing the contributions of the laity by awarding the title of Archon has been extended. This practice not only honors those who have served the Church with distinction but also serves as an example for all to emulate.

THE RESTRUCTURING OF THE ARCHDIOCESE

As he entered into the second decade of his ministry, His Eminence became aware and concerned of the need for the establishment of an hierarchal and administrative structure which would more effectively serve the needs of the Greek Orthodox communicants and insure the survival of the Church in the Americas.

The new Charter superseded the 1931 Charter and resulted from a formal request made to the Ecumenical Patriarchate in November 1973.

Having dealt with the problems of Orthodoxy in the Americas first hand for fifteen years, His Eminence made the request for a new Archdiocesan structure. His objective was to enable the Church to better serve the needs of the people and the parishes of this vast Archdiocese. At the same time it was important to maintain internal unity in the Church of America as well as unity of the Church with the Ecumenical Patriarchate.

Shortly after the formal request, a first delegation representing the Archdiocesan Council visited the Patriarchate. They presented the views of American clergy and laity as to the status, problems and needs of the Church. They described the pluralistic society in which the Church functions and explained the necessity for restructuring. The delegation was welcomed as the first such delegation ever to visit the Patriarchate. The Patriarchate, through a special committee, listened carefully to the delegation and promised prompt and thorough consideration of the needs of the Church of the Americas.

Consultations and the exchange of drafts continued for over three years. The Archdiocesan Council and its committees met in extended sessions to consider the drafts. Finally, a second delegation visited the Patriarchate in November 1977 to transmit the latest thinking of the Archdiocesan Council. They were greeted by His All Holiness, Ecumenical Patriarch Demetrios I. His most understanding and encouraging words indicated that the Ecumenical Patriarchate had indeed given serious and deep study to the problems of Orthodoxy in the Americas.

His All Holiness stated:

On the occasion of this official meeting with you, we wish to make known to you that the Ecumenical Patriarchate, throughout the course of the historical development of Greek Orthodoxy in America, as well as that of American Orthodoxy in general, with great effort, attention, understanding and constant motivation for the well-being of the whole of Orthodoxy in the Americas, is possessed by a sacred desire. This holy desire on the part of the Mother Church is to offer itself in service for the increasing growth and development of the Holy Archdiocese in a fashion so that the Archdiocese may give - together with all of the Orthodox in America - a united, eloquent and powerful witness of Orthodoxy in the New World. It is our expectation that this take place within the realities of the present world and as a Christian offering, more broadly spiritual and more broadly civilizing, as the unique contribution of Orthodoxy to the geographic and cultural American reality. All of these things are said for one basic purpose; that Orthodoxy offer its very own interpretation and solution to the primary and ultimate problem of the world, that is, the relationship of humanity with God and the relationship of human beings among themselves that arise from the divine-human relationship. It is in this broad, Orthodox view of the world and humanity that the Ecumenical Throne intends to examine the special problems of our Holy American Archdiocese.

Interpreting to you the spirit of the Holy and Great Church of Christ we wish to reiterate - even though this is known to all persons - that in reference to the important topic of American Orthodoxy (though we respect and recognize the value of special characteristics of the many ecclesiastical and other traditions of the different Orthodox ethnic groups and various ecclesiastical jurisdictions in America), we desire a single, united Orthodox presence and witness in America, always within the framework of Orthodox canonical and ecclesiastical order.

It is with this spirit that the Ecumenical Patriarchate seeks to conduct the conversations between you and those who have been appointed by the Holy and Great Synod of the Ecumenical Patriarchate for the ultimate well-being of our Holy Archdiocese in America and for the broader good of all of American Orthodoxy. Welcome.

These inspiring words were further amplified by His Eminence, Metropolitan Geron of Chalcedon, Meliton. As President of the Patriarchate Committee of Coordination and Programming, he opened the conference by stating the basic principles of the discussion and saying:

Our meeting in this holy center of Orthodoxy may be characterized as an historic juncture of the reality of Orthodox presence in America. We use the word "reality" in order to emphasize that we speak in a concrete manner.

Within the general climate which combines the ancient Orthodox tradition of the Mother Church, her canonical order, ecclesiastical ethos, and American reality, and with the service of the faithful people of God in America and the common Orthodox witness in America as our ultimate purpose, we initiate these conversations with the prayer that the results will be the best possible for the common good during this present stage of the reorganiza-

230

tion of our Holy Archdiocese. We do so, with an even fur-
ther goal, i.e., that this more perfect structuring of the
Archdiocese will become an essential, basic and concrete
contribution to the common Orthodox witness in the two
Americas.

What followed was a series of long but gratifying meetings which culminated in a draft of the Charter that the delegation believed was a reflection, not only of the hopes and aspirations of our Archbishop, but also of the clergy and laity.

When the delegation returned to the United States, they were gratified to learn that the Holy Synod had granted the Charter on November 29, 1977.

In his letter transmitting the Charter to the Church in the Americas, His Eminence Archbishop Iakovos has stated:

The main characteristics of the new Charter are: 1) the
establishment of dioceses bearing the name of the geo-
graphic location and the city in which each will be located,
replacing the present Archdiocesan districts system; 2) the
election of Bishops to each diocese, replacing the pres-
ent system of Auxiliary Bishops; 3) and the establishment
of a Synod of Bishops which will have all the rights and
responsibilities of an Episcopal Synod.

While the new Charter provides more privileges as
well as more opportunities for initiatives in their dioceses,
the Bishops will also bear greater new responsibilities to
better serve the spiritual needs of the faithful and pro-
vide for the progress of our parishes.

In accordance with the wishes initially expressed by His Eminence, the Charter specifically provides that the Archdiocesan Clergy-Laity Congress "is concerned with all matters, other than doctrinal and canonical. which affect the life of the church including its unity, uniform administration, education and financial programs."

Following the granting of the Charter, the implementation thereof began and continues to this day. In addition, studies continue under the guidance of His Eminence as to the need for changes or modifications thereof.

THE STEWARDSHIP PROGRAM

Under the paternal and loving guidance of His Eminence,

the rethinking by the communicants of their obligations to the Church eventually matured into the Stewardship Program which now functions to encourage each Orthodox to make wise and meaningful use of all that he or she may have. The Archdiocese provides materials and consultants to the parishes in order that each parish can implement a Stewardship Program. The Stewardship Program is an important factor in the present and future status of development of the Church and its expanded scope should play a major future role in the progress of the Church.

LOGOS

As His Eminence began his ministry asking for a new kind of thinking with respect to the obligations to the Church, a few laymen answered the clarion call and responded with the establishment of the Sustaining Membership Program which has now become LOGOS, the League of Greek Orthodox Stewards. Originally founded for the purpose of serving as a vehicle for concerned Orthodox to make direct financial contributions to the Archdiocese to meet the needs of administration, the LOGOS program has now developed where thousands of Orthodox are being given an opportunity to give of their talents and material things to specific projects and institutions.

THE FUTURE

The visionary and persevering stewardship of His Eminence Archbishop Iakovos has brought the Archdiocese to a level of development that could not have been envisioned at the beginning of his ministry and is characterized in a transformation of the thinking of clergy and laity of the Church. As a result, the Archdiocese can face the future with a strong administrative structure and sound financial programs involving dedicated and able clergy and laity who are aware of the need for unity.

Impact Through Public Relations

By

George Christopoulos

o other church leader in contemporary America has understood Marshal McLuhan's concept that "the medium is the message" as well as Archbishop Iakovos. He has not only been an excellent communicator of Christianity's message, and personifies it, but is also an outstanding newsworthy subject for the Archdiocese's internal and external communications—the subject of this article.

His extraordinary speech in November 1983, for example—at a mass rally before the United Nations protesting the unilateral attempted formation of a Turkish "nation" on divided Cyprus—had an electrifying impact on thousands of people present and among many more thousands who heard it on radio. He said his "dream" was not that the Turks would be destroyed but that, one day, the Turks and the Greeks of Cyprus would live together in harmony. There were echoes of Martin Luther King's similar speech in 1963 urging love and true co-existence between blacks and whites in the South, an "impossible dream" which miraculously is being realized more rapidly than anyone could have imagined. Those hearing the talk sensed the Archbishop's honesty and were moved by its "love your enemy" philosophy, an ideal of The Master that Archbishop Iakovos was simultaneously preaching and living. It is well known, of course, that because of the Turkish government's opposition, Archbishop Iakovos was not selected as the successor to Patriarch Athenagoras I. Any other leader might have been rankled and sought thereafter any opportunity for retribution—but not Archbishop Iakovos. His ideal is Christian love—which he embodies, preaches and communicates to millions through the mass media.

As Primate of Greek Orthodoxy in the Americas for 25 years, Archbishop Iakovos has succeeded in strongly communicating the Word of God and projecting in the Western Hemisphere the wisdom of the Orthodox church and the beauty of its profound glorification of God. He has been following in the steps of St. Paul, the Patristic Fathers, the Greek Christian missionaries to the Slavs and more recently, his worthy predecessors in the New World, Archbishops Athenagoras and Michael. Iakovos has carried out the dictum of St. Matthew almost literally (with his celebration of the Easter liturgy on national television and through educational films distributed by the Archdiocese): "Let your light so shine among men that they may see your good works and glorify your Father which is in Heaven."

From the day of his enthronement, the Archbishop has often simultaneously been the *message* of a TV or news story and the *medium* carrying Christ's Good News. This was caught accurately in a nationally-syndicated interview on April 1, 1959 by the religion editor of the Associated Press, George Cornell, who—soon after the enthronement—wrote:

Archbishop Iakovos. . . personifies some distinctive qualities of his Church—qualities that are giving it increasing impact and esteem. Like the church, he is urbane, cheerful, learned, cosmopolitan, warm-hearted, tolerant and immensely confident. . . A tall, handsome man of 48, Archbishop Iakovos' conversation sparkles with scholarly references and homey anecdotes. He is genial, informal, with an easy smile. He likes to fish, read poetry, writes a bit of it himself. In his new post, he holds the second highest office in his church, next to Patriarch Athenagoras I of Constantinople (Istanbul), Senior See of Orthodoxy and dominant center of ancient Christendom. With Orthodoxy now extended into the West, including its growing U.S. prominence, Archbishop Iakovos hopes to bring about some form of confederation or consolidation of all its national branches in America. "The time is ripe for this now," he said. "All Orthodox groups are faced with this transition period. We need to grow. . .and contribute the most to life in this country."

BACKGROUND OF THE COMMUNICATIONS EFFORT

The Orthodox have grown, indeed, during the quarter-century Iakovos reign, formed strong social connections and contributed substantially if not "the most" to life in America. Greek Orthodox stability, and its American parishioners' deep belief in the fundamental reality of the dogmas as presented in the Nicene Creed, have been very welcome in the United States especially in the 1960-1984 period. As we know, these last 25 years have been a crisis period. It has been socially and politicaly divisive and troubled by economic hardship and by seismic cultural shocks stemming from racial tensions, the Viet Nam war and whirlpooling technological development. During this period there have been widespread rampages of immoral sexuality, violence, extreme pornography and drug abuse. These dehumanizing trends and other forms of secularism have driven families apart and alienated and depreciated the self-esteem of individual men and women. An unimaginable number of people and families have, because of such forces, disassociated themselves from traditional values and divine worship. Also, this troubled generation has been living in constant nuclear terror and, unwisely, looking less and less to the Church and to God for renewed sanity, health and spirituality.

Fortunately, the Greek Orthodox Archdiocese through its many activities, including its growing communications effort, has been able to substantially shield many of its parishioners from such attacks. The Church—with its leaders at the national and parish levels—has inspired parishioners to hold fast, grow stronger and pass on their Christian Orthodox tradition to their children and grandchildren. As we see in other sections of this book, as the cultural capacity and economic capabilities of parishioners grew, so did their spiritual needs. To these challenges the Archdiocese and the parishes responded vigorously with increased religious fervor and celebration as well as numerous innovative programs. The struggle for Christian survival (and even perfection) goes on, even now. Therefore, it is important to see what Archdiocesan communications and public relations have contributed to this struggle and what success has been achieved. Fortunately, there is substantial evidence in historical literature and sociology studies documenting that progress has been made by the Greek Orthodox in the Americas, especially in the last 25 years. (The best history to-date, even though anticlerical, is Theodore Saloutos' *The Greeks in the United States*.)

The first major sociological study of our religious community was sponsored a few years ago by the Archdiocese at the request of the Communications Department. It found the Church's message had been effectively communicated by His Eminence, the hierarchy, clergy and lay leaders throughout the 500 parishes to the public—especially some two million Greek Orthodox. The steadfastness and dogmatic "correctness" of Greek Orthodoxy in America has borne healthy fruit. While other religious groups were being weakened and drained of their faith and spiritual content through dogmatic and ecclesiastical confusion (at the national leadership and local level) the Greek Orthodox branch flourished—materially and in spirit.

In the *Study of the Greek Orthodox Population* conducted by the Gallup Organization in 1980, the following major, positive, elements were cited:

● Levels of attested belief among Greek-American adults (18 and older) are extraordinarily high—higher not only than the levels recorded for most other major faiths in the U.S., but also for most other nations of the world as well.

● The Church is the focal point in the lives of most Greek-Americans and serves them in a great variety of ways, social as well as religious.

● Levels of religious practices among Greek-Americans and among church members are also high.

● A large majority of parents of children under 18 believe it is "very important" to them that their children become "religious persons."

● A high level of awareness of Archdiocesan institutions and organizations is found among the Greek-American community in the U.S. and particularly among church members.

● The proportion of adults who say they have read the newspapers for the Archdiocese within the last 12 months is an impressive 75% for Greek-Americans and 81% for church members.

With regard to this communications function, the Gallup study found a very large majority of those surveyed (260 Greek Orthodox members, 195 persons of Greek heritage and a supplementary survey of 98 teenage church members) indicated they knew enough about the Greek Orthodox press to explain its function and to give their opinions about it. The great majority said this press also was a very key source of information for them about Greek Orthodox affairs.

What they want to learn more about from the press (and presumably from other public media and internal print, audio-visual and other communications from the Archdiocese and their parishes) are the following:

1. Historical information, such as the roots of Greek Orthodoxy's early church history.
2. Moral and ethical guidance—how to be a better Greek Orthodox.
3. Ecumenism—developing closer relations between Greek Orthodox and others.
4. Liturgical matters, i.e., pertaining to the Divine Liturgy and the Sacraments.
5. Readings and interpretations of the Bible.
6. Greek Orthodox missionary activity in other parts of the world.
7. Controversy within the Church on sexuality, ordination of women and traditionalist liturgies.

The study included suggestions to the Archdiocese as to how the unique strength of the Greek Orthodox Church could be maintained or strengthened. Its main conclusion was the following:

If the Greek Orthodox Church can meet the spiritual needs of youth, continue to provide vital education programs, reach out to those not in the church, while giving succor to those within the church, its continued vitality in the 1980s seems assured.

Obviously the success the Archdiocese and the parishes have had in keeping the faithful close to this most ancient of all Christian denominations is not complete. The study did point out some of the impact on Orthodoxy of the destructive sources affecting all other religions in America. Imperfections that were uncovered by the study included negative feelings by some parishioners on liturgical matters, alleged restrictiveness of the Church in some areas (particularly toward "outsiders"), and a difference of opinion (by younger members) regarding the acceptability of pre-marital sex.

The significance of the Gallup survey is manifest and is in accord with the experience of many of the Church's leaders at the parish, diocesan and Archdiocesan level. The study proves the Greek Orthodox Church in America has been successful in preaching and reaching its constituents. But as the new generation further feels the impact of secular and other forces, *there are warning signs that there will be even a greater need to communicate to all two million Greek Orthodox—*

in and out of the church setting! Let us examine therefore the history of the communications efforts of the Archdiocese and determine how it is prepared to meet the future.

PIONEERING EFFORTS AT PUBLIC RELATIONS

From the founding of the Archdiocese, some efforts at public relations and communications have been made by every Archbishop. As a shepherd of many parishes, he is automatically a celebrity among the Greek Orthodox and newsworthy in the Greek American press. Also, every Archbishop is given some attention by the general religious press. He is an acknowledged leader and representative of the Ecumenical Patriarch of Constantinople (known by learned editors everywhere to be an historic See equal in honor to that of the Pontiff Maximum and Bishop of Rome). But only through special talent and success in communicating the reality and *image* of importance to the public does any religious leader receive major coverage in the general news media.

Although Archbishop Metaxakis (see biographical article First Section) was acquainted with President Hoover and moved in high Anglican circles, he was not known to the American public. Nor was he charismatic enough to raise public opinion regarding his Church and its constantly quarreling immigrant parishioners. In contrast, Archbishop Athenagoras exercised tremendous charisma and through frequent celebration of impressive Greek Orthodox Epiphany ceremonies in Atlantic City, New Jersey and in Tarpon Springs, Florida, got to be widely known. In doing so, he raised the conscious level of the public about the Church and the Greeks in America. Athenagoras was active in the Hellenic cause during World War II. He was well regarded by many Americans for representing heroic Greece at the highest levels of society during that terrible conflict. That outstanding prelate practiced personal diplomacy, mixed with notables, and in other ways became newsworthy. Athenagoras, of course, was widely hailed as National Chairman of the Greek War Relief. Along with Spyros P. Skouras, president of the organization founded a week after Italy attacked Greece, Archbishop Athenagoras was invited to the White House and received due praise in the press for that organization's success in saving thousands of Greek lives during the terrible Nazi occupation. The elevation of Athenagoras to the Ecumenical Patriarchate caught the fancy of the news media in 1947, as did the singular honor by President Truman of having the high cleric transported to his See by the Presidential airplane. This was a gracious act on Truman's part (and, of course, a political sign to the USSR that a second "Greek-American" prelate had become head of the Orthodox faithful—some 400 million people—if the Russians were counted).

Archbishop Michael (Primate 1950-1959) continued to practice private diplomacy as his preferred means for projecting the importance of the Church on the national and international scene. During his reign certain major news events took place which put him, and the Greek Orthodox Church in the U.S., in the media spotlight from time to time, but not often or systematically. The Archdiocese still required major internal strengthening; therefore, public relations and other "external affairs" were secondary as will be seen in other articles in this volume, especially by Constantelos, Hatziemmanuel, and Kourides. Michael was still building the internal strength of the Archdiocese and probably his most notable success was an internal program—the Greek Orthodox Youth of America. Nevertheless Michael became alert to the value of, and the need for, media relations. He pioneered in institutionalizing the public relations function and getting the assistance of media professionals to help in obtaining international publicity.

To this end, Archbishop Michael formed the Greek-American Press Association and held many strategic meetings with 16 members (all American journalists of the Greek Orthodox faith). They served as public relations advisers to the Archdiocese. In 1955, Archbishop Michael received national coverage in the secular press for protesting Turkey's savage attacks upon the Ecumenical Patriarchate. Much of this publicity was due to the efforts of the Archdiocese, and that handful of Greek-American journalists led by George Douris who wanted to stop the attacks. They sought to mobilize Greek-Americans (and the general public) and get them to express their anger and concern to the United States President and Congress. This *ad hoc* public relations campaign, combined with similar protests from the Greek press and AHEPA, did raise the responsiveness of Congress somewhat and probably contributed to the slowdown of the Turkish attacks, which might have—if left unchallenged by the U.S. Government—resulted in the total destruction of the Patriarchate. Also helping in the defense was Ecumenical Patriarch Athenagoras himself, a friend of President Truman. In the press he reminded America that he (Athenagoras) had been an American citizen, loved the U.S. and its ideals and expected his former second homeland to show concern for his venerable office. In 1955, the Archdiocese at a cost of $25,000 published a special tabloid newspaper showing pictures of the Turkish atrocities at the Patriarchate and photos of the murder of Greek Orthodox clergy and distributed the newspaper to members of Congress, news media and opinion leaders. For a variety of reasons, including the combined public relations campaign described above, the Ecumenical Patriarchate was saved—at least temporarily. From outright destruction of the Patriarchate, Turkey moved to the less noticeable strategy of atrophy, suppression and attrition.

In 1965 I was invited by Archbishop Michael to direct public relations for the Archdiocese. I did so for a short period. I found the state of the public relations function to be at its early stages. The Archdiocese's conception of "public" was what might be objectively called "internal affairs"—such as concern with, and events around, its parishes and institutions. Its publications were very sparse, almost limited to the yearbook—a small handbook filled with an assortment of bilingual information. There was almost no effort to impact the non-Greek community or general public with standard public relations methods. With the exception of an occasional press release announcing a conference, an appointment or a religious event, there was little media contact. Also, the Patriarchal crisis was over, at least as far as the press was concerned.

Despite a crying need for good media contact development, the leadership did not understand the need nor could it expend much of its limited resources on external affairs. (See preceeding article by Foussianes.) The Archdiocese was insular; it must be noted however, that Archbishop Michael, as an astute diplomat, was able to carry on high level discussions with the then Secretary of State and to be the first Greek Orthodox prelate to participate in the inauguration of a U.S. President (Eisenhower, 1953). The relationship between the two remained strong. President Eisenhower and other dignitaries participated in the well-publicized ceremonies of the St. Sophia Cathedral in Washington in 1958. Evidently, the federal government has become increasingly appreciative of every Greek Orthodox Primate since Archbishop Metaxakis.

PROFESSIONALISM GRADUALLY ACHIEVED

Public relations is a communications effort practiced differently by various leaders and organizations. It is an artistic-sort of social science that blends diplomacy, journalism, marketing and management. Its purpose (regardless of what form it takes) is to increase the sponsoring organization's total viability and historical impact.

As noted earlier, the current affairs and the cumulative effect of this outreach by the Archdiocese under Iakovos has reached a very high level for any organization. Today, there are many in the press and broadcast field who increasingly turn to His Eminence and the Archdiocese for their comments and reaction whenever a major church-related story or historical event takes place. The Archdiocese's authority is highly regarded and the Archbishop's comments widely sought and quoted by the press and broadcast media. This is a mark of excellent relations with the communications media—something that took more than a decade to develop. That this outreach function has been raised to a high level is attributable to the acumen of Archbishop Iakovos, and to the healthy development of the Greek-American community and Church. As stated earlier, the most recent success of communications can be attributed in great part to His Eminence. He is both the medium and the message and has helped nurture to maturity the Archdiocese's public relations structure and function. Of course public relations and communications are not expected to be an end unto themselves but to reflect the needs and abilities of "the client"—in this case the whole Church. Therefore, as it developed in strength and influence (as detailed in this volume by Messrs. Kourides, Foussianes and Charles) the Archdiocese increased its public relations capability and success. This was not automatic improvement but reflected the skill and devotion of the public relations staff.

For many years in the 1960s the public relations department was headed by Arthur Dore, assisted by Panos Peclaris; in the 1970s it was headed by Reverend Basil S. Gregory and Ms. Terry Kokas (who served as director until 1981). Through these dedicated professionals and others who served on committees or/as employees, consultants, advisers and volunteers, the Archdiocese effectively stayed in contact with the national wire services and broadcast networks, major dailies and national magazines, the religious press and Greek-American

241

media. The first half of the secret of the staff's success was in building upon the respect and friendship of many journalists for the Archbishop. The other secret of this small staff's effectiveness, then and now, lay in its ability to stay in close contact with reporters, editors, broadcasters and other media representatives, serving their information needs quickly, accurately and thoroughly. Such trusted relationships are invaluable. They have been nurtured through sincere openness by the staff and the Archbishop, even extending to journalists frequent invitations to participate as individuals in Greek Orthodox conferences, meetings and events. Through such trusted, close, liaison, the department was instrumental in the arrangement of several major productions on major TV network programs (including the presentation, live, of the Easter Agape service on NBC under the title "Festival of Love").

Additionally, the public relations staff is often called upon to coordinate Archdiocese programs and events, especially those that have a public interest content. This constitutes planning and helping carry out programs and events, and is not merely the issuance of a press release to the media. Without belaboring the staff's planning, coordination and multiple contact work, I list some of the very important events during the Iakovos reign for which the Archdiocese issued news releases and which resulted in good "pickup":

● Iakovos offers prayers at the inaugural ceremonies for three American Presidents: Kennedy, Johnson and Nixon.

● Iakovos is elected co-president of the World Council of Churches, a post he held for eight years, and undertakes trips throughout the world to WCC, ecumenical and interfaith meetings.

● Iakovos travels to many parishes; delivers lectures and speeches before various college and universities and at meetings with numerous Christian denominations and other religious groups.

● Iakovos travels abroad to participate in the Pan-Orthodox meetings in Rhodes (1961). Iakovos accompanies Patriarch Athenagoras to Mt. Athos (1963), to his dramatic meeting with Pope Paul VI in Jerusalem (1967) and subsequent visit to Rome in 1967—where the 1,000-year old mutual "anathema" of Constantinople and Rome is nullified. Iakovos travels to Athens, South America, Korea and Vietnam.

● Iakovos marches in Selma, Alabama with Rev. Dr. Martin Luther King, Jr. and expresses his own support, and symbolically the support of all Christianity, for the civil and human rights of all oppressed people.

• Iakovos and Richard Cardinal Cushing issue a joint statement protesting the banning of prayers in public schools by the U.S. Supreme Court.

• Iakovos officiates at a requiem service for Francis Cardinal Spellman in St. Patrick's Cathedral, the first such ecumenical event involving Eastern Orthodox and Roman Catholic hierarchs at a high level in the U.S.

• Iakovos has a unique, private, audience with Pope Paul VI during his visit to the United Nations.

• Iakovos and Archbishop Terence Cooke (later Cardinal) pray together in the St. Paul chapel in the Greek Archdiocese directly after being informed of the assassination of Dr. Martin Luther King, Jr.

• Iakovos strongly protests the U.S. Supreme Court's decision to permit abortion.

• Iakovos prays for world peace at the official U.S. Bicentennial ceremony in Philadelphia, alongside President Ford.

• Iakovos is presented the Presidential Medal of Freedom by President Jimmy Carter.

• Iakovos asks parishioners to protest continuous violations against the Ecumenical Patriarchate.

• Iakovos participates as a leader of the World Council of Churches convening in Vancouver, B.C. (and earlier in other meetings in other cities world-wide).

• Iakovos annual nameday dinner is marked by U.S. Secretary of Education Bell (and in prior years by other celebrities).

• Iakovos officiates at the funeral of Captain George Tsantes, U.S. naval attache slain in Athens.

• Iakovos celebrates his 25th (and prior his 20th) anniversary as Archbishop with an ecumenical doxology, invocations before the Senate and the House of Representatives and other events.

• Iakovos lobbies on Capitol Hill against Turkish Cypriot attempts to create their own "nation" in Cyprus and to win multinational recognition of that "sovereignty."

• Iakovos delivers keynote address at Clergy-Laity Congress (in 1984 and other years).

DEPARTMENT OF COMMUNICATIONS FORMED, 1980

Despite the increasing public relations presence of the

Greek Orthodox Church in America, many meetings of several Clergy-Laity Congresses surfaced the need for greater efforts in this field. A committee, headed by James Scofield and Alfred G. Venetos, studied this matter in 1978 and developed a "blueprint" for restructuring this activity of the Archdiocese. On January 27, 1979 the master plan by the committee was endorsed by the Archdiocesan Council. His Eminence, on October 1, 1980, announced the formation of the current Department of Communications and said in the years ahead he would give high priority to its activities, which "would be a major factor in the future growth of the Greek Orthodox Church in the Americas."

The Department is now headed by Reverend Alexander Karloutsos, Executive Director, and includes the offices of News and Information, director Nikki Stephanopoulos and Broadcast and Film, director, E. Phil Eftychiadis.

The Department has available a listing of nearly three-dozen videotape cassettes, suitable for VHS or Betamax, that can be rented or bought for home or other use and another list of four-dozen 16-mm Greek Orthodox films suitable for parishes and other organizations. Additionally, descriptions of pamphlets and other publications published by the Archdiocese can be obtained for personal or parish use.

The mandate of the Communications Department is to be the principal agency of the Archdiocese in the field of mass media and internal communications.

In 1984, Father Karloutsos said the primary purpose of his unit, as is the ultimate goal of each Archdiocesan Department, is to proclaim the Good News: "To do so, we must seek to communicate—that is, to inform and edify our own members and believers and the society in which we live—about Greek Orthodoxy as God's ancient, yet ever-new, vital and vibrant religious presence in the world."

He said the Department had a five-year goal (1981-86) of creating the best communications system possible and of concentrating on the following three approaches and methods:

● Stronger professional and technical relationships with the press, broadcasting, telecommunications industry, cable TV systems and others.

● Education and training in media skills for Diocesan leaders, parish priests, lay leaders and seminarians.

● Media (radio, TV and print) production including the establishment of a national media resource center.

In the last four years the enlarged department has been operating, it has increased news coverage by the media significantly and it has also carried out its above stated missions. It has been especially successful in creating a broadcast and videotape capability for the first time. This has been possible through the efforts of Professor Philip Eftychiadis, a television production specialist and academic, and currently Director of the Office of Broadcast and Film. Additionally, several parishioner's donations have been directed to media production costs, including a $100,000 donation from George Coumantaros to underwrite the videotaping of a series on the Sacraments. The Department has inaugurated many ambitious programs: parish-level training seminars in public relations techniques and videotape methods, preparation of videotapes for parish use; production of radio tapes for Ecumedia News' 1,600 radio stations; dissemination of half-hour sermons to the Armed Forces Network, and the placement of a half-hour segment for broadcast by 300 cable TV stations. Other major publicity and educational successes were the hour-long special on NBC, "The Divine Image: The Orthodox View," and a half-hour Paschal program on CBS's Orthodox Series, For Our Times.

Father Karloutsos said the traditional public relations operation (now mostly concentrated in the Office of News and Information) has continued unabated and has resulted in a steady stream of articles and feature stories disseminated by United Press International, Associated Press, Religious News Service and other wire services and appearing in major regional and city newspapers as well as in Greek-American publications. For example, during the 25th anniversary of the enthronement of Archbishop Iakovos on April 1, 1984, there were significant articles written in the New York area by the two major dailies as well as three-day coverage by the three networks.

Among the very important publications recently completed or revised by the Department are the Communications Media Handbook (to provide parishes information and how-to-guidelines in their media relations) and the Communications Directory (listing several hundred Greek Orthodox journalists, broadcasters and others involved in the mass communications media). Another that has become available in August 1984 is A Companion to the Greek Orthodox Church (a comprehensive, one-volume introduction to our Church in America and Orthodoxy worldwide including sections on its history, theology, art, administration, activities, statistics and policy statements on various matters. It was compiled and edited by Dr. Fotios Litsas in conjunction with an editorial board.

Staff members are also represented on several commissions and boards, including : Religion In American Life, Communications Commission of the National Council of Churches, the New York City Council of Churches, and SCOBA. These responsibilities involve a considerable amount of time but are considered by the Director to be essential if the Greek Orthodox Church is to become stronger, more visible and an identifiable influence and presence in America.

"Our work has increased considerably," Father Karloutsos said, "and it will continue to grow because the need for communicating with our members and constituencies is enormous. It might be impossible to reach all our target audiences from the Archdiocese; therefore, in the next phase we hope to begin to see programs and success at the diocesan and parish levels. The role of the Archdiocese is to work with the national media and to supply leadership and support for the other levels of the Church. Between us we will be having even greater impact in the future."

Reflections from Brookline

Archbishop Iakovos and the Development of Hellenic College/Holy Cross Greek Orthodox School of Theology

By
Stanley Samuel Harakas

y purpose is to record some aspects of the leadership Archbishop Iakovos has provided with his 25-year service as Chairman of the Board of Trustees of Hellenic College, Inc. and to highlight some of the most significant policy decisions taken by the Board governing the highest education institution of our Church in the western hemisphere.

In a non-exhaustive, quite subjective manner, I look back on the stream of events which have taken place over my own 18 years of service as a faculty member. My tenure includes four years as Dean of our undergraduate Hellenic College, conducted concurrently with my service as Dean of Holy Cross from 1970 to 1980. From this flowing course of events and developments, there are some experiences which strike me as significant. These, I would suggest, had an innovative and constructive impact not only upon the institution itself, but also on the life of the Greek Orthodox Church in this country and abroad.

LEADING THE LEADERS

From my vantage point of participation in the meetings of the Trustees, I saw at close range the manner in which the Archbishop exercised leadership among men of strong will and character, accomplishment and success, vigor and vision. The task of the Trustees is to set policy and to implement it by providing the personnel, the financial resources, together with the continuing oversight and evaluation to actualize the policy. At different times and under different circumstances these responsibilities wax and wane in importance and urgency. As

247

a result, identifying the issues and setting the priorities is one of the most important tasks of the Chairman of the Board of Trustees.

I always had the impression that Archbishop Iakovos was usually several steps ahead of the rest of us in discerning the future issues and directions in which we were to move. Below, the reader will notice that in nearly all of the other topics addressed, there is a certain *avant garde* character. Nearly all of them are indicators of forward movement and development. It can be stated categorically that as Chairman of the Board of Trustees, it was Archbishop Iakovos who either initiated them himself, or quickly adopted and made into his own agenda, progressive suggestions and ideas of others.

Two overarching impressions regarding the Trustees and the Archbishop's relationship with that body remain with me. The first is the mutual loyalty and regard of the Chairman and the members of the Board. The second is the embodiment in institutional policy of the general Archdiocesan perspective, which seeks to place the Greek Orthodox Archdiocese and its institutions in the mainstream of American life today.

Many of the Archbishop's suggestions for the development of the School were innovative to the point that some of his long-time colleagues on the Board of Trustees—who were in fact friends and former parishioners from the early Boston Cathedral days—often resisted his leadership. In those cases, the Archbishop used a panoply of means to overcome their opposition, including the whole range of appeals: to reason, to feelings, to the past, to aspirations, to loyalty, to anger, to persistence in the face of strong objection. He didn't always win them over, but the one thing that remained constant in all of those debates, was the mutual respect, Christian love and loyalty of the Archbishop and his co-workers on the Board.

I will never forget the financial crisis which almost closed the institution in the early 70s and the vigorous concern for the restoration of financial stability to the School. It was a challenge which welded the Trustees and their Chairman into an extremely committed and hard working body. Tribute must be paid to them all, but the vigorous leadership of Vice Chairman George Condakes must be noted and the careful legal counsel of Charles Bucuvalas highlighted. Condakes' task was to work hand in hand with the Archbishop to avert financial catastrophe and save the School for its future development.

1922: Theological School of St. Athanasios, Astoria, N.Y. — also the first Archdiocesan headquarters of the Greek Orthodox Archdiocese of N. and S. America

Graduation procession at Hellenic College/Theological School, Brookline, Mass.

Needless to say, the School was saved and prospered.

From the very beginning of his ministry in the Americas as a priest, and later, as Archbishop, there was a very special characteristic in the performance of His Eminence. He has wanted the Greek Orthodox Church not to stand on the sidelines of public life, but to assume an active involvement in the mainstream of American life. In my book *Let Mercy Abound: Social Concern in the Greek Orthodox Church*, there is adequate documentation of that fact, in particular the relation to the moral questions which we face as a Church.

No less idiosyncratic is his commitment to a forward-looking, involved and contemporized approach to education—theological education in particular. Archbishop Iakovos easily and readily keeps the ancient traditions and values of Greek Orthodoxy incorporated into a vision of the contemporaneous character of the Faith. Hence, the demand that theological education be relevant to the contemporary realities, while remaining faithful to the age-old deposit of the Faith. This perspective guided and guides today the goal-setting process of the Board of Trustees of Hellenic College/Holy Cross School to Theology.

"YOU ARE ALL MY APPOINTMENTS!"

Sometimes said in anger, sometimes said complainingly, but mostly said with pride and as a challenge, those words will be heard by the faculty members on occasion in the course of deliberations about the work of the School, and particularly, the work of the School of Theology. Technically, of course, the members of the faculty *are* appointed by the Trustees. But the process of faculty appointment is complex, involving a search and evaluation procedure by the present members of the faculties, recommendations to the deans, evaluations by the Dean together with his own recommendation to the President, a recommendation by the President to the Trustees, who dispose of the issue through an internal process of their own. Yet, inasmuch as the Archbishop functions as Chairman of the Trustees, there is a real sense of truth to the assertion that the faculty members are "his appointments."

The assertion is most often a challenge, never a possessive claim. It is articulated as a call to responsibility and to excellence. Once or twice every year, with very few exceptions,

250

His Eminence has met with the collegiate and theological faculties in a joint meeting to share concerns and to discuss the work and progress of the institution.

More frequently, and on a more intimate and informal basis, he meets with the faculty of the School of Theology. We have met at various places: at the School, the Archdiocese, Boston restaurants, and most memorably, at His Eminence's home in Rye, New York. The breaking of bread together, and a moment of shared prayer are essential elements of our meetings. The agenda consists usually of exploring situations to the current problems of our educational endeavor. The tone is set by an intense concern on the part of the Archbishop with the School and by the communication of a sense of urgency and challenge to the Faculty members to do more, to work harder, to heighten commitment, to demand the best from themselves.

For many years Fr. George Tsoumas was a symbol—the faculty member who was a bridge between two worlds, the old and the new. Born in Lowell, MA and educated at the Theological School of Halki in Constantinople, he became the first Greek-American to hold a full-time faculty position at the School. Under the chairmanship of Archbishop Iakovos, the American-born members of the School of Theology faculty have risen to 85% of the total. Nearly all have either graduated from Holy Cross or have studied here. All have earned graduate degrees in the United States or abroad, primarily in Greece and in Germany.

In many ways, faculty are beneficiaries of the direct and indirect support of His Eminence. One example is the distribution of funds for study and the publication of books by faculty members through the Taylor Scholarship Fund. A survey of the published list of recipients of the Taylor Fund shows that 28 of the recipients have gone on to become faculty members or administrators of the School of Theology over the years. At present, the focus of the fund is to publish the writings of faculty members of both schools, an inestimable encouragement for scholarship endeavors by faculty members.

Not the least of these actions of support is the endowment of three academic chairs, with his own personal funds, as well as the accumulated proceeds of the annual Archbishop Iakovos Dinners in New York City, held on the occasion of his nameday. The endowed chairs are for the Greek language, the New Testament and Patristics. In his name, as well, an annual prize is awarded to mem-

251

bers of the faculty for outstanding contributions to education in our institution.

Needless to say, such support is more than symbolic. It is a support which shows in action His Eminence's intense concern with education in general, and education at Hellenic College-Holy Cross in particular. Perhaps that which symbolizes his relationship with the institution is the name given to the message he gives at each graduation service to the students. Called the "Paterexhortatory," it captures the fatherly concern of the spiritual mentor of the School, together with the persistent call and support for growth, development, advancement and the striving after excellence. No reality embodies that spirit of challenge more than the establishment of Hellenic College.

HELLENIC COLLEGE

The ground-work for the establishment of an undergraduate college preceded the Archbishop's assumption of his hierarchical duties in the Archdiocese. The Seminary began as a two-year preparatory institute whose purpose it was to linguistically train young candidates for the priesthood, so that they could study theology in Athens, Greece. Although established in 1937, international events leading to the out-break of the Second World War made foreign study impossible.

Courses were then added to the program, concluding arbitrarily with five years of post-high school study. This pattern continued for 17 years. However, there was no graduating class in 1955, since that year marked the extension of the course of studies to six years. Until then, the curriculum mixed liberal arts and theological studies together in one program. The new six-year program divided the curriculum into a pre-theological division and a theological division. A series of deans—Rev. Dr. Nicon Patrinacos, Bishop Athenagoras Kokkinakis, Rev. Dr. (later, Bishop) Panteleimon Rodopoulos—little by little prepared the ground for the establishment of the college. It was under Dean—subsequently President—Fr. Leonidas Contos, that the plans for the establishment of Hellenic College were implemented.

It was the shared dream of Archbishop Iakovos and John and Thomas Pappas, of Boston, to establish a Greek-American University. In the early sixties, a "Hellenic University

Foundation" was established. The Board of Directors authorized extensive feasibility studies which indicated that many Greek-American academics, intellectuals and fraternal leaders endorsed the idea. Staff, organization and faculty were recruited in the middle-sixties. Formal applications to state and accrediting agencies were made. The agencies, reflecting the limited programs originally envisioned, suggested a name change: and Hellenic College came into being. Unfortunately, a severe financial recession took place in the United States and world-wide, precisely at the time that fund-raising for the college began.

The Archbishop announced his full support of Hellenic College and committed his full energies to its success at the Clergy-Laity Congress in Athens in 1968. It was a commitment which was to be deeply tried over the next few years. The financial crisis for the School worsened. It took bull-dog determination not to allow the School to close. The Archbishop's commitment was deep and unwavering. I remember long Board of Trustees meetings, punctuated by faculty and student protests, resignations by chief school officials, financial crises on almost a monthly basis—and behind it all the powerful, persistent, persuasive presence of the Archbishop who held it together.

As the hard work and sacrifice began to reduce the debt and difficult decisions about program and staff began to take hold, the most important question about Hellenic College had to be faced: leadership. Never before had it been led by a lay leader. Always before there were clergymen at the helm. It would be a true innovative step to suggest and appoint a layman for President. Certainly, it could not be done over the objections of the Chairman of the Board. But when it was his own recommendation at the suggestion of leading clergy faculty members there were none on the board to object. Yet in the light of the long tradition, the appointment of Hellenic College Board Member and Dean of Graduate Programs at Babson College, Thomas C. Lelon, as President in 1976, took both courage and vision.

If the undergraduate college could be made to work, it would need a professional administrator guiding its destinies. But the confidence and full support of the Archbishop was absolutely necessary if the graduates, and particularly the clergy of the Greek Orthodox Archdiocese were to accept him, and more importantly, cooperate with him. The appointment and the necessary support were forthcoming.

The wisdom of the appointment is now seen in the careful and deliberate development of Hellenic College in both its programs and enrollment. The undergraduate college is staffed by an excellent faculty, with a carefully planned range of courses, which attracts a uniquely mixed body of students, primarily Greek-American, but which also includes a significant number of Orthodox Christians of other jurisdictions and nationalities. President Lelon has been capably assisted by another person who enjoys the support and confidence of the Archbishop, an extremely well educated clergyman and scholar, Fr. Michael Vaporis, who has been Dean of Hellenic College since 1976.

WOMEN STUDENTS

Not the least of the innovative changes which contributed to the new and viable character of the School, was the acceptance of women students in the degree programs and as residents on the campus.

I know from personal experience that the idea of bringing women to the campus of Hellenic College and Holy Cross was an original insight of the Archbishop. Two observations are needed to set it in proper perspective, one dealing with the ethos of Hellenic/Holy Cross, and the other with the state of education at the Teachers Institute at St. Basil Academy at Garrison, NY.

The radical character of this decision is understood only in the context of the kind of educational ethos which has characterized Holy Cross School of Theology from its inception. Its spirit was formed by the Patriarchal Theological School of Halki, the traditional training center of the hierarchy of the Ecumenical Patriarchate. The monastically-inspired character of that institution was a model for the character and lifestyle of Holy Cross in Brookline, even though it was its policy to encourage graduates to marry before ordination to the Orthodox priesthood. Even after the college was established, no women were accepted, and it was totally inconceivable that women could stay overnight on the campus. It was a rare occurrence to see women on campus and this, only on special formal occasions, such as graduation.

In the face of this, was another reality which Archbishop

Iakovos had to address in his capacity as Chairman of the Board of another Archdiocesan institution, St. Basil Academy. A hybrid entity, consisting of an orphanage and child-care home, and a teacher training center for the afternoon Greek schools of the Archdiocese, the Academy had not been equipped to respond to aspirations of the women students there to earn a recognized academic degree.

Needless to say, it was a bold initiative that the Archbishop presented to us when he proposed that we study the possibilities of merging the Academy's teacher training program with the undergraduate programs of Hellenic College. A two-year study of the possibilities began, necessitating long and involved curricular planning efforts to establish the new program, the careful preparation for living arrangements of the new women students, and most of all, the preparation of both men and women students for the incorporation of St. Basil's Teachers Academy into Hellenic College. Still, it was a shock to many of our alumni when, on March 23, 1978, at the annual celebration of the joint feast of the Annunciation and Greek Independence Day, that Archbishop Iakovos announced to the public gathered in the Hellenic College Pappas Gymnasium that the merger would take place.

Many felt a sense of the end of an era. Others felt the opening of a new horizon for the life of the school. For all, it was a significant change. It probably is still too early to fully assess the impact, but the wisdom of the decision can hardly be denied. The women on our campus, in both Hellenic College and the School of Theology make a significant contribution to campus life and to the educational process. Judging only from my own vantage point, I would have to say that the dire predictions of some simply have not materialized, and that there have been many significant additions to the quality of life in the school because of the presence of the women students. Most importantly, our priestly candidates now have an environment which more nearly represents the parish situation in which they will serve as priests. The move, by and large, has had salutary consequences.

ACCREDITATION

From an administrator's perspective, achieving and maintaining accreditation for an educational institution is primarily a process of making reports, and reacting to the responses to the reports. At this time, our schools enjoy a number of accreditations. We are accredited as an undergraduate and graduate education institution by our regional agency, the New England Association of Schools and Colleges. The several degree programs of Holy Cross School of Theology are individually accredited by the national professional theological education agency, the Association of Theological Schools. In addition, our School of Theology has received official recognition as a "University Level School of Theology" by the Universities of Athens and Salonika, in Greece, facilitating doctoral candidacies by our graduates in those schools.

The story of our efforts to qualify for accreditation is long, complex, and not particularly interesting, after the fact. What needs to be pointed out in this context is the commitment of Archbishop Iakovos to the process. The Archbishop never flagged in his encouragement and articulation of the idea of achieving accreditation for our schools. His Eminence knew that accreditation was not only an issue of recognition and acceptance into the mainstream of American education. It also meant an improvement of academic standards. This very rarely had to do with the content of our teaching. It nearly always had to do with requirements associated with the structure of the educational process. Examples of this were teaching loads for faculty, salary levels, library holdings, the integration of practice with theory in programs such as Field Education, distinctions between degree programs, the development of student services, etc.

It also always meant that more money had to be spent, that more people had to be committed to the effort, that more time had to be allocated for the educational process from restricted material and human resources. It is no secret that we had several occasions when our applications for accreditation were either rejected or received responses of "postponed action." These occasions were very depressing for us. Some reacted negatively with a "who needs it!" attitude. However, never were such attitudes communicated by the Archbishop. Rather, we had persistent encouragement to keep working, to respond to the accrediting agency comments with positive steps. He never let us forget the goal.

But, secondly, the Archbishop—as much as his other duties and responsibilities as Primate permitted—responded to our requests for policy changes, material resources and the placing of church personnel in the service of the School. Though others on the Board of Trustees frequently needed extended argumentation to be convinced that the Board should act in a given way to support the accreditation process, the Chairman of the Board seemed nearly always to understand, and to support our requests. I say "nearly always" because sometimes I felt that the Archbishop supported us in the councils of the Board of Trustees not because he fully agreed with us, but because he wanted to support us in our commitment and our striving after the prize. I remember such moments vividly, not only at Board of Trustees meetings, but at the Archdiocesan Council, at the Clergy-Laity Congresses of our Archdiocese, and during visitations to Greece, for discussions with academic and governmental authorities.

As the administrator chiefly responsible for the efforts to achieve accreditation following the first rejected application, I can assure the reader that accreditation could never have been achieved without the permanent, unwavering and totally committed support of Archbishop Iakovos.

ECUMENICAL AND PAN-ORTHODOX OUTREACH

A final brief word needs to be said about the ecumenical and pan-Orthodox outreach of Holy Cross Greek Orthodox School of Theology. More in his capacity as leader of the Archdiocese, than as Chairman of the Board, Archbishop Iakovos has enlisted the faculty members of Holy Cross School of Theology to become involved in ecumenical and pan-Orthodox meetings and conferences. More often than not, these meetings have required that faculty members study, research and write on topics, which in all likelihood they would not have investigated. Many of our publications in the *Greek Orthodox Theological Review* and other theological journals are precisely the fruits of such meetings.

Frequently funded by the Archdiocese in whole or in part, faculty members have travelled the world over to participate in ecumenical events. The most striking example of such involvement was the delivery of one of the keynote addresses at the 1983 Vancouver General Assembly of the World Council of Churches by Holy Cross faculty member, Fr. Theodore Stylianopoulos.

One of the most important ecumenical involvement decisions which could not have been made without the approval of the Board of Trustees and its Chairman was the application of the Theological School for membership in the Boston Theological Institute, an ecumenical consortium of theological schools in the Boston area. In my judgment, our participation in the B.T.I. has given us much more than we have contributed. It is a remarkable fact that at least once a month, our Dean, Fr. Alkiviadis Calivas, sits down with the Deans of eight other theological schools such as Episcopal Divinity School, the Jesuit Weston Theological School, Harvard Divinity, Boston University School of Theology, St. John's Roman Catholic Diocesan Seminary, Boston College's Theology Department, Andover-Newton School of Theology, and the evangelical Protestant Gordon-Conwell Seminary. Nowhere else in the world is there a theological consortium which includes an Orthodox School of Theology. It is a proud first. But more importantly, it is a significant "plus" for theological education at Holy Cross.

The benefits for Holy Cross from this cooperation are legion. Perhaps the most important tangible consequences are to be seen in the areas of the library, the field education program, the interchange of scholarly opinion in the various disciplinary meetings, and the privilege of free cross-registration. The intangible benefits are harder to describe, but the increase in understanding among the traditions, is certainly not the least of them.

The Archbishop's pan-Orthodox concern is seen again and again in funding faculty participation in numerous meetings designed to address issues of common concern for the Orthodox Church. Here too, the fruits are many, including significant publications and the privilege of participating in the development of theological thinking on new and perplexing problems facing the Orthodox Church. The international reputation of our School is enhanced with every one of these activities, which permit faculty members to interact with theologians of the Orthodox Church the world over.

Closer to home, pan-Orthodox cooperation has been fostered by the Orthodox Theological Society in America, the majority of whose officers and active members are on the faculties of Holy Cross Greek Orthodox School of Theology and St. Vladimir's Orthodox Seminary. One of the most important activities of the Society is the organization, on a periodical basis, of international theological conferences. His Eminence has been one of the most generous supporters of these conferences through financial and moral support. The proceedings of these conferences have been published, some in the *Greek Orthodox Theological Review.*

. . .AND MORE

There is much more which could be written here to describe the very close tie of His Eminence Archbishop Iakovos with Hellenic College and Holy Cross Greek Orthodox School of Theology. The special assignments given to the School to research projects for the Archdiocese, the special attention given by His Eminence to students, his special concern for mission students and encouragement of support of their study at Hellenic and Holy Cross, the encouragement of scholarship contributions for our students, the careful attention to major donors, such as Charles and Mary Maliotis, who gave the beautiful Cultural Center Building to the School, the assignment to the Seminary of the responsibility for clergy continuing education programs, are only a few of the many different topics which could have been developed if there were more space.

But even such topical development would not fully explain the continuing and persistent identification of Archbishop Iakovos with our College and Seminary. We know that deep in his heart the School has a special place. We know that in his prayers, the well-being of the School is a constant concern.

For you see, he is Chairman of the Board. . .and much more!

Greek Education and Learning in the Iakovian Era

By

Emmanuel Hatziemmanuel

he enthronement of Archbishop Iakovos took place in the New York Cathedral on April 1, 1959, in a majestic ceremony of an ecumenical character. The new Archbishop, understanding his mission and work to be "a God-given command",[1] took over the government of the Greek Orthodox Church of America at that difficult time when the religious and cultural life of the nation was undergoing a new orientation and reordering of its moral values. This, as expected, had a serious effect on the spiritual life of Greek American Orthodoxy, which had just begun to outgrow its immigrant stage and was becoming a creative factor in all phases of the American society.

The new Archbishop, having wide intellectual horizons and liberal but deeply religious personal experience, began to move freely within the perimeter of Greek Orthodoxy—this being a mosaic composed of our liturgical life and tradition, our Patristic literature, our Greek and Byzantine literature and art, and, of course, the Greek language and heritage.

In one encyclical after another, and by erudite archiepiscopal papers presented to Clergy-Laity Congresses, the new Archbishop analyzed and exalted Greek Orthodoxy by way of ideas and texts that have become classiscs in Greek-American Orthodox literature. For His Eminence, Greek Orthodoxy became a spiritual weapon against the materialism, hedonism and social realism of the time. Our ancient religion became a source of inspiration and an imperative element in the structure and identity of the Greek Orthodox populace in American society.

The Archbishop's keen intellect discerned that the survival of the Greek Orthodox identity in America demanded a new and all-embracing program. Our community and ecclesiastical life was in the midst of a "whirlpool of the libertinism and materialistic interpretation of religion and life." Above all, our communities exhibited serious weaknesses in the area of Hellenic-Christian education.

THE FIRST DECADE, 1959-69

From the very outset of his reign, the new Archbishop specified Greek education was the "first concern of the Archdiocese."[2] He stated in his introductory report to the delegates of the 15th Clergy-Laity Congress, ". . .without an Education worthy of the history and value of Orthodoxy, the Greek Orthodox Church is not going to survive."[3] He looked forward and demanded "the resurrection of the Greek Orthodox conscience, even in the hearts of those now in deprivation."[4]

A work of this magnitude, however, was not to be easily accomplished. Our communities at that time were involved in an unprecedented building activity that continued unabated throughout the decade of 1960-70. New and spectacular churches were erected with adjoining spacious community centers, yet very few school buildings were erected. This building vitality served as a reason for "thankful doxologies to God"[5] but it also awakened deep concern in the Archbishop. He knew the true, and ultimate, goal of the community lies not so much in building expansion as in the spiritual, moral, and social cultivation of its members. The Greek Orthodox community, in the words of the Archbishop, "must place at its center and at its perimeter the Resurrection; this means for the restoration of the morally fallen, for the revival of those who are spiritually dead, and above all to make the Resurrection an ideal, and an all-embracing pursuit of its members."[6] But unfortunately, ". . . many of our great communities prefer to build large and impressive community centers for the purposes of entertainment rather than centers that would offer both, social and cultural mixing and education, that is, halls as well as schools."[7]

In pursuing these theses, Archbishop Iakovos reprogrammed the priorities of our Church in America and considered the primary one to be education, ". . .an education including the ideals and life of both America and Greece. . . It seems that a number from among us have forgotten that our heritage and our mission is education. . .,"[8] he said to the representatives of our communities at the 16th Clergy-Laity Congress.

261

By his introductory report to the representative of the above Congress, the Archbishop offered for discussion and adoption an extensive educational program. This included:

• The effective improvement of religious education to be offered by qualified Sunday school teachers aided by parents, the revision of curriculum and the composition of new textbooks;

• The reorganization of the Greek Orthodox Youth of America (G.O.Y.A.);

• The organization of a community Boy-Scout program so that our children might grow up within our community and ecclesiastical perimeter;

• The opening in the largest and more affluent of our communities of Day Schools, instead of the building of large and impressive community centers, because "Day Schools constitute a certain guarantee for the future of our Church, while hardly the same can be said for the community center";

• The advanced accreditation of our Theological School to university status;

• The accreditation by the American educational authorities of the Teachers Training Department of St. Basil Academy as a junior college;

• The revision and re-evaluation of the curriculum of our Afternoon Schools to halt and reverse the lowering of standards resulting from a lack of interest on the part of both parents and the governing boards of our communities;

• The strengthening and widening of the scope of the programs of summer camps in this country, as well as in Greece.

The Archbishop knew that hinging on the success of his education program was the survival of our Church and the materialization of his dream for Orthodoxy to grow 'into a major factor of spiritual influence in this country."[9] At the foundation of his educational program, together with the systematic teaching of the Greek language, the Archbishop placed the teaching of our Orthodox religion. This because, in his opinion "the teaching of the language alone will not suffice. We need the molding of character which is the work of teachers with Greek Orthodox convictions who would carry out their mission with faith, love, and intense personal interest."

As regards the teaching of the Greek language, the Archbishop rejected the excuse offered at that time to the effect that the teaching of the Greek language would place obstacles to the further development of our Church. On the contrary, he strenuously argued that the Greek language constitutes an education of great value because "it familiarizes our children with the spiritual world in which they are born and grow"[10] and "gives them the opportunity to understand and express themselves in a language which is theirs, of their Church, and, to a great measure, lies at the foundation of higher learning here and elsewhere."[11] Moreover, "the Greek program must introduce our children into the world of the Greek mind, Greek thought, Greek art, and Greek civilization. All this to be pursued diligently with professional competence, with love and faith."[12]

The educational program of the Archbishop was adopted by the Church and began to produce definite improvements, mainly in two areas: the opening of Day Schools and the summer camps in Greece. Thus during the ten-year period of 1959-69 there began to function in community-owned school buildings the following five Day Schools: "Argyrios Fantis" and "Rev. Thomas Daniels" of the communities of Sts. Constantine and Helen in Brooklyn, NY and Washington, DC, respectively (1963); "Soterios Ellenas" of the community of "Koimisis" in Brooklyn, NY (1966); the Day Schools of the communities of St. Demetrios, Jamaica, NY; and of "Transfiguration" in Corona, NY (1967). The following parishioners donated large amounts for the building and operation of Day Schools: Theodore Tsolainos, Constantine Goulandris, Argyrios Fantis, Soterios Ellenas, and William Spyropoulos.

During this 1959-69 period the program of summer camps in Greece reached its final state of completion. The Archbishop from the very beginning discerned the educational value of these camps: "Our children," he said at the 16th Congress, "by visiting Greece, widen their cultural horizons; they come to know the life of the country where their fathers and grandfathers were born. They make friends with children of the same national origin and the same religion, and they prove able to form a personal appreciation of what Greece really is—this immortal land of eternal light, as it is called even by non-Greeks."[13]

In support of this program the Archbishop succeeded in securing the cooperation and support of the Greek government and of some of the larger banks of Greece. From 1959-67, 313 boys and girls, 11-15 years of age, together with their counselors were guests of these camps, funded by Greek banks. And in the year 1969 the Greek government donated to our

Archdiocese 11,500 *stremmata* in the area of Eleia-Bartho-
lomeo in the Peloponnese on the condition that the Archdiocese would forest the area and erect the necessary buildings
for summer residence for American youngsters of Greek descent. Thus came into being an impressive summer camp complex, equipped with modern facilities and with the capacity
of accommodating 200 children and 50 counselors. It was
named Ionian Village and began functioning in the summer
of 1970.

SECOND DECADE 1969-79

The 20th Clergy-Laity Congress, which took place in New
York City, June 27—July 4, 1970, marked the first year of the
second decade of the Archbishop's leadership. This period
was to be vital for the survival of our Church as "a sanctify-
ing and saving institution,"[14] and also as "a center of enlight-
enment and education."[15] The concern of the Archbishop at
the time was further accentuated by the "revolution" of a sizeable part of the American youth against traditional moral, political, and social values. A large number of the young resisted
"the Establishment" by political protests, drug abuse, sexual
licentiousness and bizarre music. By that time, the Greek-
American family had become part of American society, and
our young, an inseparable part of American youth.

The Archbishop guided by the lessons of the historical experience of the Orthodox Church during crises, proposed
the need for a fundamental revision pertaining to all facets
of the life of the Church. This revision would seek to develop
"a new religious experience"[16] "in place and in time"[17]—an
experience which he considered to be the only stable factor
guaranteeing the survival and preservation of the Church as
well as a means for the spiritual progress of the faithful.

In the area of education, the Archbishop requested the representatives of that Clergy-Laity Congress to consider and
take decisions in line with the report which the Office of Education of our Archdiocese had submitted. That report included
concrete proposals pertaining to all divisions of the Archdiocese's education: its primary aim, curricula for all school
types, classroom textbooks, teachers workshops and seminars, and school finances.[18]

The 20th Congress appproved this new program of the

Office of Education by which—among other prerequisites—the teaching of the Greek language was recognized not only as an obligation but as a "privilege" as well. The teaching of Greek was to be systematically and methodically pursued according to the principles of modern linguistics as regards foreign language instruction with the aid of readers and other books that the Office of Education of the Archdiocese would publish. The 20th Congress also requested that our young be taught to personally live "the spiritual, moral, and cultural values of Greek Orthodoxy, of the Greek heritage, and the almost-now-forgotten virtues of citizenship, patriotism, and the ideals of the present-day American ethnic synthesis."[19]

The decisions of the 20th Congress regarding the Archdiocesan Greek Education program have constituted since then the foundation, the framework, and the guide for formulating a system of community education with a Greek-American substance and character. This system has become "a recognized branch of American education,"[20] as noted by the eminent linguist, Dr. Joshua A. Fishman, who stated that the Greek-Americans, "have one of the more wide-ranging and better organized community school systems from among all national groups."[21] Also, *The New York Times* noted in an article published on November 16, 1980, on ethnic languages in America, that the Greek Archdiocese has "a sizeable school network and in its 400 Afternoon and 22 Day Schools, together with religious education, a comprehensive program of teaching the Greek language, history and cultural heritage is offered."[22]

The Archdiocesan Council of Education and the Office of Education of our Archdiocese, under the unwavering determination of the Archbishop, carried out the decisions of the 20th Congress. The results have been very satisfactory and can be summarized in the following:

1. The issuing of a new curriculum containing instructions for teaching Greek as a second language and for teaching our cultural history and religion (First edition 1972, revised editions 1973, 1982).
2. The publication of a series of new textbooks for teaching the Greek languages and Greek cultural history. This series is entitled: *Matheno Ellinika—Learning Greek*. The language books of this series have been written according to principles of modern linguistics and the methodology of teaching foreign languages. The first book was enthusiasti-

265

cally approved by the special Board (KEME) of the Greek government and was published in 1971 by the Institute of Textbook Publications of Greece.

3. Teachers workshops and seminars have been permanently instituted in the larger cities of the United States and Canada (New York, Chicago, Atlanta, Philadelphia, Baltimore, Boston, and Montreal). The first of these seminars took place in New York City in 1973.

4. The issuing of a syllabus for teaching modern Greek in American high schools. This syllabus was approved by the Department of Foreign Languages of the educational authorities of the State of New York, and was cited as being "an outstanding program for teaching the Greek language."[23] As a result, the Greek language was recognized as equal to other foreign languages that are taught in American high schools. This helped to introduce the teaching of modern Greek in a number of high schools in the Greater New York area, such as Bryant High School, and Long Island City High School. Also, the Department of Foreign Languages of the educational authorities of the State of New York, granted to the Archdiocese Office of Education the right to prepare the annual examinations in Modern Greek for students of American high schools, Greek community schools and for new high school students arriving from Greece. The results of this examination are submitted to the principal of the American high school and are entered on the student's report card. The first State examination of this type was administered in June 1972. Up to now 5,000 high school students of both sexes have taken this examination.

5. The question of teachers' salaries for both Day and Afternoon Schools has been extensively considered in accordance with the financial capability of our communities. As a result, the salary distinction between teachers of the American program and those of the Greek program was eliminated. A salary scale was established comparative to that of teachers serving in Catholic, Lutheran and other non-public schools of America.

6. Brief biographies of professors of Greek descent who teach in institutions of higher learning were collected and published in a volume entitled, *Who's Who of Greek Origin in Institutions of Higher Learning in the United States and Canada.*

7. Programs of teaching Greek in colleges and universities of

the United States and Canada were financially aided. These include Columbia University, Queens College (New York), Ohio State University, University of Connecticut, Rutgers University (New Jersey), Indiana University, University of Florida, and McGill University, Canada.

8. A number of Greek-American associations and societies of Greek letters and arts (such as the Modern Greek Studies Association, the Center for Neo-Hellenic Studies, the Greek Theatre of New York and the Philiko Society) were morally supported and some were financially aided.

9. The following Greek-American Day Schools were opened in the following communities: Evangelismos in Houston, TX, 1970, Three Hierarchs in Brooklyn, NY, 1975; St. Nicholas in Northridge, CA, 1977; St. Nicholas in Flushing, NY, 1977. Also, the high school of the community of St. Demetrios, Astoria, NY, was completed in 1975.

10. Relations with the Greek government in educational matters have appreciably improved. The Archbishop from the very beginning sought the understanding and cooperation of the Greek government in matters pertaining to our educational program on the understanding that the Greek Ministry of Education "possesses the experience and the expertise on Greek education" while our Archdiocese "is solely responsible for and exclusively administers the education of the Greek-American communities in the Western Hemisphere."[24]

Working together with the Greek government in educational matters began in 1961 at an important meeting at the Archdiocesan headquarters between the Archbishop and Constantine Caramanlis, then Prime Minister and now President of the Republic of Greece, and Evangelos Averof, the Minister of Foreign Affairs and now head of Parliamentary Opposition. Since then, cooperation has been harmonious and productive. And apart from the Ionian Village, already mentioned, the following have been accomplished:

a. Our Theological School and the Teachers' Training Department of St. Basil Academy were accredited by the Greek authorities as institutions of higher learning;[25]

b. All of our Day Schools and 95 Afternoon Schools have been given the status of their corresponding schools in Greece;[26]

c. The years of service of full-time teachers in our community schools have been recognized for pension purposes;

d. Annual workshops and seminars for teachers of three weeks' duration were established in Greece and funded by the Greek Ministry of Education. The first such seminar took place in 1973;

e. Three Counselors on Education were appointed in Canada, and one in New York City in 1977.

THIRD DECADE, 1980 TO DATE

The third decade began with a definite philosophy and clearly-defined goals with regard to our Archdiocesan Education. The curricula covering all of our community schools have been tested, language books for the teachers of Greek as a second language have been published, ampler means of teaching have been procured, teachers have become better prepared, and our Community System of Education constitutes a healthy part of American educational network.

Also, new Day schools have been added to those already mentioned: the "Peter Stathakos" school of the community of Holy Cross, Brooklyn, NY, 1980; the high school of St. Demetrios, Astoria, NY, 1980; the high school "Archbishop Iakovos" of the community of St. Demetrios, Jamaica, NY, 1984; and the school "Demosthenes" of the community of St. Nicholas, Laval, Montreal, 1982. And during this year the community of the Holy Trinity in Chicago—with the support of the Greek Americans of the entire city and the spiritual guidance of Bishop Iakovos of that city—will open the first Greek high school in that city.

In April 1984, 25 years of fruitful spiritual and ecclesiastical leadership were completed by Archbishop Iakovos. The work of the Archbishop has been particularly difficult. The way he followed was one against the spirit and philosophy of the times, a period "permeated by materialism, eudaemonism, and social realism." But it was the way dictated by the experience of history and the tradition of Greek Orthodoxy. This was a way of struggle, agony, and success.

The positions of the Archbishop regarding Greek education and learning were stated from the very beginning with undoubted clarity and were followed with unyielding self-consistency:

Our Archdiocese will never be brought to its knees while ascending its destined heights. This, at any rate, is its prescribed course. It will always support to the measure of

*its strength the Greek language and learning in their bat-
tle for survival. This, because we believe that the Greek
language and learning constitute the staff on which the
lyrics of the history of our nation are intoned together with
the rhapsodies and symphonies of our Hellenic-Christian
civilization that finally lead to "Christ is Risen. . ." of our
Greek Orthodoxy.*

*Our Archdiocese will do everything possible in sup-
port of the Arts—whether music, drama, theater, paint-
ing, sculpture, literature, architecture—for them to re-
cover their place in our hearts, in our homes, in our esteem, in formulating and expressing our esthetics, an element which has been part of the life of our Nation and
our Church during the long span of our history. . ."*[27]

FOOTNOTES

[1]Enthronement Speech, Holy Trinity Cathedral, New York City, April 1,
1959.

[2]Keynote Address, 15th Clergy-Laity Congress, September 17-24, 1960,
Buffalo, New York, p. 7.

[3]15th Congress, p. 7.

[4]15th Congress, p. 10.

[5]Keynote Address, 16th Clergy-Laity Congress, June 24-29, 1962, Boston,
Massachusetts, p. 26.

[6]16th Congress, p. 31.

[7]16th Congress, p. 31.

[8]16th Congress, p. 36.

[9]Keynote Address, 17th Clergy-Laity Congress, June 28-July 1, 1964, Denver, Colorado, p. 9.

[10]17th Congress, p. 20.

[12]Introductory Remarks to the Committees of the 18th Clergy-Laity Congress, June 26-July 2, 1966, Montreal, Canada.

[13]16th Congress, p. 22.

[14]Keynote Address, 20th Clergy-Laity Congress, June 27-July 4, 1970, New
York City, p. 8.

[15]20th Congress, p. 8.

[16]20th Congress, p. 8.

[17]20th Congress, p. 8.

[19]The representatives of the 20th Clergy-Laity Congress were presented
for comment, discussion, and adoption the following two texts: *The Aims*

and Goals of Archdiocesan Education, by the Rev. Dr. Nicon D. Patrinacos and *The Present and Future System of Archdiocesan Education,* by Emmanuel Hatziemmanuel, Director of the Archdiocesan Department of Education.

[19]Decisions on Education of the 20th Clergy-Laity Congress.

[20]Papaioannou, George, *The Odyssey of Hellenism in America* (Athens Academy Award - under publication), p. 145.

[21]Fishman, Joshua A. and Harkman, Barbara R., Project No. 8-0860. Prepared under N.J.E. Grant G-78-0133.

[22]The Ethnic Mother Tongue in America: Assumptions, Findings, and Directory. *The New York Times,* Nov. 16, 1980.

[23]Dammer, Paul E., Chief, Bureau of Foreign Languages Education, The State Education Department, Albany, NY. Letter to His Eminence Archbishop Iakovos, Sept. 10, 1973.

[24]Archbishop Iakovos' Report to the Greek Ministry of Education, Oct. 15, 1970.

[25]The Teachers' Training Institute of St. Basil Academy was accredited by the Greek Ministry of Education in 1970 and the Holy Cross School of Theology in 1975.

[26]Decisions of the Greek Ministry of Education: No. 160319 of December 15, 1972 and No. 27034 of March 13, 1973.

[27]Archbishop Iakovos' Speech at the Greek Letters Celebration, January 23, 1963. See: AGONES KAI AGONIES, published by the PATRIARCHIKO IDRIMA PATERIKON MELETON, pp. 1108, 11109.

"Philoptochos'" Major Achievements

The Greek Orthodox Ladies Philoptochos Society Had Extraordinary Success in 1959-1984

By
Stella Coumantaros

he philanthropic mission of the Greek Orthodox Ladies Philoptochos Society underwent a major transformation following the arrival of His Eminence Archbishop Iakovos, who was enthroned as Primate of the Greek Orthodox Church in North and South America on April 1, 1959.

A new epoch of progress began for the Greek Orthodox Community, and particularly for the Philoptochos Society. Archbishop Iakovos made a stunning impact on the membership of the Philoptochos Society. He inspired and motivated the women to seek new horizons, to undertake programs which were meaningful and productive, to render their assistance to the Community. He had an excellent rapport with the members of the Philoptochos, who had deep respect and admiration for him. They were very much aware that Archbishop Iakovos, like the founder of the Philoptochos, Patriarch Athenagoras I, was a visionary . . . a dreamer, who envisioned a philanthropic agency capable of offering extraordinary service to the Greek Orthodox Christians of the Archdiocese.

The new Archbishop was no stranger to the Greek Orthodox community. Years before, he had come to the United States from his home in Imbros to serve as Archdeacon to Archbishop Athenagoras. He then served in many other increasingly responsible capacities: Priest, Professor, Dean of Holy Cross Seminary, and Dean of the Annunciation Cathedral, Boston, for 14 years. He was well-known as a dynamic clergyman and an eloquent preacher. He knew first hand the dedication and efforts of the Ladies of the Philoptochos. He knew, too, that the lofty goals of this charitable organization had not yet been fully realized.

After his formal investiture as Archbishop, His Eminence invited members of the Central Council to a meeting at the Archdiocese, to discuss their programs. He reviewed their past efforts and offered many suggestions for the expansion

271

of their activities. A few months later, during the National Philoptochos convention, which was convened simultaneously with the Archdiocese Clergy-Laity Congress in Buffalo, New York in 1960, some of the Archbishop's recommendations were presented and adopted. Also, the names of the executive committee of the Philoptochos was changed from "Central Council" to "National Council". Mrs. Hariclea Malamas, a prominent member of Philoptochos in New York City, was elected National President.

Several new programs were undertaken in 1959, including:

1. The founding of the Sisterhood of Saint Basil. This new entity would devote itself totally to increasing the financial support of St. Basil Academy (a major Archdiocesan institution). The Sisterhood of Saint Basil was completely separate from the Vasilopita which had begun in 1942 at the request of Archbishop Athenagoras. The Sisterhood's efforts proved to be a highly important element in the progress made by the Academy. Mrs. Anthoula Tsougros of Commack, NY, was appointed to serve as chairman, a post she held effectively for several years.

2. Two new Philoptochos programs were initiated under the high auspices of Queen Frederica of Greece and with the recommendation of His Eminence. The Foster Parent Program was created to offer financial assistance to poor children in Greece. This proved to be a very successful project. In less than 10 years the Philoptochos raised $107,000 to aid the needy Greek children. The chairman of this program was Mrs. Helen Kavrikas. Contributions were solicited to help provide dowries for poor young women in Greece, with substantial results.

MEETING THE CHALLENGING SIXTIES

With the influx of new immigrants from Greece and other countries during the 1960s, Archbishop Iakovos established an Archdiocesan Department of Social Welfare. The Philoptochos was asked to be actively involved in this new department, which would assist newcomers to America, and offer aid to the sick and disabled. Mrs. Alexander Shikar, a member of the New York City Philoptochos was named chairman. Contacts were made with City, State and Federal social welfare agencies, in an effort to offer adequate assistance to those in need. More than $53,000 was raised for this relief program.

In 1961, the Government of Greece asked the Archdiocese to become the adoption agency for Greek children, an enormous responsibility. Again the Archbishop turned to the Philoptochos Society's National Council. A study showed several persons would be needed to administer and coordinate this program. Considerable work had to be performed to process and handle all aspects of the adoptions. Three dedicated ladies were chosen for this project, including: Mrs. Lycurgus Davey, chairman; Mrs. Basil Vlavianos and Mrs. George Kertatos, co-chairmen, of the Adoption Committee.

The National Council office of the Philoptochos was fortunate to have a remarkable lady as executive secretary: Mrs. Despina Vrachopoulos, who had worked with the Philoptochos during the reign of Archbishop Athenagoras. She served with great zeal and devotion, assisting in every program undertaken, maintaining the office records, attending meetings of every committee, and communicating regularly with the hundreds of Philoptochos chapters.

In 1963, at Archbishop Iakovos' request, the Philoptochos launched a new yearly collection throughout the Archdiocese on November 1—the feast day of Saints Cosmas and Damianos, patron saints of the Philoptochos—for the benefit of the charitable institutions of the Ecumenical Patriarchate. The members of the Philoptochos heeded the request and worked diligently and with gratitude for their founder Patriarch Athenagoras. His love and inspiration still permeates all of the projects undertaken by the Philoptochos. Since then hundreds of thousands of dollars have been raised by the Philoptochos for the institutions of the Ecumenical Patriarchate in Constantinople. The recipients of these contributions by Greek-American parishioners include: the Patriarchal School for Girls, the orphanage on the Principos Islands, the welfare agency operated by the Patriarchate and hospitals, schools, shrines and churches.

At the National Philoptochos Convention in Denver in 1964, Mrs. Sophia Hadjiyanis was elected President. She had a distinguished record of achievements with the Holy Trinity Cathedral Philoptochos in New York City. She was the founder of the Philoptochos' Chrysanthemum Ball, she worked with the American Red Cross and the National Conference of Christians and Jews, she was one of the founders of St. Michael's Home for the Aged in Yonkers, NY., and an active member of the Anglo-Hellenic Bureau of Education, which brought more than 500 students from Greece to study at American universities.

A major decision was taken at the Denver Convention: The construction of three dormitories for girls at St. Basil Academy in Garrison, NY. This was a most ambitious project, which was spearheaded by Mrs. Hadjiyanis. It required all the co-operation, exceptional efforts, and resources the ladies of the Philoptochos could muster. The cost of the undertaking was $1,000,000, plus an additional $54,000 for the furnishings. Mrs. Kay Papageorge, an interior decorator and member of the National Council of Philoptochos, offered her services to dec-orate and select the furnishings for the new buildings.

Another decision taken by the Philoptochos Conference in Denver was to participate in ecumenical programs. Archbish-op Iakovos had encouraged the Society to become more "ecu-menical" in their endeavors. He suggested developing con-tacts with women's organizations that were church-related, such as Church Women United.

His Eminence was pleased that this decision was taken. He was one of the six Presidents of the World Council of Churches, and was very active with the National Council of Churches. He was known internationally as a prominent ecumenical leader. He knew the value of having contacts with leaders of other faiths.

The delegates elected Mrs. Yorka Linakis of Jamaica, NY, an attorney and member of the National Council, to serve as liaison with Church Women United, an organization of Prot-estant women affiliated with the National Council of Churches. This was the beginning of an active ecumenical program which brought the Ladies Philoptochos Society into contact with Christian women of several denominations.

The Archbishop urged that Philoptochos chapters develop contacts with Christian women's organizations on a local level to get acquainted with women of other faiths. This has proven to be a most rewarding spiritual experience for the women.

A crisis developed in 1964 when thousands of Greek Or-thodox Christians were expelled from Constantinople, Imbros and Tenedos by Turkish authorities. Archbishop Iakovos be-came very alarmed and urged the Philoptochos to help the victims of Turkish oppression. Immediately, $5,000 was sent to assist in the relocation of these unfortunate persons, who fled to Greece and the United States. Among those expelled were two high-ranking Metropolitans who were members of

274

the Holy Synod of the Ecumenical Patriarchate of Constantinople: Metropolitan Iakovos of Philadelphia and Metropolitan Emilianos of Seleucia, who came to the United States. The National Council of Philoptochos, as well as many local chapters, offered hospitality and hosted receptions for the Metropolitans during their stay in America.

In 1965, a major earthquake hit Arcadia, Greece. The Philoptochos Society offered the victims assistance by sending financial aid in an effort to alleviate their suffering.

The passage of a new immigration law by the U.S. Congress, increasing substantially the Greek quota, brought thousands of new immigrants to the United States in 1966. Archbishop Iakovos in an effort to address the many needs of the newcomers, established the Archdiocese Social, Health and Welfare Center in Astoria, New York, where most of the new immigrants settled. Bishop Philotheos of Meloa was appointed General Director, and Mrs. Stella Coumantaros, a professional social worker, was named Director of Social Services. A year later this agency was moved to the Philoptochos National Office at the Archdiocese headquarters in New York City, where it continues to offer complete social services to the Community.

Another era where Philoptochos has made a significant contribution is education. Originally, the name of the organization was "Philoptochos kai Philekpaitheutike Athelphotes", which embodied not only love for the poor but also love for education. It is a known fact that when a parish was being organized the women encouraged and participated in the formation of schools to teach the Greek Orthodox faith, and Greek language, history and culture. This was strongly advocated by the late Patriarch Athenagoras I in a 1936 encyclical, which urged the establishment of Sunday Schools and Afternoon Schools. He asked that that Philoptochos take responsibility for these schools.

Archbishop Iakovos, a proponent of the perpetuation of Hellenic education and culture, discussed with the members of the National Council the possibility of undertaking a scholarship program. In 1966, during the Philoptochos Conference in Montreal, Canada, the delegates voted enthusiastically to establish a National Scholarship Fund, to provide deserving high school students with financial support to pursue college studies. Chapters throughout the Archdiocese endorsed this educational project to grant annual scholarships of $2,000 to worthy students. Mrs. Dorothea Prodromidis of Jamaica, New York, a former Vice President of the National Council of Philoptochos worked diligently administering the Scholarship Fund.

The "Open Heart Surgery" program was an historic humanitarian activity, initiated in 1968. The program began when Father Peter Kalellis (who was serving the Saint Sophia Cathedral in Los Angeles), and cardiologists at the Loma Linda Hospital in Loma Linda, California, met with His Eminence at the Archdiocesan Headquarters in New York City, and presented him with an extraordinary proposal: They asked the Archbishop to arrange for a team of heart specialists from the Loma Linda Hospital to travel to Greece to perform "open heart surgery" on Greek children with severe heart ailments. At that time, this advanced procedure was not used in Greece. The Archbishop wholeheartedly endorsed this proposal and immediately asked the Philoptochos to adopt it as a major project. Mrs. Jennie Scourby of Brooklyn, NY, was assigned to head the program as chairman.

Since it was responsible for all financial aspects of the program, the Philoptochos raised thousands of dollars to:

a) purchase the necessary surgical equipment and send it to Greece;

b) make all arrangements for Loma Linda heart surgeons to go to Greece to examine and operate on certain children requiring this treatment;

c) arrange visits by Greek doctors and nurses to the U.S. to receive training in this new technique so that they could establish "open heart surgery" programs in Greece.

The program was immensely successful with more than 324 children receiving the special "gift of life" from the Philoptochos.

The Clergy-Laity Congress and the Philoptochos' National Conference convened in Athens in 1968. One of the most gratifying experiences for the delegates was the visit of several children who had been cured by means of open heart surgery. The children warmly conveyed their thanks and appreciation to the Archbishop and the Philoptochos for their expression of Christian love.

Several years later the "Open Heart Surgery" program was enhanced and is now known as the "National Philoptochos Children's Cardiac Program". It currently functions under the able chairmanship of Mrs. Alice Nicas of Long Branch,

*The new Archbishop Athenagoras surrounded by the ladies of the Philoptochos
Society of Weirton, West Virginia, and their pastor.*

*Past and present Presidents of the National Philoptochos Society, Mrs. Ka-
therine Pappas and Mrs. Beatrice Marks.*

NJ. As a National Project it has expanded to include donations to hospitals in the United States and Canada, that accept children from Greece who need heart surgery.

Decisions taken at the Athens Conference included the construction of a fourth dormitory at St. Basil Academy and contributions to the Ecumenical Patriarchate of Constantinople and the "Lyceum ton Ellinidon", a prominent Hellenic women's organization.

THE PRODUCTIVE SEVENTIES

The Philoptochos Chapters began the 1970s with enthusiasm and zeal. The members of the Society, women of deep faith and commitment to their Church and community, expressed their Christian dedication by launching a plethora of activities that surpassed all expectations. Inspired by Archbishop Iakovos' paternal love, the Society undertook a new and imaginative philanthropic program that included involvement in social and moral issues, the expansion of educational, humanitarian, charitable and welfare concerns, and the continuation of its efforts for the needs of their brothers and sisters in Greece.

During the Philoptochos Biennial Conference in Houston, TX in 1972, the delegates, together with the members of the Clergy-Laity Congress, were attending the grand banquet when the tragic news was announced: The beloved founder of Philoptochos, Patriarch Athenagoras I was dead! Everyone was devastated.

Deeply grieved, the Philoptochos members and others at the conference attended a Trisagion service in the Chapel celebrated by the grief-stricken Archbishop. Bishops, priests and delegates chanted for the repose of the Patriarch's soul. To honor their founder's memory, the Philoptochos decided to:

a) establish an annual Patriarch Athenagoras Scholarship to a student pursuing theological studies at Holy Cross School of Theology;

b) send donations to the Kidney Foundation of New York; the Joint Disaster Fund for victims of Hurricane Agnes; and the Sycarides Institution for Retarded Children in Greece.

In gratitude to its late founder, the Society vowed to make an even greater contribution to the needs of the Greek Orthodox community. Several months later during a meeting of the

National Board, His Eminence reminded the chapters of the Philoptochos of the late Patriarch's exhortation to support education. The Archbishop recommended that the organization undertake a national program to financially assist Hellenic College and Holy Cross School of Theology. The Philoptochos Women for Hellenic College Program was founded, primarily to conduct fund-raising events and to contribute to the successful operation of the institution. This proved to be a popular and succesful project which continues to expand annually.

In October of 1972, the President of the National Council, Mrs. Sophia Hadjiyanis announced the appointment of Mrs. Stella Coumantaros, a professional social worker, as the new Director of the National office. Mrs. Despina Vrachopoulos would continue to serve Philoptochos as senior advisor.

Mrs. Katherine Pappas of Milton, MA, was named President of the National Board during the Conference convened at the Conrad Hilton hotel in Chicago in 1974. Mrs. Pappas is a distinguished lady known for her many benefactions and indefatigable efforts on behalf of the Philoptochos of the Annunciation Cathedral in Boston, and Hellenic College and Holy Cross School of Theology. Her devotion to the Church and community earned her the respect and esteem of the Greek Orthodox community and countless civic organizations in Boston, and throughout the United States. A long time friend of Archbishop Iakovos, Mrs. Pappas accepted the post with humility and resolve. Many new and beneficial programs developed under her leadership.

At the Philoptochos Conference the assembly voted enthusiastically to undertake Hellenic College and Holy Cross School of Theology as a National Project.

The delegates decided to expand its participation in Church Women United, which now included women of the Roman Catholic Church, making it truly representative of the three bodies of Christendom. Named to represent the Philoptochos, in addition to Judge Yorka Linakis, were: Mrs. Sofia Shane of Milwaukee, WI, and Mrs. Vivian Hampers of Grand Rapids, MI.

PHILOPTOCHOS LAUNCHES MASSIVE RELIEF PROGRAM FOR CYPRUS REFUGEES

The Greek Orthodox Community was shocked and distressed by the devastating invasion of Cyprus by Turkish Armed

Forces in July, 1974. Archbishop Iakovos called an emergency meeting of the Archdiocesan Council, the National Board of the Philoptochos Society, the Presidents of the National fraternal organizations, and prominent individuals, to undertake immediately the mobilization of an immense relief program for the 250,000 Greek Cypriots who fled their homes following the bombing and invasion of their homeland. Philoptochos members were urged by Archbishop Iakovos to express their concern, love, and compassion for the Greek Cypriot people. The Philoptochos responded by launching an intensive six-point campaign to send aid to the refugees in Cyprus: 1) hundreds of thousands of pounds of food were sent to Cyprus, together with blankets, clothing, medicine, medical equipment and pharmaceuticals; 2) blood drives were sponsored to help the wounded; 3) letters of protest were sent to government officials, including President Gerald Ford, Secretary of State Henry Kissinger, and Members of Congress; 4) funds were collected for the Archdiocese Cyprus Relief Fund established by the Archbishop; 5) a national Canister Drive was undertaken in every major city in the United States and Canada; 6) a Foster Parent Program was launched to help the displaced refugee children of Cyprus.

Philoptochos' extraordinary organizational ability was demonstrated most emphatically during the Cyprus crisis. The National Office, under the leadership of Mrs. Pappas and Mrs. Coumantaros, was in constant communication with the Diocesan district offices and local chapters to apprise the members of the progress being made. A corps of volunteers, made up of ladies in the greater New York area, contributed countless hours to supplement the small staff of the Office, to effectively handle the numerous aspects of the Cyprus relief program.

Mrs. Angelica Kapsis of Albertson, NY, a member of the National Board, was appointed chairman of the Foster Parent Program, which was named "Caress". This program, which was established to offer financial aid to the thousands of Greek Cypriot displaced refugee children, was proposed by Archbishop Iakovos and the late Archbishop Makarios, President of the Republic of Cyprus. The Philoptochos was asked to implement this humanitarian program to alleviate some of the suffering endured by the children. Administered by the National Office under the efficient direction of Mrs. Stella Coumantaros, and with the cooperation and assistance of the Government and the Church of Cyprus, the Philoptochos sent $20 a month to numerous children, who urgently needed help, because of the loss of one or both parents during the invasion. More than $1,500,000 was raised during the subsequent nine years, *an extraordinary accomplishment!* The program continues to function today at the request of His Eminence, Archbishop Iakovos, and the members of Philoptochos Chapters who feel there is still a need for the Foster Parent project.

PHILOPTOCHOS LADIES APPOINTED TO
ARCHDIOCESAN COUNCIL

During a meeting in Washington, D.C. in November, 1974, Archbishop Iakovos, in an effort to express his high regard for the work done by the Philoptochos, appointed four prominent members to the Archdiocesan Council, the executive body of the Church including: Mrs. Katherine Pappas of Milton, MA, President of the National Board; Mrs. Zoe Cavalaris of Charlotte, NC; Mrs. Lila Prounis, of New York, and Mrs. Eleni Huszagh of Glenview, IL. These ladies had distinguished themselves on the local and Diocesan level of Philoptochos and were members of the National Board.

Another major project of Philoptochos was to actively participate in fund-raising events for the restoration of the historic Avero House in St. Augustine, FL, where the first Greeks arrived and settled in 1768, before the Revolutionary War. This project included the construction of a Chapel dedicated to Saint Photios, the Ecumenical Patriarch of Constantinople who served the Church in the Tenth Century. St. Photios was responsible for sending Saints Cyril and Methodios to Christianize the Slavic people. The culmination of the Philoptochos' efforts was the donation of a magnificent icon—the glorious Platytera Virgin located in the apse of Chapel. Mrs. Pappas, in announcing the Philoptochos' decision to sponsor the icon of the Virgin in the St. Photios Chapel, said: "When we saw the beautiful icon of the Virgin, we decided that the Philoptochos Society as 'Mother of the Church' should donate it. In commemorating the pioneers, we wish to remember the women who helped found our Society. Their selfless dedication has been an integral part of the growth and advancement of our Church in the Americas".

PHILOPTOCHOS CONTRIBUTIONS AID
MANY WORTHY CAUSES

The Philoptochos Society's Emergency Relief and Welfare Fund has responded generously to numerous national and international disasters, by sending contributions to aid the victims of:

- Hurricane Agnes
- Mobile, Alabama hurricane
- Cambodian warfare
- Salonika earthquake
- Italian and Greek earthquakes

The Fund has also assisted the handicapped senior citizens organization of Greece, the Kidney Foundation, and the Retarded Children philanthropy there. The Philoptochos has also participated in many U.S. national appeals and telethons such as "Stop Arthritis", "Easter Seals", and "Cerebral Palsy" programs.

ECUMENICAL INVOLVEMENT

The organization's committee on Ecumenical Relations has been most productive in planning significant projects, recommended by His Eminence, who constantly urged the Society to expand its programs. The Archbishop charged the Philoptochos to increase its contact with Orthodox women from the canonical jurisdictions of SCOBA (Standing Conference of Canonical Orthodox Bishops in the Americas). He invited several Orthodox women to a meeting to discuss the possibility of establishing an Orthodox women's organization. The suggestion was received with great joy by Orthodox women of the Eastern Orthodox Churches. Representatives were appointed by Orthodox Hierarchs to serve on a Pan-Orthodox committee to undertake a comprehensive and all-inclusive spiritual program for Orthodox women. The organization was named "Orthodox Christian Women of America", with Judge Yorka Linakis elected President.

Several years of inter-Orthodox activity followed, which included worship services, during which Archbishop Iakovos presided, interesting lectures, seminars and receptions. For complex reasons OCWA ultimately became inactive.

However, the involvement in Church Women United by

Philoptochos, increased considerably over the years with Judge Linakis serving as legal counsel, Clara Nickolson of Allston, MA, as executive secretary, and Mrs. Vivian Hampers of Grand Rapids, MI, as National Vice President. Also, three women were named to serve on the Common Council: Presvytera Alexandra Poulos, Mrs. Eva Topping, and Mrs. Sofia Shane.

Mrs. Hampers, in September 1976, was one of several Orthodox women from around the world to attend the "First Orthodox Women's Consultation" convened in Romania. The consultation, which was initiated by the World Council of Churches, took place at the Women's Monastery of Agapia. Its purpose was to discuss "The Role of Orthodox Women in the Church". Appointed by His Eminence to represent the Philoptochos Society, Mrs. Hampers participated in discussions concerning, "Education and Vocation of Women in the Church", "Family and Church Life", "Witness in Society", and "Participation in the Ecumenical Movement". Mrs. Hampers served as chairman of one of the Workshops on "Witness and Ecumenism". Four years later, Syndesmos (the international Orthodox youth movement) sponsored an Orthodox Women's consultation at St. Vladimir's Seminary in Crestwood, NY. Several Philoptochos members attended, including Mrs. Hampers, Mrs. Sophia Altin of Fort Lee, NJ; Mrs. Artemis Davey of North Haven, CT; and Mrs. Stella Coumantaros.

Another important National Project undertaken by the Philoptochos was the Archdiocesan Foreign Missions program. Initiated by the Archbishop and assigned to Bishop Silas of Amphipolis to administer. The purpose of the program is to offer monetary assistance to Greek Orthodox Christians in Africa, Central and South America, Alaska and South Korea. The Chapters raise funds to provide scholarships to educate young men as Orthodox priests and women as teachers, for service in their native lands. Funds are also donated to aid construction of churches and schools in those areas and to furnish religious and educational materials.

On Valentine's Day, February 14, 1977, the National Board launched a new Greek Children's Cardiac Fund Appeal. The appeal, inspired by the Archbishop's love for children, aimed to raise funds to support open-heart surgery for Greek children, a program which began in 1972. Several hospitals in America participate in this program. The Philoptochos is truly gratified with the response from its members, as well as from the heart surgeons who are involved in the cardiac program. These dedicated cardiologists perform the surgery by donating their services. . .an extraordinary humanitarian gesture. New York Hospital was in the forefront of the program under the direction of Dr. Mary Allen Engle, Director of Pediatric Cardiology. Over $100,000 has been donated to the following Medical Centers:

New York Hospital, New York City
St. Francis Hospital, Long Island, NY
Deborah Heart & Lung Hospital, Browns Mill, NJ
Children's Hospital, Philadelphia, PA
Children's Hospital, Chicago, IL
Metropolitan Medical Center, Minneapolis, MN
University Hospital of Alabama
Children's Hospital, Buffalo, NY
Children's Hospital, Boston, MA
University of California at Los Angeles
Children's Hospital, Pittsburgh, PA
Texas Medical Center, Houston, TX
Texas Heart Institute, Houston, TX
Toronto Hospital for Sick Children, Toronto, Ont., Canada

The founders of Philoptochos, particularly Mrs. Despina Vrachopoulos, were honored during the St. Basil's commencement day ceremonies in the spring of 1977. Archbishop Iakovos dedicated a "Philoptochos Room" at the Academy to pay tribute to the monumental contributions of the pioneers of the Philoptochos, and the dedication and tireless efforts of Mrs. Vrachopoulos who faithfully served Philoptochos for 37 years.

NEW STRUCTURE, BROADER HORIZONS

The historic restructure of the Archdiocese in 1978 included the reorganization of the Ladies Philoptochos Society. A new Constitution and By-Laws were prepared for presentation to the delegates attending the Detroit National Conference. The national structure of the organization would be expanded to include two representatives from each Diocese, appointed by the Diocesan Bishops. The Philoptochos Chapters would function under the spiritual leadership of the Bishop. Diocesan Boards would be established and the Presidents of the Boards would serve on the National Board. The policy and programs of Philoptochos would be subject to approval by the National

Assembly during the Biennial Conferences, and implemented by the National Office. This is the structure under which the Ladies Philoptochos Society functions today.

In addition to learning of the historic reorganization of the Philoptochos, the Detroit Conference delegates were informed that the Society's philanthropic endeavors had reached a new peak. In her report to the Assembly, Mrs. Coumantaros announced that *during the period of 1976 to 1978 an estimated $1,000,000 was donated to worthy causes!*

The Conference for the first time focused on several social and moral issues. Papers were presented on "Child Abuse", "Abortion", "Sex Education", "Ordination of Women", "Battered Women", "Church and Family", "Pornography", "Gambling", and the use of alcohol on church premises.

Decisions were taken to embark upon an educational program to enlighten the community to Cooley's Anemia, the disease that afflicts children of Mediterranean ancestry. Also, the delegates voted to continue to support the Heart Fund, the Arthritis and Cancer Foundations, and the Cerebral Palsy telethons.

PHILOPTOCHOS LAUDS ARCHBISHOP IAKOVOS ON 20TH ANNIVERSARY

In an expression of "deep gratitude and love", the Philoptochos President, Mrs. Katherine Pappas, announced a special gift for His Eminence, Archbishop Iakovos on his 20th Anniversary. Mrs. Pappas, speaking at the formal luncheon marking the Archbishop's anniversary at the Plaza Hotel in New York on April 1, 1979, said: ". . .we greet you today with deep gratitude and love for your spiritual leadership that has enabled the women of the Church to develop and expand their roles of service to Greek Orthodoxy. . ."

Mrs. Pappas continued, "the Philoptochos will construct the Archbishop Iakovos Athletic and Recreation Center at St. Basil Academy. It is in deep appreciation of your gifts to us, and in dedication to your abiding love for all children." The Center includes facilities for all types of sports and recreational activities. The Daughters of Penelope participated in the project by pledging to build a swimming pool at the Center. The complex, at an estimated cost of $500,000, is expected to be completed in 1984.*

An unprecedented project was undertaken under the guidance of Archbishop Iakovos, to offer assistance to the Cooley's Anemia program. Philoptochos and AHEPA, in a cooperative and united effort joined forces to sponsor a "Bike-A-Thon" or a "Ride For Life", to acquaint the Greek-American community about the generic disease, Cooley's Anemia, also known as Thalassemia. This disease affects children of Hellenic, Italian and Jewish ancestry. The results of this unified effort enabled the AHEPA Foundation to award grants exceeding $100,000 for medical research.

The increased Philoptochos involvement in the United Nation's NGO (Non-Governmental Organization) Program through Church Women United's Advisory Committee, proved most rewarding. Mrs. Sophia Altin of New Jersey, who serves as the Philoptochos liaison, arranged a "Philoptochos Day" seminar at the U.N. The "Day" began with a Greek Orthodox Worship Service at the Church Center with Bishop Philotheos of Meloa presiding. The speaker at the seminar was Ambassador Zenon Rossides of Cyprus. Mrs. Altin also participated in the U.N. International Congress, held in Milan, Italy, on the "International Year of the Child".

"PHILOPTOCHOS HAS COME OF AGE"
THE DECADE OF THE 80s BEGINS

One of the featured speakers at a seminar sponsored by the Center for Byzantine and Modern Greek Studies at Queens College in New York, was Mrs. Stella Coumantaros. The seminar, held on May 9 and 10, 1980, explored "The Greek Community in Transition". Mrs. Coumantaros' address covered the activities of the Ladies Philoptochos Society, "since its inception as a loosely organized philanthropic group of pioneer women who propagated philanthropy" to the present. "Today," said Mrs. Coumantaros, "as the largest Greek Orthodox women's organization in the world, the Philoptochos has transcended the confines of the Greek community by broadening its focus to include issues of national and international importance.

The Archbishop Iakovos Athletic and Recreation Center, as of mid-1984, is nearing completion.

Through our evolution from a parochial philanthropic organization set up to assist the Greek immigrant, the Philoptochos Society today is a relevant, world-minded, philanthropic body that has spanned five decades. Philoptochos has come of age in the Americas!"

The Philoptochos in 1980, imbued with the Archbishop's humanitarian spirit to aid children, expanded its activities to participate in the United Nation's observance, "The Year of the Child". A community handbook was created and distributed to the Philoptochos Chapters, the parishes and institutions, coordinated by Mrs. Diane Constas of Greenwich, CT. Also, a youth poster and camera contest was held with over fifty entries, which were exhibited at the Archdiocese. Among the judges were two prominent ladies: Mrs. Themis Hadges, a distinguished artist; and Miss Pamela Ilott, Vice President of CBS Religious and Cultural Broadcasts.

On the agenda of the Philoptochos Society during the 25th Biennial Conference in Atlanta in July, 1980, was the question as to its participation in the "The Year of the Family" (sponsored by the United Nations in 1981). Philoptochos President, Mrs. Katherine Pappas, in her address to the Assembly, expressed it most effectively.

"This new decade begins as we, the Philoptochos Society, are poised on the brink of a new era of loving service for our church and community.

"Remembering our dedicated founders who began in small groups almost 50 years ago, philanthropy and support of the Archdiocesan institutions continue to be the primary concern of today's Philoptochos.

"We have a special gift from the Lord, the traditions of the Greek Orthodox Family, of giving to those who need help, of respect for education, of bringing the Church family closer together.

"We are continually inspired by His Eminence Archbishop Iakovos, whose vision and strength has made our progress possible".

The Philoptochos Society in September, 1981 mourned the passing of its Past President, Mrs. Sophia Hadjiyanis, one of its most distinguished members. Mrs. Hadjiyanis had served as President from 1964 to 1974, after several fruitful years as an officer of Philoptochos. Under her leadership an important period of growth and progress had materialized. Many humanitarian programs were developed that were beneficial to the Greek Community. She was one of the founders of St. Michael's Home for the Aged, which continues to receive our love and support.

HALF-CENTURY OF CHRISTIAN SERVICE

The Ladies Philoptochos Society celebrated its 50th Anniversary with a series of festivities beginning in November 1981, and concluding in July 1982 during the 26th Biennial Philoptochos Conference in San Francisco. Several significant events marked the occasion, including:

a) A Pilgrimage to the Holy Land, Greece and the Ecumenical Patriarchate of Constantinople. An unforgettable audience with His Holiness Patriarch Demetrios I, which included a visit to the tomb of the Founder of Philoptochos, Patriarch Athenagoras I.

b) A gala Golden Anniversary luncheon at the Hotel Pierre in New York City, during which Archbishop Iakovos honored ladies from every Diocese for their years of dedicated service.

c) An Ecumenical Tea held at the Archdiocese with women from the four major Faiths attending.

d) An unprecedented World Community Day celebration: a Special Greek Orthodox Worship Service celebrated in 2,000 churches across America. This service was written by three Greek Orthodox women: Presvytera Alexandra Poulos of New York City, Presvytera Sophronia Tomaras of Tacoma, WA; and Mrs. Eva Topping of Bethesda, MD. Using the theme: "The Last Commandment: Put Away Your Sword", a beautiful and meaningful "Litany of Peace" was offered, with appropriate Gospel readings, petitions and music in the tradition of the Orthodox Church. This was a memorable ecumenical experience that will be long remembered and cherished by all who attended. His Eminence Archbishop Iakovos was particularly proud of all who participated in this notable event.

e) Dedication of the Icon of the Virgin in the Chapel in the St. Photios Shrine in St. Augustine, FL. The beautiful mosaic was dedicated to the founders, the pioneers of the Ladies Philoptochos Society.

f) A 50th Anniversary Album was published to commemorate the Golden Anniversary of Philoptochos. It contained messages of commendation, a dedication to the late Patriarch

Athenagoras I, a comprehensive History of Philoptochos, and articles on the Society's numerous programs and activities.

Also, on the occasion of the observance of their Golden Anniversary, the Society established the National Philoptochos Scholarship Endowment Fund. This fund, in the amount of $50,000 donated by the Philoptochos Chapters, provides scholarships to deserving students studying for the priesthood. The capital would remain intact and the interest it generates will be used to financially aid students attending Hellenic College and Holy Cross School of Theology, at Brookline, MA.

The President of Hellenic College/Holy Cross School of Theology, Dr. Thomas Lelon, expressed his gratitude and praised the Philoptochos for their exemplary devotion to their Church. He said, "Educating for the Priesthood and for future lay leaders of the Church is one of the most significant tasks facing the Greek-American Community. By creating this Endowment Fund, Philoptochos renews and reaffirms its commitment to this essential endeavor. All of us are very appreciative of their generosity and their consideration."

Archbishop Iakovos commended Philoptochos for a half century of devoted service with this special message:

"The blessed soul of the great and ever-memorable Ecumenical Patriarch Athenagoras would certainly rejoice and be filled with gladness on the occasion of the Fiftieth Anniversary of the life and work of your Philoptochos Society. In laying the groundwork of your organization a half century ago, the Patriarch, a Churchman of deep insight and keen foresight, envisioned a well-organized society of Greek Orthodox women striving with utmost determination to eradicate poverty and misfortune in a crusade to preserve all that is sacred and precious to our Church and all our people.

"Looking back across those 50 years, we can safely say that the Patriarch's hopes and expectations in what your organization might achieve have been fulfilled in exactly the way he had envisioned. The Philoptochos Society's outreach in the areas of philanthropy and moral and social concerns—in the local community, our Archdiocese, across the nation, in Greece and other parts of the world—has certainly been successful, perhaps even indescribable. Most importantly, it bears the seal of sensitivity and refinement that is the hallmark of your loving souls as Greek Orthodox churchwomen.

"You certainly deserve to be praised and commended. As your spiritual father, allow me to convey the gratitude and appreciation of the Church on this special Anniversary observance of your Society. I congratulate all of you on your many achievements and I pray that God will bless you in abundance and enable you to continue in your praiseworthy, dedicated service to His Holy Church and every living soul".

THE SOCIETY'S
ORGANIZATIONAL STRUCTURE TODAY

At the first meeting of the National Board, following the San Fransicso Conference, held in October, 1982, Mrs. Beatrice Marks of Lincolnwood Towers, IL, was appointed President. Mrs. Katherine Pappas of Canton, MA, who had served as President of the National Board since 1974, was named Honorary President.

Mrs. Marks, a successful business woman, has been one of the most effective leaders of Philoptochos: as president of the St. Andrew's Philoptochos in Chicago; as Diocese president of Philoptochos for the Chicago Diocese; and as a member and officer of the National Board for many years. She served as first Vice President.

A lady of immense dedication Mrs. Marks was honored for her extraordinary service to her Church and community, by the late Ecumenical Patriarch Athenagoras I, who bestowed upon her the title of Archontissa.

THE PHILOPTOCHOS NATIONAL BOARD—1984

Mrs. Beatric Marks	National President
Mrs. Katherine Pappas	Honorary President
Mrs. Anglika Latsey	First Vice President
Mrs. Mary Spirou	Second Vice President
Mrs. Velda Vasilaros	Third Vice President
Mrs. Lucy Kolovos	Secretary
Mrs. Despina Albanes	Secretary
Mrs. Argeria Logus	Treasurer
Mrs. Harriet Dodys	Assistant Treasurer
Hon. Yorka Linakis	Legal Advisor
Mrs. Stella Coumantaros	Director of the National Office

Board Members

DIOCESAN PRESIDENTS

CONCLUSION

After more than a half century of Christian philanthropy, the Philoptochos members have proven to themselves and to the world that they are women who are not only religious and Greek-oriented but also sophisticated and contemporary Americans. They are professionals embracing every field of endeavor: educators, social workers, secretaries, artists, musicians, journalists, lawyers, judges, scientists, doctors, mayors, city and state legislators, congresswomen, investment brokers and financial experts, bankers, business women, and first and foremost wives and mothers. They are dedicated laborers in the vineyard of the Lord.

In the words of His Eminence, Archbishop Iakovos, who has been the *inspiration* of the Philoptochos for 25 fruitful years: "You deserve to be praised and commended"; . . ."Patriarch Athenagoras would certainly rejoice".

Editor's Notes:

1) A comprehensive *History of the Founding of the Greek Orthodox Ladies Philoptochos Society* is included in the Fiftieth Anniversary Album, 1982; and in the historical volume *Encyclicals and Documents of the Greek Orthodox Archdiocese of North and South America - The First Fifty Years 1922-1972*, compiled by the Rev. Dr. Demetrios Constantelos, Professor of History and Religious Studies at Stockton State College of New Jersey.

2) The complete text of Mrs. Stella Coumantaros' paper on the Greek Orthodox Ladies Philoptochos Society is included in a book entitled, *The Greek American Community in Transition*, published by Pella in 1982 and sponsored by the Center of Byzantine and Modern Greek Studies of Queens College, New York.

3) The Department of Church and Society, Rev. Dr. M. B. Efthimiou, Director, has published a booklet entitled, *Put Away Your Sword* which contains the homilies delivered by Sophronia Tomaras, Eva Topping, Pearl Veronis and Efthalia Makris Walsh, during the celebration of a Greek Orthodox Litany of Peace marking World Community, Sunday, November 6, 1981.

Development of Orthodox Architecture

By

Steven Peter Papadatos, A.I.A.

eligion is a universal phenomenon. As such it develops not only man as an individual, but seeks to express its objective purpose collectively through the development of groups, nations and peoples. The projection of the religious sentiment of a group or individual is realized through worship, which manifests the faith of certain beliefs in that concept of the divinity. A necessary presupposition for the worship by such groups is the designation of a specific space within which people with the same beliefs can express themselves jointly.

In Christianity, people express gratitude and their thanksgiving to God, whom they consider the cause of their existence and of their salvation. Climatic conditions, and historical and cultural evolution deemed it necessary that worship take place in an enclosed space. The first Christians used homes for their worship needs. For example at Ephesus, St. Paul indeed wanted a house for teaching, but primarily a place to celebrate the Lord's Supper. Thus the place of worship developed which came to be called a church (ecclesia), which literally means coming together or gathering because the faithful would gather there for the collective celebration of the Eucharist. Later, and particularly in Rome during the persecutions, subterranean worship was conducted in burial places called catacombs.

With the recognition of the Christian church as a licit religion of the Roman state, the development of ecclesiastical architecture ensues. For their churches, the Christians of that time copied the architecture used for court-houses. These structures were rectangular, separated into three sectors by two rows of columns. The mid-sector was also the tallest and referred as the clearstory. On the eastern side of the midsection was situated the Holy Altar for the celebration of the Divine Liturgy. The prothesis or preparatory area and the baptistry were outside of the church proper.

A structure of this design is called a basilica. With the emergence in time of Christianity as the sole religion of the Empire, and with the further development of Christian dogma, church architecture produced a new form. This form depicts the cosmos with the heavens represented by the dome upon which Christ the Pantocrator is represented iconographically. The construction of the dome was made possible by the development of the pendentives, whose use permitted the construction of a hemisphere over a square compartment.

Lastly, the prothesis and Diaconicum (or place of storage for the holy vessels) was transferred from outside the church to the interior. This development was the result of theological considerations. The general use in the Christian East of the Divine Liturgy of St. John Chrysostom led the Church leadership to adopt one major achitectural form. This structural style is known as the Byzantine style. The entire treasure of Orthodox Architecture belongs to this style which, of course, through the progress of the centuries did develop in different ways. The Byzantine style was characterized by two elements, the dome and the cruciform shape of the church proper.

The Greek Orthodox Christians upon their arrival in the United States gathered into religious communities, with the result being naturally, the acquisition of churches. At first they did not, and could not, build churches fully in the Byzantine style. The artistic aspect of the Greek Orthodox Church of America reflected the intellectual development as well as the financial state of its flock. As one might expect, the first immigrants were not overly concerned with art and architecture. Of greater concern was their social adjustment to their new homeland and their complete dedication to maintaining their spiritual identity, faith and culture. In this difficult struggle for survival they were not able to give expression to their religious feelings through the construction of their church edifices. They set up churches to worship in and improved upon them, aesthetically and in other ways as time went by.

St. Sophia of Constantinople was always their prototype and the center of their religion, ethnic pride and aspirations. However, in the beginning they contented themselves with the purchase of old churches built by other groups which they transformed into Orthodox churches by adding icon-screens and icons, imported from Greece or made here.

The increase in the number of immigrants as well as the

improvement generally of their financial status gave impetus to local communities to embark upon the construction of new churches, faithful to the Orthodox tradition and reflecting the historical grandeur of Byzantium. This interest in Byzantine design aided the Greek Orthodox Church in the Americas to maintain its 2000-year old identity. In the psychology of the prosperous immigrant, Byzantine architecture was his greatest ethnic offering to the new homeland which had enabled him to prosper economically and socially.

Thus we can understand the willingness of many Orthodox parishes to sponsor the construction of excellent church edifices in the Byzantine style. Close to the turn of the century, two churches were erected in the Byzantine tradition to bear witness to the close ties the Greek Orthodox immigrants maintained with the Christian Orthodoxy of the Ecumenical Patriarchate. The first of these churches is situated in Central Boston, the capital of New England, and is named after the Annunciation. The second is in Lowell, Massachusetts, and bears the name of Holy Trinity. Both structures bear the basic characteristics of the Byzantine style. Following the Second World War, activity in the construction of church edifices greatly increased and most structures of that period are architectural landmarks for the cities in which they were erected. The building of these churches has proven a significant contribution to the art of Orthodox church construction. These American structures included the traditional elements which express the theology of the church and acknowledge the building is used for the Eucharistic gathering, i.e., the dome and the sides of the cross-elements uplift the worshipper to the Heavens. In addition these new churches use contemporary materials, state-of-the art church construction methods and contribute new dimensions which are pleasing to the eye and the tastes of contemporary man.

A monumental example of modern Orthodox church architecture is the Church of the Annunciation, Milwaukee, Wisconsin. According to Constantine Kalokyris, Professor of Byzantine Art and Archaeology at the University of Salonika, the building inaugurates with splendor a new era in Church design. Created by the world famous architect Frank Lloyd Wright, this house of worship was completed in 1961, three years after the enthronement of Archbishop Iakovos. At the consecration, His Eminence noted that he sanctified a structure to the glory of God which on first sight seemed different.

Yet like St. Sophia, the Milwaukee structure is for our times akin to the fabled great church in Constantinople—which for the time of Justinian was innovative and daring. Domed like Hagia Sophia, the Church of the Annunciation also rests upon a cross. This American church represents something as old as eternity itself through a very modern medium of expression which retains all of the characteristics of an Orthodox structure.

The Church of the Annunciation seats 800 people comfortably. On the lower level, seats are placed around a triangular solea emanating from a triangular altar area. The seats proceed to fill three of the triangular areas which comprise legs of the cruciform shape upon which the dome and lentilline structure rest. The fourth leg is the altar area. The gallery is completely round with five rows of seats running along 90% of the circle as in St. Sophia, and Sts. Sergius and Baccus in Constantinople. A light metallic iconostasion separates the altar from the nave.

His Eminence was so correct in his statement because although totally new and revolutionary, the new church was built upon the strong premise of Byzantine design. Wright later wrote, "The structure is not a copy of Byzantine architecture, it is something better than a copy." A visitor to the church will confirm that it is in its true form an expression of the beauty of the Byzantine architectural tradition but not a slavish copy of that period.

Professor Kalokyris also said: "This church impresses deeply in that it serves as a reminder to the viewer that Orthodoxy is not inhibited through the employment of new methods, materials and means of expression, in the very tradition of St. Sophia in Constantinople. In his use of the dome, Wright succeeds in the application of the principle of dematerialization to contemporary church style."

It was a young and progressive Archbishop Iakovos, given as he is today to looking far ahead into the future, that gave his blessing to this project amidst a spirit of conservatism and a narrow and uncreative traditionalism which clung to form more than spirit. His Eminence transcended the difficulties (and negative reactions) to grasp, comprehend and approve that which was to become a reality dear to the heart of America: A structure which conveys the spirit of Orthodoxy while reminding us that the success of the apostolate of the Church lies in the use of the media of our times to express, pure and unadulterated, the Faith of our fathers.

A classic Byzantine ensemble of frescoes and mosaics decorates the altar of the Archdiocesan Cathedral in New York.

Also worthy of mention are the Cathedral of St. George in Montreal, Canada, and the Church of the Dormition of the Mother of God, Denver, Colorado. The church of Denver is a hemispherical church which opens upon six bases. An arch in the eastern part of the church comprises the couch of the apse. In the interior, the impression of an octagon is given but on the outside the white hemisphere dominates giving the impression of an egg-shaped structure.

The first architect in America to go beyond a copy of shape to a reproduction of style of construction was Stewart Thompson. Through his design of the beautiful church of the Archangels in Stamford, Connecticut, and the Holy Cross Seminary Chapel he brought mute witness to 11th century Byzantine architecture to America.

The Greek Orthodox church of St. George in Norwalk, Connecticut, employs the basic shape and proportions of an early Christian basilica. The style of the masonry is that of the later Byzantine years, employing however, contemporary materials. The use of roman brick, stone and a crimson mortar gives the appearance without duplication of costs and ancient materials of the Byzantine period. There was much excitement in Norwalk over the Byzantine project since most of the residents were descendants from Trebizond on the Black Sea in Asia Minor. The challenge of building a large edifice whose details could be authenticated as Byzantine, was accomplished in 1977, and St. George was proclaimed an architectural landmark in the City of Norwalk the same year.

St. George of Norwalk was the basis for the Church of the Annunciation in York, Pennsylvania, which is a domed cruciform basilica. The Annunciation was the effort of many years of devoted and continuous work on the parish's part in order to recreate a true and magnificent Byzantine design. The altar windows were drawn to conform to the altar windows of St. Clement of Ohrid, Yugoslavia, which was built in 1295. The side windows were taken from Parigoritissa at Arta, erected sometime between 1283 and 1296, and they are referred to as Bilobated windows. The panels as shown on the front elevation were taken from Panagia Gorgoefekoos, at Athens, known as the "Little Metropolitan". It's perhaps the smallest cathedral in the world: its beautiful walls are studded with antique marble reliefs.

The community of St. John's, Blue Point, Long Island, wanted their new church to sustain and promote the Byzan-

Saint Barbara
Greek Orthodox
Church in Orange, Ct.
is a traditional example
of Byzantine architecture.

Wisconsin-born
Frank Lloyd Wright's
last major work,
the Annunciation
Greek Orthodox Church,
is now a famed
Milwaukee landmark.

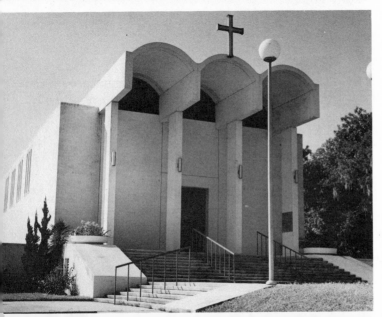

Contemporary/modern
design to fit the
climatic conditions
of the South.
St. John the Divine,
Jacksonville, Fl.

tine cultural inheritance in testimony to its dedication to the values of the Greek Orthodox church. A simple Byzantine cruciform church was built with a dome supported by pendentives to denote the contribution of this typology by the Byzantines. The proportions and style mark an adherence to classical Byzantine architecture as a tradition and reminder of the past. Details were derived from the Church of the Theotokou in Osios Loukas monastery and St. Catherine's on Mount Sinai. The forms have been simplified and the details modernized while retaining the value and scale of Byzantine ecclesiastical architecture.

A church which differs considerably from the traditional is the domed almost square church of the Dormition of the Theotokos in Seattle, Washington. The Church of St. Mary's in Minneapolis, Minnesota, is another example of modern architecture. It is a Greek cross domed without pendentives. Light enters at the base of the dome. There is no apse in the altar. Noteworthy too is the Church of St. George in Manchester, New Hampshire. It features four couches, supported by four external columns, from which arise four great arches which support a higher dome without pendentives.

Similar to St. Mary's, Minneapolis, is the Church of the Holy Cross in Lebanon, Pennsylvania. As the architect writes in his analysis, he wanted the structure to bear a striking resemblance to the Parthenon. In the design of the Cathedral of Sts. Constantine and Helen, we see an attempt of the architects to give a new variation to the ancient cruciform basilica.

Between 1960 and 1970 many rotunda-shaped structures were constructed by our parishes. Examples are to be found in the New York communities of New Rochelle and Jamestown. A modern edifice inspired by tradition is the church dedicated to St. Demetrios which was built in 1963 in Seattle, Washington. This bold structure appears to be the same from whatever angle it is viewed. Eight identical arches are placed in a circular manner, from which arises a tall and narrow octagon forming a dome. The eight sides are composed of stained glass permitting light to enter.

The Church of the Ascension, Oakland, California, follows the Byzantine tradition and when viewed from the outside appears as a lined light dome which emerges from columns which encircle the structure. Below are four semi-circular openings which form the shape of a cross and permit light to enter. The dome is made of copper on the outside and, in the

300

interior, is covered with gilded aluminum plates.

In the Diocese of New Jersey, excellent examples of traditional churches have been built during the past decade. The Cathedral of St. John the Theologian in Tenafly is of the Byzantine shape. It is a dome cruciform basilica, with an apse over the altar, and stands tall and proud on a hill.

A significant architectural design is that of the Holy Trinity Church in Westfield, New Jersey, bearing all of the basic characteristics of a dome and cruciform Byzantine church. The dome from the outside opens like a large umbrella and is supported by the walls of the church which are octagonal. In the church walls (aesthetic and dramatic semicircles) are placed stained glass windows depicting the 12 Prophets. The dome dominates and the icon of Christ the Pantocrator shines. The east apse bears the icon of the Virgin, the Platytera. The rays of the sun pierce the stained glass, and the result is a structure with brilliance, continuity and harmony. The high dome with the Pantocrator lifts one toward the heavens and gives the feeling that here, indeed, in the Eucharist heaven and earth meet.

A very modern church is the church of St. Nicholas, Flushing, New York. The structure is a church and community center. The church itself is octagonal and features a dome 20 meters high, which creates an imposing spatial statement. One side of the octagon is an apse which is used as the altar. It is separated from the nave by four concrete frames upon which are placed small icons. The altar is separated from the nave with a concrete icon screen of early Byzantine design. Scenes from the life of Christ cover the upper sections of the octagonal walls.

Another modern church is Sts. Constantine and Helen of West Nyack, New York, which is now rising above the hills overlooking the Palisades Parkway. The church has a passive solar dome and is undoubtedly the only Byzantine structure designed with this energy efficient feature. According to *The New York Times*, the cost of heating so large a space as a church has more than tripled since the oil crisis of 1973. Also, it said solar energy can often be designed into a new building at little expense.

In summary we can say that in the past 25 years we have seen a great achievement in Orthodox ecclesiastical architecture and iconography. We are fortunate to have benefitted in these advances by the enthusiasm and guidance His Emi-

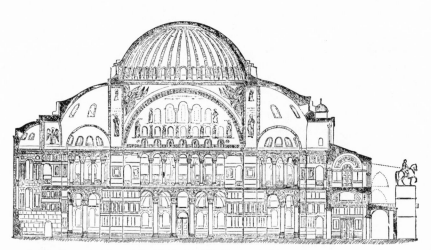

Architect's cross-section of Haghia Sophia, The Basilica in Constantinople, in Justinian's time.

nence Archbishop Iakovos instilled into the communities. The result has been the construction of many new and beautiful Byzantine structures in the Western Hemisphere. Our churches occupy a special place within our lives and communities. His Eminence Archbishop Iakovos, has our gratitude for inspiring our communities into building new, innovatively-designed churches worthy of our 2,000-year old religion. Also, I believe he will inspire the Greek Orthodox in this land to even more and greater works of art and to even higher social and religious achievements.

Growth of Liturgical Music in the Iakovian Era

By
*Frank Desby**

 eldom has an art form been subjected to as strong
a changing influence as the music of the Orthodox Church when it was brought to America. In
this century our music has been seriously impact-
ed by the art and ethos of the New World. True,
there was great influence of the Moslem world on our music
after the late 14th Century. This influence introduced some
alterations in the scales upon which the melodies were constructed and rhythmic elements as well. Yet the basic concept of a single melodic line, without a background harmony,
prevailed until the 20th Century. The Slavonic Church, on the
other hand, developed a harmonic-contrapuntal choral idiom
in the 17th Century, while the Greek and Antiochian Orthodox retained the single melodic chant line, save for the held
tone of the *isokratai.*

Harmonized choral music for the Church did not originate
in America. Greeks living in France and Germany experiment-
ed with harmonizations. Eventually, at the turn of the century
attempts at harmonized Byzantine chant were introduced by
John Sakellarides.

Outside of Greece one would expect to find choral music
without much opposition, so it is no surprise that on Easter
Sunday in 1844 in the Orthodox Church of the Holy Trinity
in Vienna John Haviaras led a choir of 24 male singers in a
four-part setting of the liturgical music. At the end of the (19th)
century Spyro Spathis settled in Paris. He offered the Greek
community harmonized settings for mixed chorus. Criticism
of this early pioneer's work lies in the fact that the subtleties
of Byzantine scales are *theoretically* not adaptable to western

---

*The author extends his gratitude to the National Forum of Greek Orthodox
Musicians for its research assistance in preparing this article and the special
contributions of Vicki Pappas, National Chairman.

major-minor tonality. It is not our purpose here to discuss the possibilities of adapting one type of musical thought (Byzantine) to another (European) except to state that compromises, once believed unworkable, are possible.

In Athens, the influence of John Sakellarides on the musical habits or customs of the Greeks was minimal; it was in America that his influence became enormous. As an opera coach and music teacher at a number of schools, and also trained thoroughly as a *psaltis*, Sakellarides came into contact with a large number of musicians of varying backgrounds. In the 1920s he came into contact with Henry J. Tillyard, of England, who was with the British school at Athens. Tillyard became quite interested in Byzantine music and secured Sakellarides as his teacher. Together they explored some of the earlier layers of melodic tradition, noting that the melodic style and indeed, the tonality, had been subjected to several transformations. Music before the Turkish conquest (1453) was different and every 150 years or so thereafter the melodies were subjected to "interpretations" which the Byzantine composers called "exegesis." The last transformation occurred during the last half of the 18th Century, culminating in 1814 with a revision of teaching methods. Also, for the first time, in 1820, it became possible to engrave and print Byzantine music.

With this vehicle of mass reproduction, it was unfortunately only the latest style that became universally established. Little of the music of the older *melourgoi* survives today except in manuscripts. Furthermore, what melodies had survived had become decorated to the point of corruption. Sakellarides studied the early examples, combed out many ornaments, and produced a kind of classical reform. This purging upset his contemporaries. However, many of his simplified versions became standard but not his harmonic versions.

With the migration of Greeks to the United States, beginning at the turn of the century, came some of Sakellarides' students. They offered their services as precentors *(psaltes)* and even as instrumentalists for community entertainment. The creation of language schools for the first generation of Greek-American children created new demands from the church. In many instances, the local precentor was also the local Greek school teacher.

The influence of public school choral music, and its existence in local non-Greek churches, was strong enough to introduce in our churches some experimentation with choral music and a participating group—the choir. Separated from the Mother Church, the Greek Church in the U.S. was not under the watchful eye of anyone objecting to innovations. So in the first quarter of the 20th Century, the Greek Church of the New World introduced choral music on a permanent basis.

EARLY CONTRIBUTORS

No one knows, for sure where and when the first choir was formed in America's Greek churches, or when organ accompaniment was introduced, but many have laid claim to being the originators. Among the early choir directors are the Sakellarides group, especially those who had their training under the Athenian master. These include Christos Vrionides and Angelos Desfis. Others who had come under Sakellarides' influence were Nicholas Roubanis, George Anastasiou, and Athan Theodores. These few were singled out because they contributed published material to our churches.

In each of the large cities the precentors organized choral groups to take part in the liturgy and occasionally other *akolouthiae* for Holy Week and major feast days. The inclusion of young girls and women seemed to raise no objection by either the faithful or the clergy. Indeed this seemed to be another vehicle for preserving the heritage, while in Greece the inclusion of women in church services is still not acceptable. The music brought to this country had been written for four-part male ensemble. It was soon discovered that with women now part of the choir, such writing was unsuitable. Rewriting was inevitable as was the influence from non-Hellenic sources. Some of the new directors had received a full formal training in western music in addition to their Byzantine training. Here are brief profiles of these leading choir directors:

Christos Vrionides had received diplomas in the U.S. as well as Greece, was thoroughly trained on several instruments and served as conductor of a number of symphony orchestras. By appointment of the federal government, he founded and conducted the Long Island Symphony in Babylon, N.Y. Vrionides was recognized in the 1951 *Who's Who of Music*. Later, he became the music instructor at the Holy Cross Seminary, teaching both Byzantine music and the formal music of the west. He wrote the first American-published treatise on Byzantine chant.

Nicholas Roubanis came to the U.S. in the twenties from Alexandria, Egypt, as a french horn player. He was active in Chicago for a number of years before moving to the East Coast, becoming involved with the musical growth of several communities. His "Divine Liturgy for Mixed Voices" was to become immensely popular due to its simplicity and effective part writing. His song *Miserlou* became an international hit in the popular music field. He was active in the church all his life; his final position was a choir director in New Jersey.

George Anastasiou was a student in Cyprus of the famous Stylianos Chourmouzios but studied the choral settings of Sakellarides with great diligence when he immigrated to the U.S. and formed a choir in Tarpon Springs, FL. Eventually he moved to Philadelphia, PA, where he was the Protopsalte, directed the choir, and became a greatly respected music teacher. Anastasiou, like others about him, realized the four-part male choir setting of Sakellarides could not be used with mixed voices. Therefore, he composed his own music for the Divine Liturgy. While the work of Anastasiou was similar to the liturgies of Vrionides and Roubanis, it introduced some additional material, contained both a major and minor version of the Liturgy, and included information for use of the organ. Most valuable was an appendix of the special hymns needed throughout the ecclesiastical year. This material, appearing in western staff notation, makes this volume indispensable. Probably no choir director even today is without one.

Athan Theodores' and **Angelo Desfis'** contributions consisted mostly of separate hymns of the Liturgy, such as the Cherubic and Communion hymns. Privately published, their circulation was not extensive. In 1947, Desfis on a visit to Greece secured the publishing rights to Sakellarides' western notation liturgical book *Hymns and Odes*, which explains how so important a transition work came to be printed in America.

Father Demetrios Lolakis studied at the Conservatory of Athens, becoming proficient in the piano and violin, which he played all his life. He came to the U.S. about 1927 and was active mostly in the Midwest. He transcribed an enormous amount of Byzantine music into staff notation and recorded much of this on tape. This contribution has proven to be quite valuable to priests and psaltes alike. After his ordination in 1922, his musical activities grew rather than become subordinate to parish duties; yet, as a parish priest he was always highly esteemed. After his retirement, he continued his musical activities with undiminished interest and vigor.

THE POST WAR PERIOD

At the outbreak of World War II most churches had a choir of mixed voices, ages ranging from 15 to 25. They also had an organ of some kind, either a harmonium with pedals to pump the bellows, or a simple electric instrument, and a choir director who might also be church secretary, *psaltis*, language teacher, or all three.

When the servicemen returned home, they displayed a new spirituality and renewed interest in their ethnic roots. This resulted in a welcome renaissance for the Church. The serviceman entered colleges and joined musical organizations there. As he participated in his local choir, he required the Church to offer more challenging musical fare. In many cases, the liturgical repertoire remained identical to the pre-war music, but standard choral music was added so that church-sponsored programs offered the community a broader selection. Many second-generation Greeks also majored or minored in music; therefore, as the original directors retired, the positions were filled by well-trained newcomers. As a result, more challenging type of music becomes manifest.

It should be mentioned that all of the Archbishops of these periods, despite their traditional upbringing, encouraged the adaptation of choral music and inclusion of an accompanying instrument. A little known but interesting piece of information concerns the introduction of a pipe organ to the church on the island of Kerkyra by then Metropolitan Athenagoras and later Archbishop of North and South America and, in 1949, Ecumenical Patriarch. Whenever a new prelate takes over there is naturally some anxiety on the part of the clergy and laity as to whether progress will be maintained. It was therefore exceptionally fortunate for the Orthodox Church in America to have Archbishop Iakovos elevated to this post in 1959. Our new Archbishop, extensively educated and cultured, was well aware of the world-wide religious situation. He was up-to-date on all facets of community life, education, spirituality, music, art and architecture. Having travelled widely, he was not given to snap decisions nor did he impede progress in any field. With his encouragement and partici-

307

pation, the musical life of the communities has flourished for many years. It was our good fortune to have a prelate that possessed experience and wisdom combined with good judgment.

MUSICAL STYLES

Some discussion of pre-war and post-war (WW II) musical style is necessary to clarify the position of both the musical contributors and the hierarchy. The single-line melody of Byzantine Chant (monophony) of the Church of Greece was rendered by two groups of precentors, explicitly guided by the *rubrics* or red-printed directions in the service books and the *Typicon*. Services in the early American churches (Annunciation, Chicago, 1893; Holy Trinity, New York, 1902; Holy Trinity, Boston, 1903) were followed by others, reaching 52 by 1962. The service music was certainly supplied by a Right and Left Psaltis plus volunteers when available, carrying on the tradition of the Mother Church. Yet, here too, a kind of melting pot was created. Because not all psaltae came from Athens or the Peloponnese, but from all over Greece and Asia Minor, a mixture of various styles was absorbed at large in the New World.

In 1902, the first edition of Sakellarides' *Sacred Hymnody* was published in Byzantine notation in Athens. This was of little use to the American based Church except that the author had managed to write, for the first time, two- and three-part harmonizations in this notation, thereby giving Athenian congregations a taste of harmony. In 1930, he published this music in European staff notation. It was this music that was used by our earliest choirs. One can imagine that in America the Sunday School children were taught the melodies at first and introduced to harmonized versions soon after. Sakellarides' music was reprinted in bootleg editions in those early days by Greek-American newspaper publishers.

What was this music like? It consisted of the melody in the top (tenor) voice and a follower a third lower, over a simple bass line, easy to sing and learn. The addition of a piano in the Sunday school or language classroom made the learning process faster. Transfer of a keyboard instrument into the church itself was inevitable. This simple style of music prevailed for at least 20 years. Everyone knows that singing a *round* is a simple form of polyphony, or counterpoint, as it is technically known. As choirs improved, a few composers began to experiment with polyphony in a limited manner.

Shortly before the second war, James Aliferis became among the first Greek-Americans to earn a Ph.D. in Music. His activity in the church resulted in the first publications of Byzantine hymns for mixed voice choir by the established American publishing house of Witmark & Sons. Aliferis' publication must be considered an important landmark in the development of our choral music. A more advanced style was on its way and we find this in the music produced by Athan Theodores and Thomas Regas. Unfortunately their work was not published and was known only locally. A priest-musician who also contributed simple choral music was Fr. E. Chrysoloras whose three-part writing was designed to be expandable to more parts.

In all these versions, the traditional melodic line was adhered to but the harmonizations were simple major-minor key associations. Even the last phase of Byzantine Chant was based on the ancient modal system, where several scales and tonalities were in use. One argument against 'westernization' of Byzantine melody was the fact that the west had only one tonality and two scales, rendering the two systems incompatible. However, modal treatment in harmonic and polyphonic music had already been achieved in the west as early as the 15th Century with adaptation of Gregorian Chant to choral music.

With the return of the servicemen in 1946, the Church was to expand in all directions, resulting in larger congregations, new churches, American-born parish and community leaders. and musical innovation. The first person to introduce music for liturgical use at this time was Anna Gerotheo Gallos, daughter of a Greek priest and married to Rev. George Gallos. Presvytera Gallos had a thorough musical education, begining in her childhood and culminating in her graduation with a Master's degree from the Eastman School of Music. Her teaching experience spans the public school and the university; in the 1960s she was for a time a music instructor at the Holy Cross Seminary in Brookline, MA. She has so far published four Liturgies, some folk songs and music for the organ. She heads the Evangeline Press, a Greek music publishing firm. At present, Presvytera Gallos is preparing liturgical music in each of the eight ecclesiastical modes.

The question after World War II was whether modality could be applied in the choral settings of Byzantine melody. Some evidence of such experimentation also appeared in one or two examples of Vrionides' first setting. When I became choir director at the Church of Annunciation in Los Angeles in 1948, this notion produced successful experimentation with Byzantine hymns. The eight modes of this music were quite easily adaptable to harmonic and polyphonic treatment. The first attempt was a musical setting for the Divine Liturgy published by the Greek Sacred and Secular Music Society in 1951. This early success at modal treatment aroused the interest of other composers including Anna Gallos whose second and subsequent liturgies are modally based. Other early successes were those of Tikey Zes, Demetrios Pappas and Theodore Bogdanos. Byzantine choral music had found its voice. By 1960 work of these Greek-Americans was beginning to find adherents and admirers even in Greece.

At this point it becomes necessary to resolve a stubborn conflict in the identification of our musical heritage. Byzantine chant, or music, is simply vocal music with a single melodic line, sometimes assisted by a particular sustained tone, also vocal. There is not, and has never been, Byzantine choral part music. In the west, Gregorian Chant is also single-line melody only vocal music without even the *Ison* or sustained tone. No one listens to a mass by Palestrina, for example, based on a Gregorian Chant and calls it Gregorian music. It is simply vocal music based on a pre-existing melody.

So it is with our choral music. It is not "Byzantine Music" at all, but a sacred music *composition* based on a Byzantine chant melody! We have heard that so and so's music is more 'Byzantine' than another writer's. Such a remark makes no sense. Some composers of Orthodox service music at present do not even use traditional chant melodies, nor do they try to imitate them. It is hoped that the public at large will understand the difference between the nature of our choral music and that of the traditional method of rendering chant.

ORGANIZATIONS AND SOCIETIES

The Seminary

Undoubtedly one of the most important events in the history of the Greek-American Church was the establishment of

Holy Cross Orthodox Seminary in 1937 at Pomfret, Connecticut, later to be moved to Brookline, Massachuessets, as part of Hellenic College.

In 1937, Ambrose Giannoukos formed the first seminary choir—a highly successful venture. Mr. Giannoukos was at the time finishing a Master's degree in music. An accomplished pianist and organist, his interest in Orthodox music resulted in 1938 in his ordination in the priesthood. Under his direction the Seminary Choir achieved wide fame and recognition throughout the New England area. After leaving the seminary, he served as professor of music at various universities and also as pastor to many parishes. At present Fr. Giannoukos is pastor of St. Nicholas Church in Corpus Christi, Texas.

At the time of Archbishop Iakovos' enthronement, the musical instruction at the Seminary was in the hands of Christos Vrionides, whose vast experience in all phases of music was invaluable to the seminarians who were fortunate to study under him. The students were brought face to face not only with Byzantine Chant in which Professor Vrionides was an authority, but also with the musical traditions of all other religious faiths, symphonic and operatic music, and a vast store of Greek folk songs.

With the passing of Vrionides in 1962, the Archbishop invited Savas Savas to teach Byzantine music at the Seminary. Born in Greece, Professor Savas was trained in Athens, and served as Protospalte and music director of the city's cathedral. He taught Byzantine music at the University of Athens, the Theological School, and the Athens Conservatory as well. Before coming to the United States, he directed choirs in Paris and Munich.

At the Holy Cross Seminary, Professor Savas felt a need for a basic aid in the English language, and in 1965, published *Byzantine Music in Theory and Practice*. This book, and another published a few years later, *The Hymnology of the Orthodox Church*, are the basic textbooks of his efforts at the School of Theology. They inspire and continue to inspire our seminarians to penetrate more deeply into the content of our hymns and their musical interpretation.

Professor Savas has also contributed articles to many important musical periodicals. His knowledge extends even to art and archeology. In 1959, Patriarch Benedictos of Jerusalem presented him with the Golden Cross of the Patriarchate. In 1976, he was knighted by the Ecumenical Patriarch as Archon

Didaskalos and selected as Outstanding Educator in America.

It seems that the musical education of our priests has been entrusted, from the beginning, to the most outstanding men in the field. Professor Savas continues his dedication by constantly updating his materials and techniques, as evidenced by his recent employment of modern audio-visual equipment in his instruction.

Byzantine music has always been the exclusive property of the male singer, although in convents some women have learned this art. Since Kassiani of Byzantine times, there has not been any recorded female contribution to the repertoire. It comes as a first for the Seminary then, to have Professor Savas train women to read and chant our music. Two young ladies, Jessica Suchy and Valerie Karras have become quite successful proteges of Master Savas. Miss Suchy has served as a *Psalta* at St. Sophia Cathedral in Los Angeles and is presently in Indianapolis serving the church there.

Choir Schools

Finding qualified musicians to train our choirs will always be a major concern. One cannot expect a young person to take a four-year college degree in music only to fulfill the requirements of a part-time job. Realizing this, Anna Gallos and Arthur Kanaracus originated the first Choir School in 1946, wherein a person could spend a few weeks during the summer in specialized subjects such as conducting, accompanying, repertoire, and some theory designed around Byzantine music characteristics. The idea worked so well that Archbishop Iakovos has supported the choir school organizations with much enthusiasm. At present, choir schools often function with the annual conference of the regional Choir Federations.

It was mentioned that Professor Savas offered the first trained women precentors, but other women have also taken part, although not American trained. Janet Christopoulos Webster migrated to the U.S. from Jerusalem after the war. She had been her father's *psalta* in the Holy City where he was an Orthodox priest. She too has served at St. Sophia Cathedral in Los Angeles.

It is no wonder then, that women as well as men in America are at present seeking to learn to perform the music as well as study it as musicologists. On the academic level, Greek Americans are also entering the musicological fields. At the University of St. Louis Dr. Diane Touliatos-Banker is contributing articles on important research in early phases of Byzantine Chant. Dr. Touliatos-Banker was trained in Thessalonica by the famous Christos Patrinelis. In Vancouver, B.C., Australian-born Dimitri E. Conomos has become internationally known for his Byzantine music scholarship and is at present editor of the *Studies in Eastern Chant*.

Choir Federations

At the time of Archbishop Iakovos' elevation there were a few choir federations throughout the country, each one covering several churches. To report on the activities of each one could require a separate article. Therefore the main purpose of these organizations will be presented, with their special contributions noted. There are now (1984) eight choir federations—one in each diocese and in the Archdiocesan District. Four issue periodic newsletters containing information on activities and articles of interest by a wide range of contributors. Some have the means to publish and sell choral music. Each of the regional federations sponsors an annual conference, where the focal point is an Hierarchical Liturgy, offering the music of a special composer. New music is also reviewed and sung. In addition, lecturers offer seminars and workshops. These are interspersed with social events and a formal banquet, sometimes a well-prepared concert is included. The annual affairs have done much to help the cause of Greek Orthodox music in this land. The influence has reached the Sunday school classroom and the fostering of younger groups, the Junior Choirs. Most important, the federations have encouraged a higher standard of performance by the individual choirs. The federations have been the major factor in getting new musical settings of our music to the public. They have also been responsible for preserving some valued traditions, such as Byzantine Chant of earlier periods. The organ has become a required item in the choir loft, with expert organists giving workshops through federation sponsorship that has borne fruit in the form of a growing number of fine players.

Other Choral Organizations

Specialists who are conductors, composers, and performers always desire a vehicle of expression that is of the highest calibre. Therefore, when a parish can afford it, a professional group may be retained.

313

THE ARCHDIOCESE

Such a choir was always available to the Archdiocese Cathedral in New York City. For thirty years its conductor was Nicholas D. Iliopoulos. This group has also presented special programs, oratorios and has featured music of several composers. In 1976, the choir was taken over by Dino Anagnost, whose vigorous leadership has not only carried out the fine work of Mr. Iliopoulos, but has added to the repertoire and in other ways contributed to the prestige of the Cathedral.

METROPOLITAN CHORALE

A choral group of professional ability to represent the Archdiocese was organized in 1962, the first director being James Stathis. In 1968, Dino Anagnost assumed its directorship and brought it to national attention. Since 1978, the Chorale has been under the directorship of George Tsontakis, who has continued to enlarge both its scope and its audience. Works of exceptional perception and prominence are being written for and presented by this group, which promises to be an important musical showcase for our Church.

Both Anagnost and Tsontakis are representatives of the finest musical development of this generation. Both have been professionally groomed, and have already caught the attention of a discerning musical public.

CHICAGO CHORALE

The midwest has recently organized another such group in Chicago at the suggestion of Archbishop Iakovos, with Georgia Mitchell as director. Mrs. Mitchell has been very active and influential in this area for many years. She is an accomplished pianist of virtuoso status, and is equally accomplished on the organ.

NORTHWEST

In the northwest, Panos Vlahos at Lewis and Clarke College organized a group that not only served the Holy Trinity Church in Portland, Oregon in the late 50s, but also presented concerts throughout the area that brought enthusiastic comments from critics. Unfortunately this group was disbanded.

After taking part in the World Council Ecumenical Liturgy in Vancouver in August 1983, plans for a Northwest Byzantine Chorale are under way. Organizers are George Lendaris of Portland; Rose Munson, choir director, Jerry Mulinos, Fr. Homer Demopoulos of Seattle as spiritual advisor, and myself as musical director at the initial stages. Repertoire will consist of Byzantine chant of various periods, and both sacred and folk choral music.

For a number of years, two professional-calibre groups existed in the state, the Bay Area Chorale, with Pericles Phillips as conductor, and the Byzantine Chorale of Los Angeles, where I am the conductor. The Bay Area group has been reorganized and is now known as the Dorian Singers with Tikey Zes, as conductor. These groups have performed extensively, recorded, and toured. Members of both groups were part of the 1974 Archdiocesan Chorale that toured Greece.

The Archdiocesan Chorale was organized by Ernest Villas and Pericles Phillips in 1972. Membership of the Chorale was drawn from a large area, and the Chorale met in Fresno, California once a month for rehearsals to prepare a major pilgrimage-concert tour of Greece and to present a Memorial to the late Patriarch Athenagoras in his home town of Vasilikon in northwest Greece. In July 1974, after a year's delay due to political events in Greece, the Chorale was on its way, giving concerts on the Islands and in Egypt, and liturgies in Athens and Salonica. Father Spencer Kezios was spiritual advisor; Thomas Lappas and Thomas Pallad, tour directors; Frank Desby, conductor; Tikey Zes, assistant; and Xenia Anton Desby, accompanist. On hearing a tape of the chanted (men's voices) portion of the memorial music, Simon Karras in Athens was amazed that Americans were able to perform this very difficult music with its tricky intonations, in such an expert manner. Karras is an undisputed leader in Byzantine music in Greece and has trained a large number of psaltae.

The Byzantine Chorale in Los Angeles has issued a number of recordings where both choral music and ancient chant are offered. The oldest extant Christian Hymn, transcribed by Egon Wellesz, appears on one of the disks.

The National Forum of Greek Orthodox Church Musicians

It soon became evident that the various federations and independent church choral organizations needed an integrating "national voice." For many years, the National Choir Committee of G.O.Y.A. served such a purpose, but increasing requirements led to the formation of the National Forum. In this way, an organization with a full staff of its own represents our choral organizations and musicians, including choir personnel, clergy, psaltae, and music educators.

The growth of musical activities on local and national levels was due largely to Archbishop Iakovos' desire to see the musical situation up-graded in all aspects. The need for an organization was on the minds of certain individuals as early as 1973, and representatives from various choir federations periodically met in the early 1970s to plan the organization.

In 1976, the Forum became official. Its first chairman was Niki Kalkanis. Mr. Kalkanis, a graduate of Wayne State University (Detroit) in Science and Business Administration, is also a talented organist and conductor, having served not only churches in the Detroit area, but as guest lecturer and administrative leader of the Mideastern Choir Federation. His critical four years in office established the Forum as the major organization of church musicians the Archbishop envisioned.

Dr. George Demos succeeded Mr. Kalkanis as National Chairman in 1980. In addition to his training to become a distinguished otorhinolaryngologist, he took advanced courses in music and conducting. His involvement in the Church stems from a very early age and is evidenced by an enormous enthusiasm for its musical heritage. He has conducted in various parts of the country and was instrumental in forming many musical organizations for the Church.

While chairman of the National Forum, Dr. Demos inaugurated the National Choir Music Endowment fund, assisted by Niki Kalkanis. He also initiated the Xenia Desby Memorial Scholarship. He has encouraged many musicians to contribute compositions to the liturgical repertoire. With the Western Federation-East, he has spearheaded the publication of the works of our outstanding composers.

Since 1982, Dr. Vicki Pappas, educator and researcher at Indiana University, has chaired the National Forum. Dr. Pappas has excellent organizational talents as well as being a well-trained musician and choir director. The early work in originating a charter and the organizational plan for the Forum was accomplished by her. At present she is able to be in constant contact with all her committee personnel, oversee the various projects, stay in frequent contact with the Archdiocese and the Holy Synod, and still continue her local church and choir commitments. During her tenure, she focused national attention on the work of the Forum, culminating in the publication of musical education materials for youth and a Symposium on Church Music co-sponsored with Holy Cross School of Theology.

Projects of the Forum include the following:

1. *The Liturgical Guidebook*, issued annually to choir directors at a nominal cost. This booklet contains the order of Sunday services and some special feasts from January to the beginning of the moveable season in the following year. There is also an annotated bibliography of related materials. The Guidebook was introduced in 1956 as a G.O.Y.A. Choir committee project at the request of Anna Gallos. Since its inception, I have been the Manual's author. When the Forum assumed publication, an editor was needed that knew the material well and could oversee the printing and up-dating of new materials; this important task is now undertaken by Peter Vatsures of Columbus, Ohio.

2. *Publications*. In addition to the Guidebook, this committee oversees many of the Forum's educational materials, and makes available reports on surveys. The preservation of the music of early contributors is another area the committee is considering. Peter Vatsuras (chairman), Thomas Lappas, Dr. James Maniatis, Harry Booras and Gerald van de Bruinhorst comprise the committee.

3. *Cantorial Training Manual*. This is a complete treatise with recorded examples for clarification, aimed towards the training of future psaltae. The text is by Frank Desby, and recorded examples by Theodore Bogdanos, both of California.

4. *Children's Hymn Series*. Mary Jo Cally of Chicago and Vicki Pappas have selected musicians trained in composition from our ranks to prepare selected liturgical hymns in both Greek and English versions. Included also are teaching suggestions regarding the meaning of the hymns, their spiritual roots, their relation to a selected icon, and meanings of key Greek words. Aimed for Sunday School curriculum, the Series promises to have even broader application to retreats, camps, and even adult education. In a similar vain, Dr. Maragos and his wife Connie of Rochester, Minnesota, are compiling an Orthodox songbook for use by the youth at informal gatherings.

5. *Archives*. Tina Vratimos of California compiled and maintains a library of historical and biographical archives. Tina, originally a very vital part of the music situation in the Midwest, is now active in choral development in the Bay Area in California.

6. *Other Activities*. Forum members from across the country handle other administrative activities of the National Fo-

rum. Publicity is in the hands of Ida Trakadas of Oak Lawn, IL. Niki Kalkanis serves as treasurer, and also manages computerized mailing lists of the Forum and federations, Lee Mellen of Atlanta serves as Membership Chairman, and John Tsokinos (Cleveland) and Dean Limberakis (Boston) coordinate the presentation of the St. Romanos the Melodian Award to outstanding church musicians each year.

PUBLISHERS

Early publications were often the product of the composers themselves, privately printed and distributed. Unfortunately, such organizations tend to be one generation firms, resulting in the eventual loss of creative work. One of the goals of the National Forum is to create a music archive of these out-of-print works.

At present, a few publishers exist that supply liturgical music and some folk music. The Greek Sacred and Secular Music Society, Inc. (1324 S. Normandie, Los Angeles, CA 90006) publishes Orthodox Service music and folk songs; several composers are represented. Thomas Lappas serves as its President. Evangeline Press (9 Kings Court, Annapolis, MD 21401) publishes, in addition to liturgical and folk, organ music based on Byzantine hymns. Anna Gallos is publisher and editor of Helicon Press (63 Gates St., Worcester, MA 01610). Arthur S. Kanaracus as publisher oversees the printing of liturgical and folk music. The Federation of Western States East Publishers (4030 S. Hudson Way, Englewood, CA 80110) represent several composers of liturgical music, especially hymns of feast days. George Demos is its publisher and editor. And finally, the Holy Cross Orthodox Press (50 Goddard Ave., Brookline, MA 02146), as the official Archdiocese Press, publishes music, didactic, theological and historical material.

MUSICAL COMPOSITION IN AMERICA

At the end of World War II, there was a mere handful of composer-arrangers of choral literature. At present, the roster has grown to more than two dozen, producing printed music as well as recordings. This post-war renaissance was especially aided by two factors: professional, advanced training sought by our new generation, and the encouragement of the Archdiocese, especially in the past twenty years.

Compositions are by nature three types: a.) based on traditional chant melody, modal in concept; b.)traditional chant

318

Sacred Music in the Byzantine and Western Musical Notations

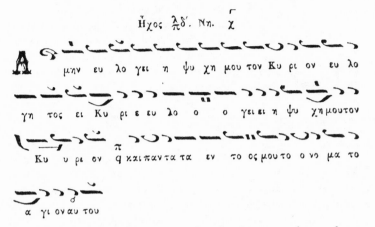

Byzantine music has its own system of musical scales, its own laws and canons, its own modes of composition, its own system of notation. Above is an excerpt from a composition by Petros the Peloponnesian (18th century), one of the most remarkable cantors and composers of Byzantine chant after the fall of Constantinople. It is written in the reformed Byzantine notation, developed by Chrysanthos (d. 1843), Gregory (d. 1822), and Hourmouzios (d. 1840). The symbols above the words do not give the pitch of every tone in the melody, but indicate how many tones a certain note lies above or below the preceding one, or whether it is a repetition of it.

ΕΙΣ ΑΓΙΟΣ

a cappella for mixed voices

George S. Raptis

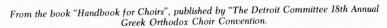

From the book "Handbook for Choirs", published by "The Detroit Committee 18th Annual Greek Orthodox Choir Convention.

in free harmonization; and c.) entirely new melodic material. Most composers have indulged in all three methods; some prefer one idiom for most of their work.

First and foremost of the post-war contributors is Anna Gallos, whose background as a thoroughly-trained musician and conductor was influenced by a family involved in the Church. Above, all Mrs. Gallos' interest in fostering the music of her colleagues is unmatched in this country. Her own output is considerable, and her untiring efforts in establishing choral organizations, schools and events has pushed our music forward at a greater pace than would have been otherwise possible.

In California, the liturgical lyre has been tuned modally. Contributors here include myself, Tikey Zes and Theodore Bogdanos. Zes and I were both trained concurrently at the University of Southern California, where a sacred music department exists. Both of us have made trips abroad to seek out new material. We have been in a sense, an influence on each other. Inspired by the 'new' modal and polyphonic technique, Theodore Bogdanos began contributing to this type of repertoire. The efforts of these modal "pioneers" surfaced because of the effort of several people on the scene at the time. They must be remembered because without their assistance any significant changes would never have become known.

An organization was born to promote the works of the new composers. This was the Greek Sacred and Secular Music Society. The original staff consisted of Thomas and Anthony Lappas, George and James Bonorris, Pericles Caiopoulos, Helen Kostelas (now Georgilas), and later on, Constantine Lappas and Xenia Anton Desby. The first president of the company was George Bonorris; at present, Tom Lappas heads the organization. Mr. Lappas and Mr. Bonorris secured the services of Dr. Charles Hirt and Dr. Pauline Aldermann of the University of Southern California, and Dr. Lauren Petran of the University of California at Los Angeles as advisors for the group. There exists an editorial board that examines all music submitted.

At this same time, Pericles Phillips and John Reckas, both of Oakland, California were the first to recognize the quality of the new music. On the East Coast Anna Gallos, almost single-handedly, created a demand for this music and encouraged the budding composers to continue producing. Working in the East at this time was Demetrios Pappas. He introduced a liturgical setting with occasional separate organ part, not merely reproduction of the vocal parts. Mr. Pappas' knowledge of traditional Byzantine melody was unusual, having grown up listening to his father, a Protopsaltis for the Archdiocese. He began early, starting to direct at the age of seventeen. For thirteen years, he sang with the prestigious Robert Shaw Chorale. In addition to two liturgies, a number of recordings with his Amphion Choir are to his credit. Their two Yassou recordings have reached international prominence.

More recently, a new wave of composers have come to the fore. Some adhere to traditional melody and modality, but many write in an 'implied' melodic and modal style, using more modern techniques. This is not to be condemned, for no art form should remain static; there is a certain freshness to these sounds that still manage to capture an essence of Byzantium, much as Bartok had done with much of the ethnic music he collected throughout his travels.

Ernest Villas has devoted a lifetime to the Church. He has taught music, and in the 1950s was musical director of the Hormel Caravan, a choral and instrumental ensemble that toured the country and made weekly broadcasts. During these tours, he visited as many Greek communities as possible to talk to the younger generation and form youth clubs. He was the originator of the G.O.Y.A., and headed the organization in its initial stages. He was responsible for forming the Metropolitan Chorale, and for a time directed the group. He published a liturgy for unison voices that is now used in many communities as total congregational participation of the Sunday Liturgy. It was this work that was presented to the congregation at the 1983 World Council of Churches Ecumenical Service.

In Madison, Wisconsin, Professor Mike Petrovich, conductor of the Greek Orthodox Church of Assumption, not only introduced some exceptionally fine music of his own, based often on traditional melodies, but was quite an inspiration to the young people in his choirs. Nicolas Maragos, one of his proteges, has become an important contributor to Orthodox service music and has written a complete liturgy and many other hymns in Greek and in English.

I mention here others who have recently contributed music of interest and value: Steve Cardiasmenos, San Francisco area; Neal Desby, Los Angeles; Dimitri Futris, Skokie, Illinois; William Harmand, Syracuse, New York; Christopher Kypros, Norfolk, Virginia; Georgia Tangeres, Baltimore, Maryland; Peter Tiboris, Plymouth, New Hampshire; Mike Pallad, Northridge, California; Steve Phillips, New York; Alex Lingas, Portland, Oregon; John Revezoulis, Milwaukee, Wisconsin, and George Raptis, Detroit, Michigan. While this roster is admittedly incomplete, it points out the fact that fully trained musicians now supply our Church with first rate choral music.

THOSE WHO MAKE IT HAPPEN

We have singled out the composers, their contributions have given them celebrity, but their "workers" at the scene are seldom honored. It is they who have the genius for recognizing talent, the faculty for promotion, and the stamina for administration and implementing.

Perry Phillips of California, previously mentioned, is an excellent conductor, has been a professional musician, and is an impresario second to none. He organized the Archdiocesan Choir that toured Greece in 1974. While president of the Western Federation from 1951 to 1954, the organization was enlarged to take in all of the Western States. At present, he is entertainment editor for the *Oakland Tribune* and director of the Ascension Choir in Oakland. He founded the Bay Area Chorale whose performances all over the west have received praise from critics. Most important is his recognition of talent and the ability to promote works of new composers.

George Dimopoulos was the most influential leader in the musical department that took place in the midwest. In Chicago, he was a psaltis, conductor, composer, and teacher. Archbishop Iakovos conferred on him the highest ecclesiastical musical honor in 1955, that of Archon Protopsaltis.

George Raptis in Detroit, and John Tsolainos of Cleveland have spent most of their lives in furthering the cause of our music, both on the organizational level and as conductors. They took part in the early stages in the formation of the Forum and have been the vital spinal cord of the Mideastern Choir Federation. They have discovered singers, conductors, composers, established a valuable archive of sacred and folk music, and in general have been surging forces in the growth of Orthodox music.

In the Southern states, John Demos, active in the Georgia State University instrumental and choral departments, has used his expertise in the Atlanta Church of Annunciation and

322

for the Southern Federation. Choral groups under his direction always achieve a high degree of musical presentation.

At the Conservatory of Cincinnati, Ann-Marie Koukios is an instructor in choral directing and a doctoral candidate at the University. She has formed performing groups of the University students to present Orthodox choral music, Byzantine chant and Greek folk songs. She directs the choir of Holy Trinity and has introduced this group to music of various composers. Her talent in developing a flowing, very expressive sound in a choral group is (to this listener) unsurpassed. A brilliant future is predicted for this unusually talented young lady.

Also behind the scenes, but very important are Evelyn Mickles, Joan Petrakis, Maggie Bovis, George Lendaris, George Haikalis, Jeffrey Economou, Rose Munson, Cathy Zarbis, George Georgantas, Agi Grigoriadis, Bill Bobolis, Paul Pronoitis, Bonnie Lozos, Adrianna Kolandranos, Dimmie Efstathiou, Athena Tsougourakis, Connie Speronis, Harry Booras, Nick Chimitras, Jane Patsakos, John Douglas, Tom Pallad, Jim Economou, Anna Marakas Counelis, Mike Hadgis, Chris Calle, James Counelis, Steve Bournos, Lois Pappademos, and Dean Limberakis. There are many others dotted throughout our continent. On a local level, these people keep up the enthusiasm, find supporters, raise funds, make arrangements for travel, housing, transportation, and countless other necessary details to keep choral organizations "afloat," but get little recognition. Many others have not come to our attention. It will be up to local organizations to see that they are eventually rewarded. One day a complete roster, possibly through the Forum, will become available.

In closing, one vital force should be recognized as having made such tremendous progress possible. This force is the Holy Synod, with its Diocesan Bishops and their clergy. In each area of our country, our Bishops have given the best possible support, encouragement, and spiritual guidance. This is real trust, and the hierarchy apparently has always had confidence in the musical leadership. Such an attitude, of course, stems from the top, our own Archbishop Iakovos. It is from him that the course of choral development has been charted and expanded. It is from him that a network now extends—of dedicated, trained, talented, and energetic church musicians working locally, regionally, and nationally to preserve and extend the rich heritage of Byzantine music and Orthodx worship.

A Vibrant Youth Ministry Since 1959

By
Angelo Gavalas

t the time that Archbishop Iakovos was enthroned in 1959, the Archdiocese was at the beginning of a new period of growth and expansion. New parishes were being established and older parishes were doubling and tripling in size. Because of "growing pains," the Archdiocese was very limited in funds and could not provide sufficient budgeting for youth work on the national and district level. This was part of the frustration felt by both clergy and laity in the youth committees who, many times over, at each succeeding Congress expressed the need for funding and personnel for the Archdiocese Youth Office.

Fortunately, Archbishop Iakovos was a spiritual leader of determination and the subject of youth continued to be a priority in all discourses and in committee meetings. His personal dynamism and spiritual presence inspired youth of all ages and imbued in them the pride of sharing the Christian faith of their forefathers. The Archbishop's personal achievements both in the Church and on the national and international scene gave the Greek Orthodox Church a new prominence that was noted by a majority of Americans. No longer was the youth of our Church asked to explain what "Greek Orthodox" meant.

Interestingly enough, the pursuit of a youth ministry began a good many years before the Greek Orthodox Youth of America (GOYA) became the official youth movement of the Archdiocese in 1950, under the aegis of Archbishop Michael. "Its inception actually took place in the minds and hearts of Greek Orthodox young people in all parts of America as far back as the early and middle 1930s. Local youth groups, many of which in later years expanded to city-wide and district-wide youth organizations, all helped set the stage for the year 1950, which ushered in a new era in the history of our Holy Orthodox Church in the Americas. . . . In November 1950, youth representatives from 29 communities in the United States convened in St.

The late S. Gregory Taylor presenting award to distinguished Youth Leader, Angelo Gerzel. Seen in picture is Bishop Polyzoides, Nick Iliopoulos, Serena Yale and Fr. Theophilos.

In San Francisco, over 20,000 attending youth festival "A Greek Orthodox Tribute to America".

Louis, MO on the weekend before the opening of the 10th Biennial Clergy-Laity Congress."[1] Representatives of the Federation of Greater New York, the Orthodox Youth of Chicago (O.Y.), the Upper Midwest Hellenic Orthodox Youth Federation (UMHOYF), and the Orthodox Youth Organization of the West, joined with numerous independent youth groups in St. Louis.[2]

The idea of uniting youth across the Archdiocese was also an effort that Archbishop Iakovos attempted very soon after his arrival in the U.S., in 1939. "In July of 1941, Iakovos met with youth in Danielson, Connecticut and formed a national youth organization and gave it the name E.O.N. (Elliniki Orthodoxi Neolaia, Greek Orthodox Youth)."[3] Unfortunately, the movement did not receive the support of the mainstream of the Church and the effort failed to materialize into a united youth movement." But the idea was kept alive. Approximately ten years later, Archbishop Michael resurrected the idea by initiating one of the most successful youth movements in the history of religion in America in organizing G.O.Y.A."[4]

PROGRESS THROUGHT THE 1960s

The best source of material on the youth ministry is to be found in the archives of the Archdiocese. The youth reports delivered to the Clergy-Laity Congresses from 1960-1982 are a resume of the youth work, its successes and its frustrations as experienced in the parishes and dioceses.

The main emphasis of the Youth Department's report presented in 1960 in Buffalo during the Clergy-Laity Congress was placed again on the development of the senior sector of GOYA. Manuel Scarmoutos, the National Chairman of GOYA proposed the motion declaring GOYA an even more integral part of the Archdiocese as follows: "That GOYA be incorporated into the By-Laws of the Greek Archdiocese of North and South America as an integral organization thereof. . . it shall follow the direction and adhere to the program of the

---

[1]Ernest A. Villas, GOYA Manual, Greek Archdiocese, New York, May, 1958.
[2]GOYA Manual, May, 1958.
[3]"A Glorious Spiritual Odyssey," *Orthodox Observer*, 1984.
[4]"Spiritual Odyssey."

Greek Archdiocese. The Parish Priest shall be the spiritual leader of GOYA and not merely religious advisor."

A further illuminating comment was made by the then Archimandrite Silas Koskinas that had significant importance indicating as well that GOYA must be treated as part of the Church. He said, 1)"The Church is an Organism and not a social or community organization. 2) Our Church should not be divided into the Old and New Generation." The above two statements made it quite clear that GOYA was not to be treated as a separate entity within the Church but as an integral part within the already established Archdiocese. Before that, the Minneapolis GOYA Conference (1954) proclaimed the Annual National Meeting *international*. GOYA was being given an official status by both the Clergy-Laity Congress and the GOYA Conference in the hope of strengthening the efforts of the faithful Greek Orthodox Christians who comprised this age grouping (ages 18 - 35).

GOYA had developed many exceptional programs that had won wide respect, and developed leaders among the youth who serve the Church to this day as parish council members, Sunday School teachers, youth advisors and Archdiocesan and Diocesan officers. GOYA sponsored the Ionian Earthquake Relief Drive which raised $113,000 in 1954; the "Books for Brookline" Drive, which netted $33,000; filming of the Divine Liturgy in sound and color; publishing a quarterly magazine, the *GOYAN*; provided the initiative for the formation of the Council of Eastern Orthodox Youth Leaders of the Americas; accepted the responsibility of the Junior GOYA Program; initiated several choir projects and raised $150,000 toward the construction of the Memorial Chapel at Holy Cross Seminary."[5]

As further indication of the foresight of the GOYA leaders at that time, the Congress of Buffalo in 1960, made an official effort to establish a form of GOYA that would meet the needs of older and married young adults to be called *St. Paul Fellowship* and *St. Andrew Fellowship* respectively. Both efforts, though well meant, never materialized nationally.

At the Boston Congress, 1962, Scouting became an official youth program of the Archdiocese. In his address His Eminence stated: "Boy Scouting is not only a highly benefi-

---

[5]GOYA Manual, 1958.

cial International Youth Movement but also one of the most beautiful activities in our communities. Those communities which are able, but do not have a Boy Scout Troop, are guilty of an unpardonable act of omission. I fear that this is at the expense of our children. The Archdiocese not only anticipates the creation of new scout troops but has conscientiously undertaken certain responsibilities in cooperation with the National Council of Boy Scouts of America, where we are represented by a clergyman of the Archdiocese. As you know, scouting offers a program to boys and girls. In our particular situation, we have need of scout troops for another reason— so that our children may grow up in the community and in ecclesiastical surroundings. Scouting could become the medium for the enactment of a missionary movement among the youth and consequently, I consider it necessary that each community adopt a Scouting Program and introduce it systematically and with great care into the life of its youth."

At that time, the Eastern Orthodox Committee on Scouting (E.O.C.S.), was formed in New York City and represented the scouting troops of all Orthodox jurisdictions. Although at present, only thirteen troops are active and are of the Greek Archdiocese, the organization continues to flourish and welcomes troops from any canonical Orthodox jurisdiction. Under the guidance of the Archdiocese Youth Ministry Office it has developed the Alpha Omega Religious Award for Boy and Girl Scouts; the Chi-Rho Religious Award for Cub Scouts; and the Prophet Elias Award for worthy Adult Scout Leaders. At one time it offered the Theotokos Religious Award for girls, but this has been dropped for the Alpha Omega Award. These awards are being administered by Fr. Joel McEachen and the Orthodox Committee on Scouting Commission. Fr. Gavalas is now preparing a *Guideline for Scouting* in the Eastern Orthodox Church. At the Denver Congress in 1964, a first Boy Scout Report was made by Frs. Charles Sarelis and George Gallas and the listing of Jr. Youth grouping was made as follows: Acolytes, Boy and Girl Scouts, Jr. GOYA, Jr. Choir.

At the Denver Congress (1964) the constant hard work of the national youth leaders and the youth committees began to formulate a clearer structure. The following description of GOYA was propounded, "the term 'young adults' is best described as a category which begins after high school (usually at age 18). It is a most dynamic and challenging age level of religious life and action in as much as it combines a keen affinity for idealism, with the rejection of many standard patterns and norms of accepted living. In this classification we find the following general grouping: the college student; the young working person or short term student; young married couples." We note at this Congress, the determination in the Archdiocese and at most of the parishes to find good and profitable solutions to youth issues. It was reported that in some Districts there were a greater number of Jr. GOYA Chapters and a request for a full-time clergyman as Director of Jr. GOYA had been made as far back as 1960. The motto, "Learn Your Orthodox Faith" was adopted as a national theme and Jr. GOYA was expanded into an exciting religious movement. The program encompassed a varied and interesting calendar of activities of the 13 - 17 year old age level, including sports, social, religious education and cultural events.

A combination of religious retreats and sports/social outings became very popular with teenagers. To meet these growing needs of youth on the Archdiocesan level, the following decisions were made: 1) His Eminence appointed a Commission composed of clergy and laity to serve as special consultants to youth. 2) A short-term course for training of professional youth workers be implemented at Holy Cross. 3) The Archdiocese Department of Laity was established in 1961 with Ernest Villas appointed as Director. 4) It was recommended that a Youth Office be established in each Diocese District.

At this same Congress extensive discussion was held about the problem of youth on campus. A very definite vacuum was reported and an urgency for action was expressed. SCOBA, the Standing Conference of Bishops in America (Council), was asked to assume the responsibility of appointing a Pan-Orthodox College Committee. In 1965, SCOBA assumed all campus work and James Couchell was employed as Executive Secretary of member jurisdictions. His office issued *Concern* magazine and *Guide to Campus Work*.

At the Boston Congress in 1962, His Eminence recommended camping as an excellent opportunity to teach our youth the Orthodox Christian principles in a most conducive environment. Living the Orthodox experience at camp had to be provided by the Church and by the right leadership. The report about camps was met with great interest and all Districts were encouraged to establish such camps. At that time only four camps were recognized, located in Detroit, Minnea-

329

polis, Gastonia and Georgetown, MA. Today, there are also four camps in the Diocese of New Jersey, two in Chicago, three in Atlanta, two in Boston, one in Pittsburgh and seven in San Francisco. There are also four foundations whose main endeavor is camping for youth. Also the Ionian Village Summer Camp at Bartholomio, Greece, was instituted in 1970, at the Congress in New York. *(See Ionian Village article in this book.)*

At the Congress of New York, 1970, the following report was given: Since 1964 campus ministry (and now campus work) has been done in 100 college campuses. National conferences, retreats and Bible study groups have developed. Relations have been established with other campus Christian groups in this country, with the World Student Christian Federation and with SYNDESMOS (the World Fellowship of Orthodox Youth groups).

With all these developments in the Youth Department, it became increasingly apparent that Archbishop Iakovos had great concern for the problem of youth. At the Athens, Greece, Congress in 1968 he stated, "Our first concern must be to instill a feeling of belonging in our children before it is too late." In addition to this statement, Youth Director, Chris Demetriades stated, "If no other business were to come before this Congress but the question of the needs of our youth, then it would be time and effort totally justified."

OVERCOMING THE CRISES OF THE 1970s

In the Seventies, the hard work of the Archdiocese Youth Department and Youth Committees was producing results. Although Senior GOYA showed signs of decline, youth work among teens and on campus was showing vitality and progress. The Archbishop's "fireside chats" continued to enjoy a popularity among the young adults and teenagers. Meeting in a more intimate setting with the youth, through these chats His Eminence was able to answer many of their queries concerning the Church and society, pluralistic morals, the nature of the society in which all Americans live, and our Greek-American heritage and lifestyle. At this time, the Archbishop proposed a four-point program that met with general approval: 1) scholarship fund, 2) cultural program, 3) national athletic program, and 4) series of inspirational lectures and

330

symposiums (and the publication of such talks).

The 1970s were the times of rebellion of youth, the Vietnam War, the cults and the drug scene. The Youth Committee reminded all and especially the youth of America, "that a better way of life has been the result of the countless and immeasurable experiences of the past. The Committee rejects the idea that our society as we know it, the home, the school and the church is not meaningful or relevant to us today."

In order to deter the rising youth rebellion among our Church youth, the Archbishop showed foresight by appointing a National Youth Commission composed of laity and clergy to discuss all aspects and problems of youth and young adults. With myself as Chairman and Fr. Nicholas Soteropoulos as National Youth Director, the Commission met for two years and laid the foundations for the youth program that was in effect in the 1970s and is still bearing much fruit in the 1980s.

The Archdiocese Youth Office headed by Fr. Soteropoulos created many fine publications during this period when youth problems were intense throughout the country. The publication *You Are Something Special*, gained the admiration and gratitude of parishioners because it provided effective education on drugs, abortion, teenage problems and on moral and social concerns.

The *Youth Calendar and Planbook* was published for the first time in which a guideline of activities for teenagers was set forth for the whole year. The *Acolyte Instructional Manual* was prepared to train altar boys and to give them the proper recognition for their services to their respective Churches. Since then, a new acolyte manual has been prepared by Fr. Michael Kontogiorgis under the auspices of the Diocese of Atlanta. It is a workshop manual for altar boys to be used in a week-long religious retreat.

These publications and the youth programs that came out of the Youth Office gave guidance and direction throughout the Archdiocese. A "grass roots" total youth program was being realized in most parishes with each parish adopting the program to reflect its own, unique situation.

In the late 1970s the Directorship of the Youth Office was assumed by two very vital young clergymen, Fathers Alex Karloutsos and Costas Sitaras. Their exciting rapport with youth and their innovative and updated programs met with great success. Teenage retreats, camping, Ionian Village and athletic activities took on a vital quality that was emulated by all the districts of the Archdiocese. In the short time they were in the Youth Office, they were able to enact all the recommendations of the 22nd Clergy-Laity Congress (Chicago, 1974). They developed a close rapport with the students of Hellenic College/Holy Cross and with the counselors and youth who attended the Ionian Village and other Greek Orthodox camps. Many of these youth have become the backbone and life of the new organization known as the Greek Orthodox Young Adult League (GOYAL).

Eight years ago GOYAL began to become a reality as its structure and direction were strongly established. Guided and inspired by Bishop Athenagoras—the Reverend Athenagoras Aneste—Chancellor of the Diocese of Chicago, GOYAL began to formulate a program that would be useful and relevant to the young adults. At the Philadelphia Congress of 1976, a small group of young adults held a conference and determined that socials and sports were not enough to hold the interest of a sophisticated young adult. The booklet, *The Young Adult League Guidelines,* already written by sòme young adults, was refined, approved and printed. The spiritually-oriented book embraces the very heart of GOYAL. It provided "the magic" of the success of the young adult movement in 1983-84. It is divided into four sections as follows: 1) *Diakonia-*Stewardship; service to the Church and the community. 2) *Koinonia-*Communion; fellowship with Christ the Savior and fellowman—including all social, athletic and cultural programs. 3) *Martyria-*Witness; learn our Faith and witness to it in the Gospel and Holy Tradition. 4) *Liturgia-*Worship; attend the communal services of the Church faithfully and observe your private devotions.

But something more than GOYAL was necessary. It was becoming obvious to our Bishops and youth leaders that the secularism of parishes and their very narrow attitude toward youth programming was not achieving the desired results. Youth needed the stimulation and enthusiasm that interaction with other youth could provide. Many of the Dioceses were offering such programs and were giving the opportunity for youth across the Diocese to get to know each other. The San Francisco Diocese was well into such a program when Bishop Anthony arranged a Youth Conference, Youth/ Young Adults, a few days prior to the San Francisco Clergy-Laity Congress, 1982. The Youth Conference (especially its spiritual workshops) had an excellent attendance and its success gave impetus to the developing young adult movement. The Conference also determined that in the future, the conference for young adults would be held annually and apart from the teenage conference. In this way, more attention could be given to the needs and concerns of young adults.

At that time, because of the untimely death of Fr. John Geranios at St. Basil Academy, the Director of Youth, Fr. Sitaras was appointed Director of St. Basil Academy and the Archdiocese Youth Ministry Office remained vacant for two years. Nonetheless, the San Francisco Congress Youth Committee was very determined to see that youth work continued. Although the Youth Committee was composed of an extensive age range from young adults, college students, to the elderly, the most verbal were young men and women in their thirties. They expressed their regret that somehow their generation had been lost to the Church as active members. The Vietnam War, the cults, mixed marriages, the "latch key child", and drugs had taken their toll on Greek Orthodox youth. They requested the immediate appointment of a National Youth Director and explained that without a strong National Center, the newly formed Dioceses would function at a disadvantage. One complements the other. His Eminence appointed me Youth Director in June, 1983. The purposes of the Youth Ministry Office are to: 1) serve as a resource center, 2) coordinate Diocese youth programs, 3) project the Orthodox Youth Ministry on the national and international scene, and 4) to be in the mainstream of the ecumenical Orthodox movement.

THE OFFICE TODAY

Since June of 1983, the Youth Ministry Office has issued numerous publications and achieved the following: seminars in every Diocese; seminars and dialogue with the students of Holy Cross/Hellenic College; a summer camp at St. Basil; helped organize the SYNDESMOS Movement in America and the plans for the next World Meeting to be held in America; organized the structure of a viable teenage youth activity between the three districts of the Archdiocese of N.Y.; organized a GOYAL Federation of New York; initiated inter-jurisdictional relations with youth offices of sister Orthodox Churches for GOYAL; encouraged the activities of the Eastern Orthodox Scouting Committee (E.O.C.S.), and served as the

333

director of the Second Archdiocesan Young Adult Conference (New York City, June 29 - July 2, 1984).

Also in 1983 at approximately the same time that the new Director of Youth Ministry was appointed, Fr. William Gaines was named Archdiocesan Director of Campus Ministry. A professor and youth specialist at Tulane University and highly skilled in matters of concern to college students, Fr. Gaines is providing leadership and guidance in the campus ministry for the entire Archdiocese. With his monthly publication *Campus Review*, he is in communication with over 15,000 collegiates of our parishes and with other jurisdictions.

Fr. Gaines participates in the Youth Ministry Office Task Force. He, I and Fr. Kerhulas—an expert in teenage and J.O.Y. (Junior Orthodox Youth, 8 - 12) work— have led seminars at Hellenic College/Holy Cross Theological School twice, and also in the Diocese of Boston. The Task Force plans to lead seminars in every Diocese in 1984-86, with the youth directors and youth leaders of each Diocese. It is the intent of the Task Force to hold these seminars at the Diocese Clergy-Laity Assemblies, where most of the leadership of the Diocese is usually gathered.

CONCLUSION AND PROGNOSIS

Although youth work throughout the Archdiocese in the last 25 years was often trying and even frustrating, one detects a continuity of effort with new programs and new ideas. When a new program became successful in one part of the country, such was not the case in another. Oftentimes, it was a parish that assumed leadership and on occasion an Archdiocesan district. It became obvious to the astute that different parts of the country had different youth problems, as did every parish. It has also become evident that the Archdiocese's best function was to serve as a resource center. Also, the parishes and dioceses had to use their ingenuity in providing unique programs for their local youth. It is believed widely that the combination of Diocesan and Archdiocesan efforts will provide meaningful and rewarding youth activities throughout the Church in the years ahead.

Ionian Village: A Unique Success

By
George Poulos

ne of the many bridges, as real as those that link the Florida Keys, is a tiny span in history known as Ionian Village. It links the young of Greek ancestry today with their glorious past, assuring the unbroken lineage of an ethnic heritage whose identity is clearly defined in the modern era in America. The architect of the Ionian Village is the venerable Greek Orthodox Primate, Archbishop Iakovos, but for whom the luster of Hellenism would be considerably tarnished and the bright light of Christianity a shade dimmer.

A song says that in summer the living is easy, but for the young it is a prolonged recess from school when they unleash their energies in street games, playgrounds and in summer camps. Archbishop Iakovos, ever alert to the needs of Orthodoxy and Hellenism, had the ingenuity to conceive a so-called summer camp where the adventuresome nature of youth could be directed and their energies channeled wisely. One can learn to swim from a book but there is no substitute for water. One can read about the glory of Greece in any number of books but there is no substitute for standing on its cherished soil and absorbing all of Greece that the five senses will allow, if only for a short period in the summer. It was thus that the idea of getting the young back to the land of their forbears was born. Through resourcefulness found only in the heart, together with the soul, this idea gradually became a reality. Today, the Archbishop's dream for Greek-American youth is proudly referred to as Ionian Village.

As noble a concept as it was when conceived by Archbishop Iakovos, the prospect for the successful establishment of a summer retreat for young Greek people some 5,000 miles away from America's shores was looked upon with some skepticism in a substantial cross-section of the Greek-American community. No one actively opposed this ambitious undertaking but by the same token there wasn't any groundswell of support for it either. The undaunted Primate, in a reversal of an adage, saw that invention was the mother of necessity, the necessity being a summer place for children in Greece and invention being the means by which to do it. Several partners in this charitable undertaking were drawn in, not the least of which was the Greek government which was as anxious, to bring its progeny back home for at least a visit. Greece stood to benefit by playing host as much as the young guests because of the need for the bond between Greeks and Greek-Americans to remain as strong as their character. But politics being what it is, the government didn't exactly jump with both feet from the outset.

At the insistence of the Archbishop, the program got underway, albeit somewhat feebly, in 1960 with the cooperation and assistance of the Greek Consulate, a smattering of private donors and a group of the major banks of Greece, which made available campsites which were their own private property. The first group of youths to make the journey were pilgrims in every sense of the word for they not only were forging a way for the rest to follow but they did it, by today's comparisons, the hard way. First of all, they had to travel by boat and because they ranged in age from 11 to 15 there was an adequate number of escorts on hand to guide them and see to their needs. These pilgrimages by the young Americans were carried out in scattered bank campsites on an annual basis, increasing in popularity despite the fact that the voyages were interrupted in 1964 and 1965 by the volatile and lamentable conditions in Cyprus following the Turkish uprising and subsequent invasion. By 1967, 313 children and 35 escorts had made the summer event a permanent aspect of the far-reaching undertakings of the Archdiocese, even though it was not in any spectacular fashion. But with the entry of the Greek government, the summer camp program became a highly viable function for those concerned on both sides of the Atlantic.

Convinced by the Archbishop that the youths of summer were as natural to Greece as its own residents, the government of Greece granted the Archdiocese of the Americas in 1966 a tract of land on the Ionian shore, some 3,000 acres in area, for a period of 99 years. On the western edge of the Peloponnesos, the camp is centered in the town of Bartholomio; it has been formed from a neglected tract into a seaside resort that rivals *the* Riviera. Built at a cost of over $850,000.00

it now boasts of a comparatively plush complex of some 30 buildings attractively arranged over about one fourth of the area, the remaining three-fourths flowered, gardened and forested with trees, with an access road worthy of a magnate's estate. Lumber is not the material of this area, as a result of which the structures are typical Greek limestone, with granite and marble everywhere in evidence.

The summer camps of America are usually in remote, tree studded areas, with barracks little more than tents that make for that great outdoors feeling, with pine needles, laden paths and an occasional pond or lake. Not so with Ionian Village. Even though it has an Olympic-size swimming pool, as do some of the most plush hotels on any seaside, it has the sparkling crystal clear waters of the Ionian Sea and a stretch of sandy beach, reminiscent of a millionaire's playground than a tanning site for Greek-American children. The facilities are, in a word, superb. Since it opened in 1970, 6,500 youngsters have come together in a cluster of buildings that include cabins, bungalows, an infirmary, an amphitheater, dining room, five athletic fields, a large air-conditioned lecture hall, and last but not least, a charming Byzantine Chapel where young hearts are filled to the brim. The camp facility has inspired some to call it a veritable paradise. There is nothing quite like it in any other country—a Church-state sponsored haven for descendants of a mother country and mother Church.

Open for the months of June through August each summer, the Ionian Village is no computerized center for young robots, but a carefully outlined and well-planned program in which each individual youngster absorbs Hellenic culture and religious experience which might otherwise elude him. Trained, experienced personnel supervise all of their activities and a dedicated clergy sees to their spiritual fulfillment. Nothing is spared and no corners cut to fill their limited time in the most rewarding experience than can possibly go towards enriching their lives. It is certain that their children will follow in their footsteps on the Ionian shore.

The active but silent partner in these undertakings, without whom the Ionian Village would be less than what it has become is the Byzantine Fellowship whose President and benefactor, Captain Nicholas E. Kulukundis together with its Executive Director, Fr. George Poulos, pooled their efforts out of their Stamford, Connecticut base and sponsored the Ionian Village during the first five years of its inception

(1969-1974). Considerable credit must be given to Olympic Airways not only for its benevolence but for its complete co-operation in every phase of the operation. The Directors appointed by His Eminence have been as follow:

The First Director, Fr. George Papadeas (1970-71); Fr. Nicholas Soteropoulos (1971-73), and George Patsios (1974). For the past 10 years the very capable director has been Fr. Constantine Sitaras (1974-84). Volunteers too numerous to mention have been of incalculable assistance in the camp's development.

A special tribute belongs to the counselors whose volunteer services provide direction for the youth with an expertise that increases with the years. Out of the 500 applications annually only 50 are selected after careful screening by experienced clergy, thus making as certain as possible that the young people will be placed in the direct care of only the most capable and dedicated young adults. Parents are thus assured that the care of their children is entrusted to an elite corps of counselors all of whom are of Greek descent and are eager to guide their youth charges as they are eager to serve God, Greece and America. It is to their everlasting credit that the Ionian Village gains in stature with each passing year—they add a cultural luster. Elaborate buildings, beaches and countryside in themselves are not enough if the visitors are to fully realize the purpose for which they have travelled thousands of miles.

In the years since its inception, the Ionian Village and its lofty purposes have not failed to leave an indelible mark on all concerned. The young adults who have served as counselors have, in many instances, and at their own expense, chosen to remain in Greece for an extra month or more. The net result has been the formation of a kinship among counselors not to be found elsewhere, to the extent that more than 30 couples have been united in marriage and 25 engagements have come out of the association found in the corps of counselors. Not to be outdone, the young who have been to the summer camp have seen fit that their experiences not stop with the departure from Ionian Village. Not wishing to let memories fade, they correspond and, like some high school graduating class, have since 1970 organized reunions throughout the country under the auspices of the Greek Archdiocese. This is absolute testimony to the success of the work of the summer program in uniting the young in a common cause for

Ionian Village, aerial view.

New campers arriving at Ionian Village, the summer of 1971.

Christianity and Hellenism. So long as they remain in Greece these young visitors see at every turn Grecian beauty and Christian love and they are not about to lose those images when they return to America.

There are three divisions of the summer program each of which lasts for 19 days. They comprise the Summer Travel Camp for those 12 to 15 years of age; the Byzantine Venture for those 16 to 18 years of age, and last (but no means least) the Spiritual Odyssey for young adults 19 years of age and up-wards. This revolving chain of young Greek-Americans has been forged to provide an unbroken line of those who are fortunate enough to receive the Christian Orthodox and Hel-lenic messages on the sacred soil of Greece and return to share it with those less fortunate than themselves.

The youngsters spend the first of the 19 days in travel from New York to Greece. Arriving in Athens, they transfer to buses which take them across the rugged Greek countryside, stop-ping off at the Corinth Canal area for refreshments and then on to the quaint town of Bartholomio, site of Ionian Village. The second day is one of acquaintance with the camp, coun-selors, clergy, a period of orientation, but mostly swimming and general recreation. The fourth to the fifteenth day is a period for activities. At that time the resident program is complemented with side trips to ancient Olympia with its sacred games site, the island of Zakynthos, the medieval for-tress of Chlemoutsi, Patras, the heralded City of Delphi with its famed oracle, Arachova, Osios Lucas and ultimately to Athens. The 16th, 17th and 18th days are devoted to tours in which are crammed a visit to the renowned Acropolis, shop-ping at the Monastiraki flea market and a full day cruise to the islands of Aegina, Poros, and Hydra. Before departing Athens for home these leg-weary but very happy youngsters will have sights that tourists save up a lifetime for.

A typical day for the first group at the Ionian Village is as follows:

8:00 a.m. - morning prayers and breakfast

9:00 a.m. - morning participation in aquatics, athletics, arts and crafts, religious discussions (rap sessions) and Greek culture.

11:00 a.m. - a community swim in the Olympic size pool or a dip in the clear Ionian Sea.

1:00 p.m. - lunch and a rest in shade or the sun

3:00 p.m. - a choice of afternoon activities

4:30 p.m. - social projects and a winding down of activities

6:30 p.m. - general cleanup time

7:00 p.m. - prayers

7:15 p.m. - dinner

8:00 p.m. - fun time in too many ways to be counted but which do include Grecian glendi, talent shows, Greek folk dancing, pool parties, music festivals, bonfires on the beach and on the last evening a farewell dinner-dance.

For the Byzantine Venture group the tour is escalated with less time devoted to Ionian Village and more to the extensive travel compressed into 19 days. For the third and final group, the Spiritual Odyssey, this trip is a traveler's *tour de force* wherein there are experiences crammed into 19 days not recommended for the weak of heart but which delight young adults who give off a glow for days after they have finished what best can be called a whirlwind tour. The only sad aspect of this summer project is that not every youngster of Greek extraction can participate, but the glow can rub off on them from those who have taken part. There can be no doubt that the Ionian Village has greatly benefitted the boys and girls who have been privileged to share its cultural and religious offerings. If they have learned nothing else they have learned the Christian message. They have learned to love one another. None who attend can fail to grasp this and that in itself is justification for this magnificent project.

The year 1984 is an historic one for Greek Orthodoxy because this one year marks three milestones; the 60th anniversary of the establishment of the Archdiocese of North and South America, the 50th anniversary of the ordination of Archbishop Iakovos into the sacred priesthood and the 25th anniversary of His Eminence as the Greek Orthodox Primate of the Americas.

While Archbishop Iakovos can look back over a quarter of a century of leadership with a deep sense of satisfaction, one of his achievements which ever looms larger on our horizon, and shall ever remain so, is the institution he envisioned and successfully developed for the perpetuation of Orthodoxy and Hellenism. It is called IONIAN VILLAGE.

Achievement and Transition
at St. Basil Academy

By

Constantine L. Sitaras and Allen Poulos

ne of the oldest, most successful and beautiful institutions of the Archdiocese, St. Basil Academy is a philanthropic and educational institution located on 220 acres in the Catskill Mountains and holding a commanding view of the Hudson River, opposite West Point. Since its founding in 1944, St. Basil has served as home and school for some 4,000 troubled, orphaned or disadvantaged Greek-American children. The Academy has prevented them from getting "lost in the crowd" and has enabled many to grow up as responsible adults. In addition the Academy has trained, in Greek and college studies, more than 300 young women who have become teachers at the parochial schools of the Archdiocese or secretaries and choir directors at many of its 500 communities.

Every one of the three Archbishops involved with the Academy has contributed significantly to it—Athenagoras as founder, Michael as supporter and Iakovos as stimulus for its growth in size and professionalism throughout his 25-year reign.

The institution is located on the main sector of the beautiful estate of the famed brewmaster and New York Yankees owner, Colonel Jacob Ruppert. After his death and because of tax considerations, the estate remained unoccupied and became one of several large tracts considered for an orphanage by Archbishop Athenagoras and the Greek Ladies Philoptochos Society. By appealing to all communities, funds were collected by the Society for the purchase of the estate and operating the children's shelter. It was named after Saint Basil, the 4th century bishop in Asia Minor who was one of the greatest philanthropists of Christianity.

The Academy is a non-profit organization. Income for the operation of the institution is obtained from the following sources: the Philoptochos Society, Greek Orthodox Archdiocese, guardian donations, state aid, and individual donors (who contribute over 50% of the annual operating budget).

In its early years, the Academy could occupy only a half dozen buildings because six years of vacancy had taken its toll on the estate. Operating funds were at a minimum and the Academy existed "by the grace of God." The Archdiocese supported only by the Monodollarion (or one-dollar-dues system) could not afford to support the Academy and the Philoptochos Society assumed responsibility of providing necessary operating funds. At its inception, children of both sexes were accepted but after a few years, the Academy became an all-girls school. Under the auspices of the Greek Orthodox Archdiocese, the Academy was also established as a junior college for women.

During the 1950s, under the leadership of Archbishop Michael, the economic situation of the Academy improved somewhat. Supported by the Decadollarion ($10-dues system), the Archdiocese was better able to assist this institution. The fifties was an era of improvement and renovation of existing buildings. The mansion of the estate was initially used for administrative offices and the women's college. The second and third floor rooms served as sleeping quarters with three or four girls assigned to each room. The first floor and basement were used for classes, study hall, library, dining room and kitchen. The young girls of the orphanage were housed in the former guest house of Jacob Ruppert. The building, which at one time housed members of the New York Yankees and the Ruppert Brewery employees during their vacations or overnight visits, became home for these young charges of the Academy. Other structures were converted for the purposes of the institution. The potting shed became the home of the resident clergyman, the estate manager's home became the schoolhouse and two residences belonging to the stablemaster and groundsman became residences for numerous staff personnel. A sole spring had supplied the estate with its water. It was pumped into a water tower located near the stable and from there distributed as needed. This supply was inadequate for the needs of the Academy and it was decided to tap Indian Brook which lies within the northern boundaries of the grounds. The Philoptochos Society and the Daughters of Penelope supplied funds to make many of these necessary changes.

The year 1962 saw the construction of the first addition to the existing physical plant. In that year, the Ahepa School was opened. Now included in this 1983-renovated building are nine classrooms, a science lab, health room, computer room with two Apple computers and administrative offices. This building was a gift of the American Hellenic Educational Progressive Association, a men's fraternal organization founded in 1922 to deal with the discrimination experienced by the first Greek immigrants.

In 1963 the Pan-Arcadian Gymnasium was built. Attached to the school by a covered walkway, this building also serves as an auditorium with stage. This ambitious building program, inspired by the energetic and progressive Archbishop Iakovos, saw the addition of four new dormitories for girls in 1969. They were built by and named in honor of the Philoptochos Society, the Daughters of Penelope, George Spyropoulos and the late Bishop Germanos Liamandis—who had been the resident clergyman and one time Director of the Academy. In one of these residences, a fully equipped dental office was installed.

For nearly 30 years, the teaching academy was part of St. Basil Academy. However, in 1973 it was merged with Hellenic College in Brookline, Massachusetts. Rather than meet the New York State requirements for accreditation, which would have required prohibitive financial expenditures, it was deemed more logical to unite these two educational institutions. In 1981 the Nicholas J. Sumas Library for Children was dedicated; it serves as a multi-media learning center. Also, three Ahepa-financed dormitories for boys were opened in 1982. The boys had been housed in the renovated stable of the estate. This building, which at one time had also housed facilities of the Education Department of the Archdiocese, is now slated for renovation into a retreat house with accommodations for 80-100 people. Scheduled for the forthcoming year are the completion of the first stage of the Archbishop Iakovos Athletic Field (donated by the Philoptochos Society), the construction of a swimming pool (donated by the Daughters of Penelope, the female counterpart of the Ahepa) and a new Saint Basil chapel (donated by William Chirgotis) to replace the edifice destroyed by fire in 1983.

Over the many years and under the dedicated leadership of numerous directors, the Academy has become one of the best institutions of its kind. The composition of the student population has changed somewhat from the early years. The following is a breakdown of the primary reason for student placement during 1983-84:

344

Orphaned	18%	Abandoned	5%
Divorce	46%	Chronic Illness	9%
Poverty	13%	Other	9%

Approximately 75% of the entire student population is on some type of public assistance (social security, welfare, etc.). The educational program which is accredited by the New York State Board of Education, includes grades 1-8. Included in the curriculum are courses in Orthodox Spirituality, Modern Greek, Byzantine and European Music and Art. Two computers assist in an educational concept emphasizing individualized and, in some cases, remedial instruction. The children come from all walks of life. Although each child is a unique individual, the students share one basic trait—they are there because of a tragedy in their young lives. Being a residential childcare center, the Academy assumes roles normally undertaken by parents or guardians of children. Programs facilitating healthy and wholesome physical, emotional, spiritual and intellectual growth are essential to Academy life. The Academy *is* their home!

In addition to performing its current basic function as a shelter and school, St. Basil Academy carries out an active visitation schedule. During these past almost three years, more than 3,000 people have visited the Academy including the Patriarch Athenagoras Conference and Retreat Center, Camp St. Basil, and the Scout House. The visits include the following:

a. *Group Visits* - There have been hundred of visitors to the Academy—Philoptochos Chapters, youth groups, AHEPA chapters and many others who come for a day visit or a weekend. This program of visits not only gives excellent exposure of the Academy and its work to those visitors but also brings the real world into the Academy life of the children. With exposure to so many people, they have grown accustomed to socializing and feeling comfortable with people, not intimidated and frightful.

b. *Patriarch Athenagoras Retreat and Conference Center* - In an attempt to meet the needs of our Greek Orthodox faithful in a more comprehensive way, the Academy is planning to broaden the scope of its mission by establishing a retreat center designed to accommodate up to 100 people. The retreat center would be available for retreats and conferences and be equipped with a chapel, sleeping facilities, kitchen and dining accommodations, administrative offices and conference rooms. Although the retreat facility is not in full operation, the Academy has hosted many retreats and seminars. This can become an excellent service to the faithful of our Church while producing an income for the Academy.

c. *Camp St. Basil* - For the past three years, Camp St. Basil has offered an exciting camp program for young people ages 9-15. In 1984 there will be three one-week sessions, with major support coming from the Archdiocesan district and the New Jersey Diocese.

d. *Scout House* - The Eastern Orthodox Committee on Scouting has renovated an older building and put it to use for their scout troops. Hundreds of Greek Orthodox Scouts now utilize the Academy grounds for their annual retreat and camporee. They tent and camp out in Eagles' Rest forest and are very well supervised.

This article has given a definition of the Academy and highlighted some of the measures taken to improve the quality of life for the children at St. Basil. Unable to prevent the circumstances bringing troubled children to its doorstep, the Academy and its supporters face the mission of intervening in family crises. It succeeds in salvaging tragic lives and providing for victimized children a loving, healthy, wholesome, Christian environment. Through the help of its many supporters, the Academy is able to be more than a school for the young, more than a church for Sunday services. The Academy is a loving, warm haven for young hearts, a learning center for the fullest development of young minds and a caring, concerned Orthodox institution for inculcating a Christian lifestyle among young souls entrusted to the institution.

Graduation Ceremonies at St. Basil's Academy, Garrison, N.Y.

St. Michael's Home and Other Programs for the Aged

By
Spiro Pandekakes

t. Michael's Home for the Aged was barely one year old when His Eminence Archbishop Iakovos became Primate of the Greek Archdiocese of North and South America. The Home was dedicated in May, 1958 by the late Archbishop Michael. At that time, it was named "Pioneer House", in honor and loving memory of the heroic first generation of Hellenes, who founded our communities and Churches in this country. "Pioneer House" (or as it was known in Greek "To Spiti tou Protoporou") was the fulfillment of the last wishes of Archbishop Michael. At the Clergy-Laity Congress in Montreal, Canada in 1966, Archbishop Iakovos spearheaded a resolution which formally changed the name to "St. Michael's Home for the Aged", in honor of Archbishop Michael. Thus began the determined and dedicated efforts of His Eminence to carry on the legacy of his predecessor and bring hundreds of men and women under the protection and shelter of their Orthodox Church.

His Eminence serves as Honorary President of the Board of Trustees of this important institution and is, in the strict sense of the word, its Steward. Although the Board Members have changed many times over the past 25 years, His Eminence has constantly and diligently maintained a vigil over the lives of the beloved men and women who truly call St. Michael's their home.

For a quarter century Archbishop Iakovos has communed with them, supped with them, prayed *with* them and *for* them— and they have become as one, in mind and spirit! Through such personal involvement he has been able to provide the residents with the best care and services. He has carefully guided a dedicated and concerned Board of Trustees and facility staff, who have been inspired by His Eminence's visionary expression.

Through a close liaison with the President of the Board of Trustees, His Eminence is constantly in touch with the problems facing the institution and, thereby, has been able to act in concert with the Board to implement rapid changes and improvements in the physical environment and quality of life enjoyed by the residents.

During his initial tour of the Home upon assuming his responsibilities as Archbishop, His Eminence noted that the "Taxiarchai" Chapel was situated in the basement. He immediately gave instructions that the Chapel be relocated to the uppermost floor in the facility so that the residents could worship in an environment where the light of the heavens could "touch" them and also provide a more conducive atmosphere for their spiritual devotions. Through regular pastoral visits to the Taxiarchai Chapel, His Eminence has been able to bring together the community of the area in spirited cooperation and support of the continuing programs of the Home.

His visits on Holy Saturday and St. Michael's nameday have been a focus of community unity, bringing unbounded joy and gratification to the residents. It is their welfare and contentment that inspires him continually to seek improvement for the everyday life in the Home. Cognizant that an individual, at this stage of life, needs spiritual enrichment and support, he assigned a fulltime Priest to minister the residents' daily needs. This not only helps the resident community in their devotional pursuits, but emphasizes the importance His Eminence places on everyone's spiritual needs no matter his or her social or economic status or chronological age.

Aware of the general reluctance by Greek-Americans to place aged parents in an institutional setting, His Eminence has utilized the Home's annual family picnic as a vehicle to project the most positive apsects of life there.

Some years ago, he quietly suggested that the event include the participation of all senior citizens throughout the New York Metropolitan area. This fostered a greater awareness on the part of the seniors and their families that St. Michael's truly offered a viable alternative for those whose circumstances warranted admission.

Although the above example may seem somewhat basic, nonetheless, it illustrates vividly the perceptivity and grasp His Eminence has of his subject. Well aware of the increasing demand placed on the existing 40 bed facility in Yonkers, NY, His Eminence and the Board of Trustess some years ago purchased a beautiful 12 acre site in Hartsdale, NY, just a few minutes away from the present facility. Their dream was to

348

The main entrance in to the new structure.

The new St. Michael's Home for the Aged of the Greek Orthodox Archdiocese, Hartsdale, N.Y. (Architectural rendering)

build a new 146-bed modern structure incorporating the most advanced concepts in independent living accommodations. The Home would emphasize supportive services conducive to maintaining our frail elderly as useful citizens of the Greek Orthodox Community.

Coincidently, during His Eminence's 25th year, this dream *will* become a reality. Construction of the new facility has begun and completion is targeted for the summer of 1985. Thus, His Eminence will establish a "National Center" dedicated to the mind and spirit of our aged brethren who did so much to establish our Church in the Americas.

And as he so aptly stated at a recent fund-raising affair, "Circumstances do not always permit us to make a meaningful difference in the quality of life for others, but this is one of those occasions. You can be a pioneer in a different sense, responding to the challenge of creating a National Center for the elderly that will be a showplace. Hopefully, the successful completion of the new Home will inspire other similar facilities throughout the Archdiocese."

Ultimately his goal is to build a structure, not to be merely physically admired, but one set in the foundations of hope, held up by the pillars of understanding, having walls of warmth and love and illuminated by windows reflecting compassion.

In 1982, His Eminence established the Ladies Supervisory Committee to assist the Board of Trustees in their efforts to provide a more comprehensive social and recreational program at the Home. The ladies also participate by acting as hostesses for the many organizations which regularly visit the institution. The committee provides community involvement for the residents and in addition, sponsors charity functions to assist in raising funds for the programs of the Home.

From the first year of his tenure as Archbishop, His Eminence instituted a wonderful tradition that continues to this day. Each year during Christmas week, he visits the residents of St. Michael's and in an intimate gathering with members of the Board of Trustees, they share together the celebration of the birth of Our Lord and Savior, Jesus Christ.

Amid the singing of Christmas hymns, His Eminence distributes a symbolic gift to each resident signifying the high regard and respect which we hold for the elderly men and women who sacrificed and endured untold hardships, so that we might share in "the promise of America."

St. Michael's Home, under the aegis of Archbishop Iakovos,

has evolved into a "Home" in the strict sense of the word, where one can enjoy a life filled with happiness, fulfillment and expectation in an environment firmly rooted in our ethnic, cultural and religious beliefs.

We pray for His Eminence's continued foresight and wisdom to guide us in the years ahead when the challenges will be greater, the problems more complex, but the rewards for our elderly brethren well worth any difficulties which may be encountered.

As he proceeds with his ministry, His Eminence continues to look forward! Increasingly aware of the growing number of elderly throughout the Archdiocese, he knows additional facilities will be needed. He envisions the construction of other Homes in each of the Diocese. This will not be an easy task! But, he knows that as a major Faith, we *must* provide institutions to serve *all* the people, rich and poor, healthy and frail, young and old, near and far.

Paramount is his desire to make the community aware of these needs, knowing full well that the plight of the elderly often becomes a secondary issue in our youth-oriented society. The example set for us by His Eminence is that we truly must follow the Commandment: "Honor thy Father and thy Mother." His example will serve as our beacon to light the way into the 21st Century and beyond—always reminding us of our duty. Indeed the plight of our aged brethren is in reality our own plight. Therefore it must be addressed, now, if there is to be a tomorrow where we Greek-Americans can grow old, gracefully and in dignity. Part of the answer for each of us includes the success of the National Center for the aged and probably the establishment of similar facilities in each major geographical sector of the Archdiocese.

Editor's Note
ESTABLISHMENT OF SCOBA
by Rev. Dr. Miltiades B. Efthimiou

The election of Archbishop Iakovos in 1959 was hailed by the Orthodox faithful in America as a much needed step toward uniting Orthodox Christianity. At his initiative, all the major canonical Orthodox groups in America were brought together in an organization known as SCOBA (Standing Conference of Canonical Orthodox Bishops in the Americas). The first meeting took place at the Greek Orthodox Archdiocese in 1960. SCOBA immediately resolved to solve problems of common concern among its member jurisdictions, such as the: Ecumenical Movement; the chaplaincy of the Armed Forces; canonicity; cooperation among Orthodox jurisdictions; relations of the Orthodox churches with other denominations; campus ministry; Boy Scouts; Christian education; and the attempts to create a local Orthodox Synod in America which would be a prologue towards autocephaly as was demonstrated in a decision of May 9, 1968, to submit the questions of jurisdictional divisions in America to the Inter-Orthodox Committee, at a meeting held in June, 1968, in France. (Unfortunately, the SCOBA request was never put on the agenda for that meeting, while at the Clergy-Laity Congress of that year, in Athens, Greece, Archbishop Iakovos cited the achievements of SCOBA as a body!) He requested, from all the Orthodox Sees worldwide the "sanctioning and blessing the creation of a local Orthodox Synod in America, under the spiritual jurisdiction of the Ecumenical Patriarchate". (From an address by Archbishop Iakovos, in Athens, to the General Assembly of the Clergy-Laity Congress).

Since that time, SCOBA, under the chairmanship of Archbishop Iakovos, through much trial and tribulation, has been seeking ways and means of bringing its objectives to fruition.

Archbishop Iakovos as Ecumenist

By

Robert G. Stephanopoulos, Ph.D.

his is a brief review of Orthodox ecumenism in America from the perspective of 25 years of service of one of the foremost ecumenists of our times. Indeed, the lifetime of Archbishop Iakovos spans the creative years of ecumenical development in the 20th century. The ecumenical and interreligious achievements of the Archbishop must be seen against the general background of the time and place in which we live. We cannot really appreciate Orthodox ecumenism without an understanding of his many lasting contributions. His work in this area is a study in the *praxis* of ecumenism.

In general, most people welcome ecumenical action. They feel through dialogue and cooperative efforts Christians and their churches can undergo significant transformations in order to reach a *rapprochement*. There is a feeling that the "old way" of polemics and massive opposition to each other cannot lead to the universalism promised in the Christian Gospel. Inquiries into the various respective traditions have already indicated that there are many, and as yet unexplored, riches that would emerge for the mutual edification of all Christians. The Orthodoxy of the Gospel would become the norm for all our historic traditions and churches. Through ecumenical involvement, study, collaboration and common efforts a reorientated Christianity could transcend its present divisions.

The ecumenical attitude pervades significant and wide sections of society, particularly the western technological nations. Even though the ultimate goal of ecumenism—true unity and reunion—is a distant possibility to many, numerous successes may be noted. Ecumenism has become an agent of transformation of the Churches' historic reality. In North America the secularization of the public culture is being modified by ecumenism; the struggle of the Church with the pluralistic world has been joined. The task of transforming the secu-

The first S.C.O.B.A. meeting of the Hierarchs of the various Canonical Orthodox Bishops in the Americas; Archbishop Iakovos, Chairman; Greek Orthodox Archdiocese, Fall 1959. (S.C.O.B.A. — Standing Conference of Bishops in the Americas)

Praying in unity — 1969.

At the Vatican ceremonies nullifying the excommunication of 1054 A.D. At the right of Pope Paul VI is Metropolitan Meliton of the Patriarchate; Archbishop Iakovos and Metropolitan Chrysostomos of Austria. At the Pontiff's left is Metropolitan Athenagoras of London and other dignitaries.

lar order into the Kingdom of God, of elevating and humanizing and giving transcendent meaning to our present "ecumenopolis" is a priority on many Christian agendas. Religious people are speaking of the need of our society to identify, reinforce and celebrate the signals of the eternal. The call of God to life is being interpreted and applied in our modern culture.

Thus, ecumenism is a fact; it is undeniably and progressively a part of our public mission as Christians here and now. We Orthodox Christians are participants in it and our Church has been actively involved in the ecumenical movement for many decades now. Our experience as Orthodox Christians in the ecumenical movement and our reflection upon it has made certain facts evident and clear:

1. The ecumenical movement is very profound, having far-reaching proportions and is "an ecumenism of time and space;"

2. Reaches to the ends of this earth and fathoms the depths of our ecclesial life;

3. Challenges our notions and conceptions of self-interpretation and our relationships to one another as individual Christians and as Christian communities;

4. Aims to bring all people to a realization of the truth of the Gospel and to the knowledge of God's will for humankind in unity, witness, service and renewal;

5. Relies on the resources of the Scriptures and the Tradition to comfort and minister to a rapidly changing, urbanized, technological world order.

Orthodox Churches have responded creatively to the lofty ideals of the ecumenical movement, chiefly on the initiative of the Ecumenical Patriarchate of Constantinople. They have committed themselves to a consistent ecumenical policy, forged out of the ecclesiological experience and the theological teaching of the Orthodox faith. Difficulties and challenges come at two points: in discovering the true meaning of God's will for the unity of His Church and all humankind; in discerning how the revelation of the Triune God faces the issues of a secularized world community which does not know Christ. The ecumenical task is particularly relevant and even crucial here in the U.S. (our own "turf", so to speak).

What of Greek Orthodoxy in America? Are we meeting the challenges of our time in our given arena of witness and service? Have we "arrived" in America sufficiently, in order to share our vision of God's salvation with others in the Western

Hemisphere? Have we mobilized our life as a Church around the great commission of our Lord (*"Go ye, therefore, and make disciples of all nations, baptizing them in the name of the Father, and of the Son, and of the Holy Spirit, teaching them to observe all I have commanded you; and lo, I am with you always, unto the close of age."*)? Our purpose as a Church is summarily expressed in those beautiful evangelical Greek words: Kerygma (Proclamation), Didache (Teaching), Leitourgia (Worship), Diakonia (Service), Koinonia (Fellowship).

In his interreligious endeavors Archbishop Iakovos has manifested a spirit of openness and acceptance rare among even the most dedicated ecumenists. In this he was influenced by the catholic vision of the Ecumenical Patriarchate of Constantinople, which has faithfully adhered to the universal and transcendent nature of the Church of Christ throughout the centuries. The mode of "catholicity" has inspired the unifying mission of the Patriarchate, guiding its international relations in the fulfillment of its role as the preserver of unity and harmony within Orthodoxy. The Ecumenical Patriarchate has given vigorous leadership by cementing the bonds between the various Orthodox Churches, while simultaneously reaching out with solicitude and charity to the non-Orthodox Christian communions. An encompassing ecumenical vision inspired its policies, based on the famous Encyclical Letter of 1920. This and other statements form the general theoretical foundation for the active participation of the various autocephalous Orthodox Churches in the ecumenical structures of our time, most notably the World Council of Churches and in the U.S.A. the National Council of Churches.

It was at the renowned theological center of Halki that Demetrios Coucouzis (baptismal name of Archbishop Iakovos) first realized the potential for dynamic universality and ecumenism. It was there that he learned of the visions of such outstanding ecumenical pioneers as Ioakim III, Meletios Metaxakis, Athenagoras I, Germanos Strinopoulos, Gennadios of Heliopolis, and many others. The Patriarchal School by its very constitution was an ecumenical and profoundly Orthodox institution, dedicated to the policies of the Mother Church of Christendom. Archbishop Iakovos' brief early ministry in Constantinople was a helpful prelude to the American experience encompassing nearly all of his years as a clergyman.

In America, Orthodoxy is discovering unprecedented opportunities for mission, and gradually coming to realize its potential, through the efforts of leaders like His Eminence. While serving with distinction as Dean of the Boston Cathedral, the Archbishop learned to love America and its unique way of life. He led in seeking ways to apply Orthodoxy to the American reality and to transform the fabric of society into conformity with the Kingdom of God. He applied himself to important areas of cooperation, explaining the Orthodox faith and participating in a variety of ecumenical activities. Also he conducted an exemplary parish ministry, showing what a priest can do ecumenically in his own local jurisdiction. He learned to love and appreciate the diversity of peoples and their varied answers to the riddles of our common human existence. Here he saw the true potential for unity in diversity and the possibilities of genuine human community.

His years as formal liaison and personal representative of the Ecumenical Patriarchate, at the headquarters of the World Council of Churches in Geneva, sharpened his considerable talents of interpretation and mutual understanding. Those years were crucial to the WCC and to Orthodox witness; a time of maturation and mutual enrichment. His crucial role in the events of that period guaranteed the success of the ecumenical enterprise, particularly in guaranteeing permanent pan-Orthodox membership in the WCC and in resolving the thorny issues of proselytism, Christian witness and religious liberty. It placed him in the forefront of the major proponents of the ecumenical movement. Even after his election as Archbishop of the Americas, he continued to serve as one of the Presidents of the WCC (from 1961 - 1968). In America, no significant ecumenical endeavor was undertaken without prior consultation with Archbishop Iakovos.

The past 25 years since he assumed the office are a model of ecumenical leadership. When it became difficult for him to attend directly to the many ecumenical activities, due to his increasing pastoral responsibilities, he established an Ecumenical Office to organize his Church's involvements. This Office monitors and represents the Archdiocese in numerous endeavors. He continues to serve as its chief officer and representative, ably assisted by a dedicated director and staff. There have been several directors of the Ecumenical Office in the past years, including Bishop Maximos Aghiorgoussis, Fathers George Bacopulos, Leonidas Contos, Robert Stephanopoulos, Nicon

Presidents of the World Council of Churches, elected at the 3rd Assembly, l. to r. — Sir Francis Ibiam, Nigeria; Dr. Martin Niemoller, Germany; Archbishop Iakovos of North and South America; Dr. Ramsey, Canterbury; Dr. David Meses, India; Charles Parlin (Meth.), New York.

His Eminence and Catholic Archbishop Weakland of Milwaukee serving as co-chairmen when Eastern Orthodox/Roman Catholic Consultation convened in Jamaica, N.Y. to discuss 1982 issued Munich Document.

359

Patrinacos, James Couchell and Alexander Doumouras, and Mr. Arthur Dore.

The Office attends to a host of programs, agencies and ecumenical activities, under the Archbishop's careful supervision. *Ecumenical Guidelines* have been produced to provide direction in the pluriform involvements of the Archdiocese and the application of its principles and policies. Contacts with heads of communions, unity concerns, cooperative agencies, mission and evangelism outreach, theological conversations are avenues toward the solution of the problems of unity, renewal and service. The Jewish-Greek Orthodox Christian dialogue is a "first" in our relations with the Jewish tradition. Our commitment to the NCCC is deep and continually improving.

The Archbishop was the first Orthodox prelate to be received in a private audience with a Pope of Rome (John XXIII) in 350 years! Since then, of course, the relations between the two great Churches have been even more significant and are progressing steadily. The Orthodox-Roman Catholic Consultation in the U.S.A. was initiated in 1965, soon after Vatican Council II, and continues into the present. The interest and the involvement of the Archbishop has kept it on a course which provides a model for similar bilaterals, such as those with the Anglican, Lutheran, Reformed, Baptist, and Evangelical Orthodox traditions.

Archbishop Iakovos has taken a deep interest in the theological community, as such. His concern for the Hellenic College/Holy Cross reached a point where he endorsed strongly Orthodox participation in the Boston Theological Institute, so that students from the various theological schools in the Boston area may study together with our own students the various theological disciplines in an ecumenical atmosphere. He has encouraged research, discussion and publication in many theological and ecumenical subjects, encouraging especially the work of the Orthodox Theological Society of America. The Third International Conference of Orthodox Theologians met in Brookline, MA, in 1978 dealing with the topic of the forthcoming Holy and Great Council. This meeting served as a catalyst for theological reflection. He has sponsored a number of our theologians and churchmen to attend international conferences relating to the many issues confronting the Church in the modern world.

A special consultation was held under the auspices of the Archbishop at Brookline in 1980 to explore the question "The

Future of the Ecumenical Movement." The papers were prepared by notable Orthodox and non-Orthodox participants and the final communique was a formal endorsement of the commitment of the Orthodox to the ecumenical movement.

Archbishop Iakovos is an exceptional ecumenical proponent. Wherever he goes, in whatever capacity, the ecumenical dimension is never missing from among his interests. He has never shirked his ecumenical duty. Indeed he never saw it as a duty, but rather as a privilege to be celebrated. The record proves him to be a true and dedicated ecumenist of historic note. He had the opportunity, as St. Paul says, and he has done the good for the welfare of all, especially those in "the household of faith."

With the greatest appreciation for the leadership of Archbishop Iakovos we can affirm our place in America and its promise for world community.

Mission in service and proclamation, to be sure. But what of unity? Did not Christ pray for the unity of His disciples: "that they may be one?" And not unity for its own sake alone, but "that the world may know and believe" (Jn. 17:21-23). We are sent into the world like Christ, sent in His name and guided by the Spirit into all truth, a truth which saves and heals. There is one body and one Spirit, one hope that belongs to our call, one Lord, one faith, one baptism. Therefore, ultimately there is hope for unity.

Orthodoxy teaches unequivocally that the Church of Christ cannot be divided; that it is united in the one Apostolic Faith, in the same sacraments of grace and in the one unbroken Tradition; and that it is fully present in the historic Orthodox communion of churches. The unity of the Church is of paramount concern to the Orthodox Churches and all efforts at reunion, such as dialogue, cooperation and renewal, are regarded as part of the essential witness and testimony of the Church to its oneness under the Gospel and to the preservation of the catholic heritage as a whole.

The principles and policies of Orthodox ecumenism are framed against the background of an unshaken conviction that Orthodoxy, that is, the unity of Orthodox doctrine, church orders, sacraments, worship and spirituality, is essential to the unity of the Church. "The Ecumenical problem for us is the problem of the *disunity* of Christendom and the necessity of the recovery of the biblical patristic synthesis of faith which

361

is constitutive of the one Church."*

Furthermore, our collaboration with the many aspects of the ecumenical movement are conditioned by the Confessional principle and the Ecclesiological principle. Christian unity is expressed in the deposit of the Apostolic Tradition and before there can be reunion among Christians there must be full agreement on the faith universally confessed and celebrated by all. The whole Tradition in its fullness is constitutive of Orthodoxy upon which all churches must be united. It is our conviction that the Orthodox Church is the visible, historic realization of the one church founded by Christ and confessing the fullness of Orthodoxy. From these basic precepts we draw certain inferences and derive policies which have proved useful and proper to our responsible participation in the ecumenical movement.

In fact, then, we cannot and must not 'go it alone.' The so-called "Lund Principle" is as important today as it has ever been: "We must do together everything except what irreconcilable differences or sincere conviction compels us to do separately." Defining the theological and ethical issues at stake in our pluralistic and secularized society; overcoming the scandal of brokenness and presenting a united, reconciling witness of the Christian Gospel to our political life and to our national culture; exercising a responsible stewardship of our limited resources; overcoming our ethnic and class differences; sharing our particular insights in a common treasury of worship, service and dialogue; and coordinating our relief and development programs for the benefit of others—all these challenges make our continued involvement in conciliar ecumenism imperative.

Currently the NCCC is rethinking its role and service to the churches and to the nation. Surely improvements are in order. The vision of a wider conciliar organization which includes the Roman Catholic church and the conservative evangelicals can become feasible at the national level. The concerns of the clergy, of the student Christian movement, of the laity, of the local and regional levels of church life, of the theological community, of the media, of ecumenical scholarship should be included in a total effort to serve the churches in their task of proclamation and salvation. We can all contribute

---

*Guidelines for Orthodox Christians in Ecumenical Relations, SCOBA (1972), p.3.

ecumenically to a theology of culture and politics which really serves and enhances our national life rather than condemning it.

In the final analysis, ecumenism is concerned with the whole human family, not according to secular and even humanistic criteria, but according to the New Heaven and the New Earth revealed by Jesus Christ—the true life of the world. The problems which confront us today are of global significance and consequence: the threat of nuclear destruction; the danger of a collapse of the world economy; the resurgence of totalitarian ideologies, both secular and religious; the impact of relentless poverty and overpopulation; the depletion of natural resources; the destructive threat of technology; the fear of multiple wars, and the limitation of human potential. These terrors are felt poignantly in our communities and elsewhere. The U.S. has a leading role and a major responsibility to act in this global setting for a positive and decent world order. Our duty as Christians and as citizens demands that we be informed about such matters and that, in turn, we inform our fellow citizens and our government about the hope that is in us. A mobilization of our political will (enlightened by the precepts of the Gospel) is a necessary ingredient of our Christian activity. Surely we have something to say; surely we have something to learn; surely we have ample room for cooperation and joint action with all people of good will!

We need not make apologies for our involvement in the ecumenism of this age. Our record is always subject to analysis and review, but it is a record of significant achievements in cooperative action and genuine dialogue. Our Greek Orthodox Church in the United States is a regional church, but it is *the Church* as it exists here and now, in a definite and special culture. As a lived reality it must be concerned for the oneness of Christ's Church and the renewal of the universal church. Its reality is certainly greater than the expressions in ideas, words and symbols which we employ to break the silence. To the degree that we seek ways of expressing the truth and the reality of the Church and of applying that truth to the brokenness of the world; to the degree that we continually seek ways of recognizing in the other church (not so much the reflection of ourselves, but of the one, True Church!) to that degree we remain faithful to the call of our Lord "that all may be one." We sin against Christ and His Church if we build or retain dividing walls of hostility and indifference.

363

St. Paul's words (Rm. 15:7) now, as then, remind us of the necessity of extending ourselves to create a universal Christian fellowship, "Welcome one another, therefore, as Christ welcomed you, for the glory of God."

His Eminence meeting with his fellow Bishops at an Archdiocesan Holy Synod meeting.

The Orthodox Observer

By
John Douglas

 nother major communications medium that emanates from the Archdiocese is the bi-monthly bilingual newspaper, *The Orthodox Observer*, which traces its origins to 1934 when it was established as a theological magazine by the then Archbishop Athenagoras. This format was continued until the late 1960s when Archbishop Iakovos and his advisers determined a more direct method of communication was needed—one that would communicate widely and rapidly the Church's message and news developments to all parishioners. The Archdiocesan Council met in Boston on June 4, 1971, and decided that the magazine be converted to a bi-monthly newspaper of the same name and that it be sent to every parishioner in good standing. The Council also determined that a publishing board be appointed by His Eminence and that a professional staff be hired. Also, a new unit was created and incorporated as the Greek Orthodox Archdiocese Press. On October 20, 1971, Archbishop Iakovos presented the first issue of the newspaper to journalists attending a press conference called to mark the commencement of the *Observer* in tabloid format. To show it was a continuation of the publication started in 1934, the first edition carried on its masthead, *Vol. 37. no. 619.*

During the years that have followed, the newspaper has become part of the life of the Greek Orthodox faithful in America. The paper is now in its thirteenth year and has a circulation of over 150,000, more than forty times that of its predecessor.

Over the years, the *Orthodox Observer* has published editorials and commentaries which have attracted the attention of lay and religious leaders of other churches and faiths. Religious publications all over the world have printed excerpts from articles in the *Observer* or commented on them.

The goal of the paper is to report significant news of interest to the Greek Orthodox Community in America and to articulate the Greek Orthodox position and morality in current events. Because it is bilingual, it serves all parishioners.

The *Orthodox Observer* helps bridge the information gap between the headquarters of our Church (the Archdiocese) and the parishes as well. The publication helps create a feeling of identity among parishioners and keeps them informed about the activities of the hierarchy and other parishes.

Obviously, the *Observer* as it is presently structured has limited sources of income. Advertising revenues total a little more than $100,000 annually. At the same time, the revenue from individual subscriptions is minimal since the publication's aim is not to sell subscriptions but to encourage the faithful to become registered members of their parishes. Thus, the main source of income for the paper is the Archdiocese which, by paying the subscription for each member of the parish and by providing space at the Archdiocesan headquarters, subsidizes its publication.

A few months after the introduction of the new *Orthodox Observer* as a bi-monthly newspaper on July 6, 1972, its founder, Patriarch Athenagoras died. The paper launched a fund raising campaign for the erection of a statue of the Patriarch. Thousands of Greek Orthodox faithful and friends of the Patriarch responded and a statue of the former Patriarch was placed on the campus of Hellenic College/Holy Cross School of Theology in Brookline, MA. Also a bust of Athenagoras was placed in St. Basil Academy and in the lobby of the Archdiocese.

The extent of the influence of the paper was indicated after the Turkish invasion of Cyprus. Kept informed by the *Observer*, Greek Orthodox faithful in North America united and gathered large sums of money and quantities of medicine, foodstuffs, clothing, and blankets and sent them to our Cypriot brothers and sisters.

Its first Editor-in-Chief and Publisher in its new format was Takis Gazouleas, who continues in that role today. Mr. Gazouleas is a veteran observer of the political arenas of both Greece and the United States and brings this unique insight to the *Observer*. He was a journalist for several Greek newspapers before joining the *Observer*.

The *Orthodox Observer* is a newspaper for all Greek Orthodox faithful and continues to play a dominant role in enlightening and informing our populace.

Winning Major Faith Status

Incorporation of the Archdiocese and Subsequent Governmental Recognition of Greek Orthodoxy as a Major Religion

By
Nicholas Kladopoulos

he first Greek Orthodox parish was established in New Orleans, Louisiana in 1865 by a few Greek sailors and cotton merchants. According to reference material in the Greek Archdiocesan Archives, there were 120 parishes in the Americas between 1865 and 1920. These communities were established in several States to meet the religious, educational, ethnic, historical and social needs of their parishioners who had emigrated from Greece or Asia Minor. However, probably the first major historical event for the Greek Orthodox in America took place in 1921—the legal founding of the Greek Archdiocese.

In 1918, the bishop of Athens and all of Greece, Meletios Metaxakis, came to America and stayed 82 days. On October 4, 1918, he visited Woodrow Wilson. In New York on October 20, 1918, Metaxakis established the Synodical Trusteeship and appointed Bishop Alexandros of Rodostolon to administer the parishes. On October 29, 1918, the Archbishop of Greece left America and returned three years later. In February 1921, he sent an encyclical to parish priests, the Board of Trustees, and all the faithful, in which he explained certain ecclesiastical events in Greece, such as his election as Archbishop of Athens and how he had been obliged to relinquish the throne of the Archdiocese of Greece, without his canonical resignation.

The Synodical Trustee, Bishop Rodostolon, sent an encyclical to the priests and parishes on February 26, 1921 in which he explained to the Greek Orthodox faithful in the U.S. that the head of the Church in America has all canonical privileges because his position was established. by the Canonical Synod and was under the jurisdiction of the Ecumenical Patriarchate and not of the Church of Greece.[1] The Ecumenical Patriarchate has this right according to the 28th Canon of the Fourth Ecumenical Synod.[2] Bishop Alexandros of Rodostolon indicated clearly in the said encyclical that the Church of Greece had improperly administered its responsibility to the parishes in the United States.[3] He knew very well differences probably would appear among the Greeks in America and tried to unify them.[4]

In the meantime Archbishop Theocletos, who was elected in place of Metaxakis, sent a telegram ordering Bishop Rodostolon to return to Athens and to appear before the Holy Synod. However, the Bishop of Rodostolon did not obey this order, but continued his ecclesiastical functions in America under the protection of the Archbishop of Athens Metaxakis, who was then living in America. In response to the tension, Metaxakis called a conference of all the representatives of the parishes and clergy. This was the first Clergy-Laity Congress, held September 13-15, 1921, in the Holy Trinity Cathedral of New York City. The first subject on the agenda was the legal organization of the Church.

The Synodical representative Bishop Alexandros of Rodostolon made the motion that "the Greek Orthodox Church in America be organized as a corporation, according to the laws of the State of New York and in accordance with the Holy Canons of the Church."[5] Rev. Father Christos Angelopoulos seconded the motion and its was passed unanimously. After that, the president of the Congress asked about the name of the organization. Rev. Father Methodios Kourkoulis made the following motion: "I recommend that the corporation be named—Greek Orthodox Archdiocese of North and South America." This motion also passed unanimously.[6]

Thus on September 15, 1921, Meletios Metaxakis incorporated the Greek Orthodox Archdiocese of North and South America, by virtue of Chapter 15, Article 2 of New York State under the Ecclesiastical Organization Law.[7] The official 1921 charter of the Greek Orthodox Archdiocese of North and South America is as follows:

CERTIFICATE OF INCORPORATION
OF GREEK ARCHDIOCESE
OF NORTH AND SOUTH AMERICA

WE, the undersigned, MELETIOS METAXAKIS, Archbishop of Athens, residing at the Hotel Majestic in the Borough of Manhattan in the City of New York, and GERMANOS POLY-ZOIDES, Archdeacon, residing at 339 East 88th Street in the Borough of Manhattan in the City of New York, being respectively the presiding officer, namely, the president, and the clerk of Greek Archdiocese of North and South America and members thereof, for the purpose of procuring the incorporation of the same pursuant to the provisions of Section 15 of the Religious Corporation Law of the State of New York, execute and acknowledge this certificate as follows:

1st—The said Greek Archdiocese of North and South America is an unincorporated governing and advisory body of the Greek Orthodox Church having jurisdiction over and relations with several Greek Orthodox Churches, some of which are located in the State of New York.

2nd—At a meeting of the said body, at which a quorum was present, duly held in the Holy Trinity Hellenic Orthodox Church located at 153 East 72nd Street in the Borough of Manhattan in the City of New York on the 15th day of September, 1921, it, by a resolution duly adopted by it, determined to become incorporated under the laws of the State of New York under the name, GREEK ARCHDIOCESE OF NORTH AND SOUTH AMERICA.

3rd—At the said meeting the said body, by a plurality vote of its members, also elected nine persons, whose names and post office addresses are hereinafter stated, to be the first trustees of such Corporation.

4th—The name by which such Corporation is to be known is:

GREEK ARCHDIOCESE
OF NORTH AND SOUTH AMERICA
Incorporated

5th—The objects for which such Corporation is to be formed are:
To edify the religious and moral life of the Greek Orthodox Christians in North and South America on the basis of the Holy Scriptures, the rules and canons of the Holy Apostles and of the seven Oecumenical Councils of the ancient undivided Church as they are or shall be actually interpreted by the Great Church of Christ in Constantinople and to exercise governing authority over and to maintain advisory relations with Greek Orthodox Churches throughout North and South America and to maintain spiritual and advisory relations with synods and other governing authorities of the said Church located elsewhere.

6th—The number of trustees of such Corporation shall be nine.

7th—The names and post office addresses of the first trustees of such Corporation, as elected by the said body, as aforesaid, are as follows:

369

ALEXANDER, Bishop of Rodostolon, 140 East 72nd Street, New York City.

Rev. METHODIOS KOURKOULIS, 1030 Beverly Road, Brooklyn, N. Y.

Rev. DEMETRIOS KALLIMACHOS, 64 Schermerhorn Street, Brooklyn, N. Y.

Rev. STEPHANOS MAKARONIS, 359 West 24th Street, New York City.

Rev. GERMANOS POLYZOIDES, 339 East 88th Street, New York City.

LEONIDAS CALVOKORESSI, 11 William Street, New York City.

PANAGIOTIS PANTEAS, 3904—3rd Avenue, Brooklyn, N. Y.

GEORGE KONTOMANOLIS, 100 West 38th Street, New York City

ALEXANDER ALEXION, 152 East 22nd Street, New York City.

8th—The principal office of such Corporation is to be located in New York County in the City of New York in the State of New York; but branch offices thereof may be maintained and its functions may be exercised in any part of North and South America.

IN testimony whereof, we have hereto set our hands and affixed our seals in the City of New York on this 17th day of September, 1921.

MELETIOS METAXAKIS, Archbishop of Athens
President
GERMANOS POLYZOIDES, Archdeacon
Clerk

STATE OF NEW YORK
NEW YORK COUNTY:

On this 17th day of September, 1921, personally appeared before me MELETIOS METAXAKIS, ARCHBISHOP OF ATHENS, and GERMANOS POLYZOIDES, ARCHDEACON, each to me known and known to me to be the persons described in and who executed the foregoing instrument, and severally acknowledged to me that they executed the same.

DEMETRIOS E. VALAKOS
Notary Public for New York County, No. 15
My commission expires March 30th, 1923.

(Certification by William F. Schneider, Clerk of the County of New York and Clerk of the Supreme Court of the said State for the said County, dated September 19, 1921, at 3:43 p.m.)

COUNTY CLERK DOC. FILE T650

Meletios Metaxakis, while in the U.S., was elected Ecumenical Patriarch on November 25, 1921. He took the name Meletios IV. One of his concerns was the *annulment* of the Patriarchal Tomos promulgated by Patriarch Joakim III, March 8, 1908. Through this Tomos, the Orthodox Churches in the Diaspora (Europe, America and Australia) came under the temporary spiritual jurisdiction of the Holy Synod of Greece.

The Ecumenical Patriarchate from 1908 until 1922 studied the situation of the Churches of the Diaspora and concluded it would be best to put them under the jurisdiction of the Great Church of Christ.[8] The reversal (annulment) of the 1908 Patriarchal Tomos was decided and approved by the Holy Synod of the Patriarchate in March 1922 and was the subject of a Patriarchal and Synodical encyclical.[9] The Patriarchal and Synodical Tomos of May 17, 1922, elevated the Greek Orthodox Church in the Americas to the grade of Archdiocese. The Patriarchal encyclical appointed Rodostolon Alexandros as first Archbishop and announced this decision to the Greek people.

At the same time, three dioceses were established in Chicago, Boston and San Francisco. Unfortunately, the period 1923-1930 was the worst period of strife for the Greek Church in America. Political frictions brought over from Greece were dividing the congregations. In 1929, the newly elected Ecumenical Patriarch Photios II had, as his first concern, assistance for the Greek Orthodox Archdiocese of North and South America. In May, 1930, he sent to America his representative Metropolitan Damaskinos of Corinth. Damaskinos made strong efforts for the unity of Greeks in America. As a result, the Ecumenical Patriarchate on November 14, 1930 announced the election of Metropolitan Athenagoras of Corfu as Archbishop of North and South America. This event truly enthused the Greek people of America. After paying his respects to Patriarch Photios and the Holy Synod, Athenagoras arrived in New York City on February 24, 1931, and was enthroned there two days later in the Church of St. Eleftherios.

For years after, the Greek Archdiocese sought official recognition of Orthodoxy as one of the Major Faiths in the Federal and State governments. On March 25, 1943, under the State Law, the Governor of New York, Thomas Dewey, before the two legislative bodies in Albany, N.Y., recognized the Confederation of the Orthodox Churches of America. Ecumenical Patriarch Benjamin I congratulated Governor Dewey for this recognition of the Orthodox faith. This was an historic event for the Greek Orthodox Archdiocese of North and South America because it empowered it to invite the other Orthodox Churches to a conference such as: Antiochian, Syrian, Russian, Ukranian and Carpathorussian.[10]

Archbishop Athenagoras reconstructed the communities in the Americas and reconciliated the faithful by various means, including the regular holding of Clergy-Laity Congresses.

On November 1, 1948 Athenagoras was elected Ecumenical Patriarch and was enthroned in the Phanar January 27, 1949. After nine months, the Holy Synod of the Ecumenical Patriarchate elected as Archbishop of North and South America the Most Rev. Metropolitan of Corinth Michael. He continued the search for official recognition. The following listing shows the recognition of the Greek Orthodox Church in the Americas by numerous States and by the Federal government during the epoch of Archbishop Michael:

The State of Wisconsin in June, 1953, recognized the Eastern Orthodox Church as a Major Faith with a special bill.[11]

The Governor of Massachusetts on July 2, 1943, signed Official Bill 592, in which the Orthodox Church was recognized as an Official Religious Body in the State of Massachusetts.

The State of Louisiana, through Assembly Bill 226, recognized the Greek Archdiocese on July 1, 1954.

The U.S. Government recognized the Greek Orthodox religion as a Major Faith on April 29, 1955.

The State of Texas on March 9, 1955, recognized the Eastern Orthodox Church.

On May 6, 1955, the State of Connecticut with Assembly Bill 89 declared the Greek Orthodox Church a Major Faith.

The State of Indiana on May 8, 1955, recognized the Greek Orthodox Church in America as a Major Faith.

The State of Illinois recognized the Orthodox Church on June 15, 1955 and with Bill 864.

The State of New Hampshire in 1955 recognized the Eastern Orthodox Church with Bill 318.

The State of Missouri on May 23, 1955, recognized the Eastern Orthodox Church as a Major Faith with Bill 161.

The State of New Jersey recognized the Eastern Orthodox Church as a Major Faith on May 23, 1955, with the Assembly Concurrent Resolution 29.

The State of Rhode Island recognized the Eastern Orthodox Church as a Major Faith, January, 1956, with Bill S-148.

The State of South Carolina recognized the Greek Orthodox Church as a Major Faith on March 15, 1956.

The State of North Dakota on January 8, 1957 recognized the Greek Orthodox Church as a Major Faith with House Resolution C-1.

The State of Minnesota, with Concurrent Resolution 346, recognized the Eastern Orthodox Church as a Major Faith on February 11, 1957.

The State of Maryland with Resolution 11, recognized the Eastern Orthodox Church as a Major Faith on February 15, 1957.

The Senate Concurrent of California recognized the Greek Orthodox Church as a Major Faith, April, 1957.

The State of Florida recognized the Greek Orthodox Church as a Major Faith on April 29, 1957, the House Concurrent and with Bill 211.

The State of Michigan, with Bill 560, recognized the Eastern Orthodox Church as a Major Faith on June 18, 1957.

The State of Arkansas, with Bill 18, of the House Concurrent Resolution recognized the Eastern Orthodox Church as a Major Faith on February 6, 1959.

STATE RECOGNITION OF ORTHODOXY AFTER THE ENTHRONEMENT OF ARCHBISHOP IAKOVOS

It is generally understood that the recognition of the Greek Orthodox Church by many States as a Fourth Major Faith besides the Protestant and Roman Catholic denominations and the Jewish religion culminated with the efforts of His Eminence Archbishop Iakovos, who was enthroned on April 1, 1959.

In January, 1965, the press office of the Archdiocese announced the following States had recognized Greek Orthodoxy as a Major Faith:

Alaska	Massachusetts	Ohio
Arkansas	Michigan	Pennsylvania
California	Minnesota	Rhode Island
Connecticut	Mississippi	South Carolina
Delaware	Missouri	South Dakota
Florida	Nevada	Tennessee
Georgia	New Jersey	Texas
Kansas	New Mexico	Washington
Louisiana	North Carolina	West Virginia
Maryland	North Dakota	Wisconsin

Archbishop Iakovos during his 25 years of pastoral tenure visited and blessed almost every parish of the Archdiocese. He has known Presidents of the U.S. and many State political leaders. As a result of these relations, the Greek Orthodox Church has been recognized by those States which had not yet recognized our Church. For example:

The Governor of the State of Kansas signed a personal proclamation which officially recognized our Church on March 8, 1960.

The State of Tennessee recognized the Eastern Orthodox Church as a Major Faith with Bill 75.

The Commonwealth of Massachusetts recognized the Eastern Orthodox Church as one of the Major Faiths on June 19, 1962.

The State of West Virginia recognized the Eastern Orthodox Church as a Major Faith on February 6, 1963, with Resolution 13.

The State of South Dakota recognized the Eastern Orthodox Church as a Major Faith on March 4, 1963.

The State of New Mexico with Bill 2, recognized the Eastern Orthodox Church as a Major Faith in 1964.

The State of Mississippi recognized the Eastern Orthodox Church as a Major Faith in 1964.

The State of Nebraska, with Bill 19, recognized the Orthodox Church as a Major Faith on February 17, 1965.

The State of Idaho recognized the Orthodox Church as a Major Faith on February 27, 1965.

The State of Nevada recognized the Orthodox Church as a Major Faith in 1966.

The State of Kentucky House of Representatives with Bill 6, in 1966, recognized the Eastern Orthodox Church as a Major Faith and equal to the Protestant and Roman Catholic denominations of the Jewish religion.

The State of Iowa, with a Senate Concurrent Resolution, added the Eastern Orthodox Church to the list of Major Faiths with Protestant, Catholic denominations and the Jewish Religion on May 17, 1967.

After 13 years of effort for official recognition of our Church, on June 18, 1979, it was announced by the Department of Army, that the Orthodox Church was recognized equally with the Protestant and Roman Catholic denominations and the Jewish religion. The following U.S. Army document speaks for itself:

"DENOMINATIONAL MINISTRY

Religions and moral coverage is provided for distinctive faith groups, such as Protestant, Catholic, Jewish and Orthodox. Specific services within the Protestant Churches may be held, but only after general Protestant services have been provided."

Following is the text of one of the States (New York) typifying the legislative action required to gain state-by-state recognition of Eastern Orthodoxy as a Major Faith.

STATE OF NEW YORK

The Legislature

ALBANY

LEGISLATIVE RESOLUTION
Assembly No. 406

IN ASSEMBLY
By Ms. Newburger

Legislative resolution honoring and commending the Eastern Orthodox Church for their contributions to the quality of life in this great State and Nation and to give recognition to this religious body as a major faith in the State of New York.

Whereas, The Eastern Orthodox Church has enhanced the quality of life of people of this great State and Nation through its dedication and service to the community; and

Whereas, The Eastern Orthodox Church is one of the major faiths in this State; and

Whereas, The Eastern Orthodox Church is entitled to and should receive general public recognition as one of the major confessions in this State; therefore, be it

Resolved, That this Legislative Body pauses in its deliberations to express recognition of the Eastern Orthodox Church as one of our major faiths; and be it further

Resolved, That a copy of this resolution, suitably engrossed, be transmitted to Archbishop Iakovos, 10 East 79th Street, New York, New York, 10021.

By order of the Assembly,
Catherine A. Carey, Clerk

ADOPTED IN ASSEMBLY ON
April 14, 1980

This official recognition of the Greek Orthodox Church as a Major Faith by the President of the United States, the House of Representatives, the Senate, the Army, State governments, by other Christian denominations and other religious groups has been a sixty-year successful campaign by the Greek Archdiocese.

FOOTNOTES

[1]Encyclical of the Holy Archdiocese of America, February 26, 1921. From the Archives of the Greek Orthodox Archdiocese of North and South America.

[2]Stephanides, Vasilios, *Church History* (Athens, 1948), p. 258.

[3]Encyclical, February 26, 1921.

[4]Zoustis, Basilios, *The Greeks in America and Their Activity,* History of the Greek Orthodox Archdiocese of North & South America (New York, 1954), p. 131.

[5]Zoustis, p. 131.

[6]Zoustis, p. 132.

[7]Zoustis, p. 133.

[8]Stephanides, p. 639.

[9]Patriarchal Tomos, March 1, 1922. From the Archives of the Greek Orthodox Archdiocese.

[10]Efthimiou, Vasilios, *History of the Immigrant Greek in America* (New York: Cosmos G/A Printing Co., 1949).

[11]All the resolutions and recognitions of the States have been taken from the Archives of the Greek Archdiocese.

The Future of the Greek Archdiocese

By
George J. Charles

rchbishop Iakovos can look back with satisfaction on 25 years of continued growth in the Church he has served so well. However, as it is with all leaders, and His Eminence is no exception, he is concerned about the future growth and stability of the Church in the Americas.

No one can foretell the future of the Archdiocese in the Americas. Nevertheless, there is no reason to doubt that the progress our Church has made since 1864 will continue in the years to come. To forecast the future, let us glance a bit into the past and analyze the present.

Lacking the organized structure of the Church in Greece, leaders of our early immigrant communities were responsible for the evolutionary development of our parish structures. Priests were brought from Greece and functioned generally under lay authority. Newly organized parishes competed with one another to obtain priests, to the obvious disadvantage of small, less financially-able parishes.

It was difficult, in the beginning, for a proper balance to be achieved between the priest, who is responsible for the life of the religious community he serves, and the lay leaders who often sought exclusively lay authority in the administration of the parish and its religious and language schools. It was not until the Archdiocese was established that some semblance of order was developed in the assignment of priests and an effort was made to assert the proper authority of the priest as the leader of the flock.

Many parishes still do not accept the leadership of the priest, in all aspects of parish life. Partly to blame are some of the priests who are unwilling to accept the responsibility that goes with leadership. Partly to blame are the leaders of some parishes who fail to understand that the leadership of the parish must be exercised by the priest if he is to function in the true, spiritual, canonical and historical framework of our Church.

The future of our parishes, the lifeblood of the Archdiocese, lies with those parishes which accept the concept of shared responsibility: That is, that the priest as the leader of the parish, is responsible for the spiritual well-being of his flock, while the laity provides the financial and administrative *support* that enables the broad needs of the parish to be served.

The successful parish, that is, one where the spiritual needs of the people are met, and where the fullest range of community services is provided, is one where the clergy and the laity understand their respective roles.

Sometimes the laity, in its anxiety to serve and to improve the administration of the parish, tends to equate the Church with a business and often, perhaps unconsciously, downgrades the Church's spiritual mission in the quest for more efficiency and better management.

"Shared responsibility" on the part of the parish priest means acceptance, in the fullest sense, of the responsibility of leadership in every area of parish life, not just those liturgical or philosophic areas which may appeal to his own religious, aesthetic or sociological values. He must lead, as Christian conscience dictates, even in unpopular or distasteful areas.

The future of the Church depends on the fullest acceptance by clergy and laity together of the concept of shared responsibility. The challenges of today's society do not permit our Church the luxury of the kind of "leadership" quarrels that have embittered some of our parishes. First of all, such conflicts have no place in our churches and in a Christian fellowship because they mock God and His Commandments. Second, they are destructive and self-demeaning. Third, they cause our people to lose faith in the *Ecclesia*, the gathering of Christian men and women to worship God together with love and mutual understanding. Equally regrettable, these conflicts deny to the parishes the services of talented and able clergy and lay persons who want no part of such squabbles.

More than anything else, it has been the wise and patient leadership of our Archbishop that has brought peace to most of these regrettable situations. The institutions of the Archdiocese, and the Diocesan and Archdiocesan Councils, have all benefitted from the Archbishop's determination to bring to the service of the Church the highest possible caliber of persons, men and women alike.

These dedicated and selfless Greek Orthodox Christians give much of themselves and their means generously in serving the

Church. They give of themselves without expectation of recognition. While those who serve their parishes are generally well known in the community, the average parishioner would be hard pressed to name more than a half-dozen of the laymen who contribute so much in the service of our Church on the national level. It is upon all of the servants of the Church, clergy and laity together, whether on the parish, Diocese or Archdiocesan level, that the future depends. If the Church continues to grow and prosper, it is because of their present and continued efforts.

The Archbishop has labored mightily to make the Clergy-Laity Congresses truly representative of the whole Church so that the needs of our Church, present and future, may be better served. There are those who view the Clergy-Laity Congress as an institution that resolutely supports pre-determined Archdiocesan policy. This is partly true, because it is the responsibility of the Archbishop to report on the state of the Church to the Clergy-Laity Congresses and to request such support as is necessary based on need and past experience.

For many Greek Orthodox Christians, moving their sights beyond their parishes constitutes the most difficult part of sustaining a strong and united Archdiocese. Heartening progress has been made during the past 25 years in strengthening the ties between our over 550 parishes and the Archdiocese. It is in the united strength of our parishes across the Hemisphere that the strength and future of Greek Orthodoxy lies, a future that requires priests and laymen alike to accept total responsibility to one another, to the Archdiocese and to their Faith.

There is no such thing as parish autonomy. We are all part of *one* Church. The needs of the parish are as one with the Archdiocese. Neither can survive without the other. The most outrageous exercise in which some parishioners indulge is that the Church is some kind of anarchistic society in which they can demonstrate parish individualism and independence. Irrespective of how such flights of fancy take form, the ultimate absurdity is the same. Moreover, the absurdity becomes dangerous and reckless when thoughtless and irresponsible expressions, or actions, challenge the unity and authority of the Mother Church.

The greatness of our Church in the years ahead requires that we strengthen our parishes and improve community services. A growing, well-coordinated parish within the fabric of the Archdiocese helps to strengthen and stabilize the whole of the Church. Most priests do not allow time following the celebration of the Divine Liturgy to comment on matters of concern to the Archdiocese. Such discourses would develop on the part of our parishioners an appreciation that the Archdiocese is more than just an hierarchal institution. Our parishioners need to be informed about the vital and dramatic movement ahead of the Archdiocese in recent years so they may share with pride in its growth, its responsibilities, its areas of service, and its contributions to religious understanding and Christian unity.

Attacks against the Archdiocese and its leadership are heard from sources which have not particularly distinguished themselves in the service of the Church. Many of these attacks are selfishly and mischieviously motivated. The dangers of such attacks is that answering them exaggerates their importance, while not answering them gives them credence in the minds of the underinformed. They must be answered, but in doing so we must differentiate between criticisms which are mindless and those having merit and which mirror our errors.

High on the list of our priorities for the future is the essentiality of meeting our environmental challenges by intensifying the spiritual life of our parishes. There exists among some of our communicants the feeling that our Church must accommodate itself to the sociological needs that beset us, by changing the Church to meet these needs. The Church is not a political party, or an ordinary organization, rushing to change its beliefs, or its rules to accommodate the winds of social change. While the Church by its very nature is intended to be the "yeast" that changes society, that same yeast will make those changes in the rules and disciplines of the Church that satisfy the needs of humanity without conflicting with Christian doctrine and faith. The Fathers of our Church are slow to make changes. This is as it must be.

We cannot be eclectic, and take or leave Jesus Christ on our own terms just because we have problems in our society, or in our families. *We can adjust the institutions—the organizations—or the agencies that serve the Church*—and in doing so we can help to deal with the problems of our society—the needs of our youth, the challenge of caring for our aged and the entire spectrum of human rights. But we must strive constantly to bring the spiritual essence of the Church closer to our people. It is true that our Church must be relevant, but relevant to what? Relevant to the changing habits of society, or relevant to its Christian-Orthodox belief and tradition! The former constantly changes, the latter is essentially unchanging, except when tradition fails to provide for the needs of Christian humanism. *It is in this context that Orthodox Christianity will face its greatest challenge in the years to come*—the challenge of resisting those forces which would have Orthodox Christianity accommodate itself easily to changing conditions and circumstances, rather than for all of us to change ourselves so that we will be better able to live our lives according to Christian teaching.

Each generation, clergy and laity together, has its own destiny to fulfill and its own contributions to make in the perpetuation of our Faith. The sacrifices and contribution of the past are wasted unless each new generation accepts its own responsibilities and shoulders its own burdens. We cannot rest on past accomplishments because the work of the Church for the Glory of God is never done. One of the most critical priorities in the Archdiocese today, and for the future, is the *encouragement of young men to enter the clergy, and for young women to enter into the service of the Church.* As the Church in the Americas is a unique institution with its own history and developing character and tradition, it is necessary that our clergy be drawn to the greatest extent possible from our own midst. Otherwise, we run the risk of confusing new generations who may be unable to identify with priests brought from abroad who find our ways and the problems of our society difficult to comprehend.

It was one thing for the pioneers of our Church in the Americas to bring over priests from Greece. They spoke the same tongue, possessed the same background and understood the traditions and culture of "the old country." A priest brought from Greece today speaks a language few of our second and third generations speak and represents a totally different cultural background. There is a gap that religion alone finds difficult to bridge. In so many of the current problems of community life, counselling by the priest is an essential and important part of his spiritual function. Such counselling requires educational background and cultural understanding. More often then not, this is lacking in a priest not born, reared or educated in the United States. Without such a background, community services by the priest becomes a hit and miss proposition.

381

How about those members of our Church who are recently arrived from Greece? Perhaps the majority adapt rather quickly to the new American society. The Church welcomes them and embraces them as it does all of its children. Their spiritual and cultural needs can be satisfied now and in the future with a measure of give and take if they do not demand too much and if the Church does not offer too little.

The Archdiocese is committed to the preservation of the Greek language and to our Hellenic traditions with every means at its disposal. However, the position of the Church in these areas is not to be doubted even by those who believe they know best how to serve the needs of the new members. For this and other reasons, the *strengthening of Hellenic College and Holy Cross Theological School are critical to the future of our Church.* This will not happen by itself. It requires the continuing attention and interest of everyone.

Archbishop Iakovos has been one of the most influential ecumenists in the world religious community. He has encouraged the participation of our parishes in local religious groups which seek a higher level of inter-religious understanding. In this connection, clergy and laity alike have an important role to play. Too many of our parishes tend to be parochial with respect to the churches of other faiths. Unlike most other Christian churches, ours are located away from where their members reside. Consequently, the members have little to do with the residents of the areas in which the Churches are situated. This tends to prevent the kind of neighborhood familiarity that other denominations enjoy. Nevertheless, some intra-neighborhood involvements are necessary so that our Church neighbors will come to know our Churches and their members as more than the places where the "Greeks" hold their bazaars. *The future dictates that our priests and lay leaders become more involved in civic activities* and in the problems of our neighborhoods in a more committed fashion. Drugs, alcoholism and other such concerns are not limited to groups outside our own environment. Hunger, discrimination, lack of decent housing, jobs and other social ills are our business, too. If most of our own people are better off and do not suffer the blight of underprivilege, all the more reason for compassion for the less fortunate and for a higher degree of involvement on our part in the affairs of our communities. We are no longer an immigrant Church. We are a Church tending to the needs of a more prosperous and more enlightened people who need to become a more integral part of the communities in which they worship, or live.

We must seek a better-educated and better-prepared clergy who can deal with an increasingly complex society. Lawyers, doctors or architects, and so forth, essentially prepare themselves for professional careers and the rewards that such careers are expected to produce. Clergymen on the other hand, prepare themselves for a life-time of service to the Church, without any expectation of enrichment. To attract candidates to the clergy, it is necessary that the parishes offer the highest possible standards of remuneration and benefits to their priests. It serves no useful or practical purpose to say, as many do, that having chosen to serve the Church, a priest should be prepared to make whatever financial sacrifices are necessary. Such views are unrealistic. Priests are no different from anyone else in their desire to have families, to feed and clothe them decently, educate them and protect them from economic adversity. Nor should a priest be denied sick benefits or a decent pension for his old age and that of his *presvytera* because he chose the priesthood as a calling. We have made progress in providing for our clergy, but too many parishes still skimp in providing for the needs of the priest and his family.

The education of our priests on the undergraduate and graduate level is the responsibility of the Church, for it is our parishes which will be the principal beneficiaries of a better educated clergy. A priest with an advanced degree in theology, in philosophy, or in education, for example, will not be serving himself, he will be serving the needs of his flock. Yet, the Archdiocese cannot easily meet the task of attracting young men to the priesthood without a program of undergraduate scholarships and grants. It cannot provide a better-educated clergy without the financial support of its parishes. Without assistance from the Church, many worthy young men must forgo a career in the priesthood for lack of means. The preferred method of providing graduate education is obviously the awarding of full scholarships for graduate study, after finishing theological school. When this is not possible, or practicable, parishes should at least provide an opportunity for their priests to engage in part-time graduate study by paying the tuition and by allowing the priest time for his studies. Each parishioner will benefit directly in some way. The benefits to our younger people, who themselves will be thus encouraged in the direction of higher education, will be immeasurable.

How much more equipped a better-educated priest is to deal with the social problems referred to earlier. How much better for our youth to be able to relate to and identify with a priest who has the educational and intellectual capacity to deal with today's problems. How much better for the priest to have the same education background as many of his parishioners. These comments are not intended to denigrate or demean those members of the clergy who have not had the benefit of higher education, or who come from a different culture. They serve faithfully and well. My references are intended to be only a realistic appraisal of our future needs and how best they can be met.

The future of the Church in the Americas will depend on our ability to make those changes that will assure the unity and stability of our Archdiocese. There are those who insist that the language of the Church be changed, that the liturgy be celebrated in English so that everyone will understand. There are those who argue that nothing should be changed, that the liturgy and the language of the Church, as the language of the New Testament, is of a sacred nature and should remain all Greek; that those who want to know what the liturgy is all about should study Greek, or follow the liturgy in a Greek-English booklet. Neither of these positions is absolute. The continuing evolutionary development of our Church will determine how the question of the language of the Church of the Americas will eventually be settled. This development will be affected by geographic and regional influences and the exigencies of time and circumstance. Change, or refusal to change, will not be determined by fiat.

The future must deal with the matter of *achieving the unity of Orthodoxy in this Hemisphere.* It must also decide whether ancient canons and established practices can withstand humanistic and social criticisms that are not without merit. We sometimes overlook the fact that the Church defines the canons and not the other way around. Consequently, our Church in the future, will need to *examine whether some of our canons,* adopted and relevant in an earlier time, are *relevant* or applicable in today's world.

It is surely no secret that many of our parishes themselves, on legitimate religious grounds, have serious doubts about the pertinency on the American scene of many of our canons and religious practices and the necessity of addressing them realistically. Nevertheless, any changes will be determined in

384

accordance with the needs of the Church but always true to Orthodox Christian beliefs. The early practice of our Church was to ordain even married men to the episcopacy. The permitted practice at present of electing only celibate clergy limits the Church to an enormously restricted and diminishing resource. While these concerns cannot be dealt with on an Archdiocesan level, they must most certainly be dealt with conclusively in the not too distant future in councils in which our Archdiocese must play a deliberative part.

Another problem that must soon be dealt with realistically by our Church in the Americas is that of the divorced clergy who are not permitted to remarry. A divorced priest is an anomoly in Church life. Canonically he is denied the sacraments, yet he receives, prepares and administers sacraments. He faces much human suffering. He is denied a human being's natural functions and the fullest enjoyment of all of life's gifts and rewards. In a pluralistic society such as ours, such a state of existence is intolerable and religiously unjust.

The Church must address the *question of abortion* more forcefully in the future. There is probably more ignorance on abortion among our communicants than on any other religious or moral issue. This can be seen by the position on abortion taken by some politicians of Greek Orthodox persuasion who obviously either do not understand their Church's position on abortion, or do not care what it is.

Likewise, the Church must become more involved in the nationwide *concern about nuclear war*. The Archbishop has been vocal in his support of efforts to lessen nuclear arms proliferation and has called for a nuclear freeze with appropriate mutual verification. If the Bishops and our clergymen have raised their voices on this issue, very little has been heard about it. There is no greater danger facing humanity than the threat of a nuclear catastrophe. The position of the Greek Orthodox Church, like that of other faiths, should become better known on the national and local levels.

The Archdiocese *lacks an effective college ministry* because its success among university students depends on the willing assistance of the clergy in college areas. This assistance has been withheld or sparsely provided. Most of our parishes have youth programs that provide spiritual learning and guidance until our youth go to college. Then, for four or more years, our youth have only a sporadic acquaintanceship with any Orthodox Church. This is a period fraught with danger for young minds which, for the first time in 12 or 13 years, are without spiritual guidance of any kind. This situation has been of major concern to the Archbishop. Hopefully, the priests will rise to the challenge and institute programs in their areas to take the Church to the campus or, conversely, bring the students to the Church.

The relationship of our Archdiocese with the Mother Church, the Ecumenical Patriarchate, in the future is a matter of critical concern for all of us. The continued harassment of the Patriarchate and the Orthodox faithful in Constantinople, and elsewhere in Turkey, by the Turks is one of the most serious violations of religious and human rights anywhere. Only Turkish-born clergy are eligible under Turkish law to be elected to the hierarchy and even then the Turkish Government enjoys the power of veto. Consequently, the problem of diminishing resources among Greek-Orthodox males in Turkey is becoming alarmingly critical. Despite pronouncements and declarations by world and national religious bodies, the situation continues to deteriorate. That is why the Archbishop over the years has again and again directed attention to the plight of the Patriarchate. That is why the Order of St. Andrew, the Archons of our Church, strives constantly to inform our Government, our Church friends and the American public of the increasing efforts of the Turks to drive the Patriarchate out of Turkey by outrageous acts of persecution.

If the Archbishop has sought to strengthen the Church in the Americas against the misfortune that would occur were the Turkish Government to cut off our ties with the Mother Church, it is only to provide our Church with the spiritual and moral strength to continue its holy mission. In view of what the Turks have done to the Patriarchate and our Greek Orthodox brothers in Turkey in the last thirty years, and in Cyprus since 1974, we would be reckless and irresponsible were we not to consider the consequences of some further dastardly anti-Christian acts on the part of the Turks.

The decentralization of the Archdiocese into Dioceses imposes upon the Bishops the responsibility of developing stronger' relationships between the parishes. Well planned, well executed Diocesan meetings should be organized annually, as well as regional parish conferences. The Archdiocesan Clergy-Laity Congresses serve a different purpose as they are intended to develop and fulfill national agendas. Diocesan meetings should deal with local problems through workshops, seminars, and debates with full delegate participation. We have too many meetings where papers are listlessly read, poorly listened to and barely digested before the program moves on. The programs of tomorrow must be organized so that parish representatives can become truly involved and become more intensively familiar with all aspects of Church life.

The Bishop's leadership of his Diocese can be more effective if he utilizes the Diocesan media to the fullest extent. His itinerary and activities should be published in advance whenever possible and he should make frequent pastoral visits. His best work is *not* done at Diocese headquarters. He should make use of his Presbyters' Council imaginatively and constructively. At the core of our Church's leadership, present and future, will always be the priests. If they are fully involved, the future stability of our Diocese and Archdiocese are more assured. If they are left out of the councils of our Church, if the value of their potential intellectual contributions is disregarded, the Church will lose.

It is not enough that our clergy come together at the Congresses every two years. More opportunity must be provided for them to give of themselves to the Church. Despite the shortage of priests throughout the Archdiocese, a means must be found to permit priests to visit Holy Cross during the year as guest lecturers as well as to perform short tours of duty at the Archdiocese. Likewise, there should be regular exchange visits on the parish level between priests. The benefits of such visits in utilizing the rich talents among us and in terms of broadening spiritual experiences and perspectives, are obvious.

The objective of all Diocesan activities is that the Bishop should do everything he can to *emphasize the need for strong Diocesan activities* as part of overall Archdiocesan unity. Bishops must be advocates of a strong Archdiocese in terms of Church unity and work at it. There should be a reexamination of our enthusiasm for the organization of new parishes in the future. New parishes should not usually be permitted to organize unless detailed analysis is made of the ability of a prospective parish to exist. The original enthusiasm for the organization of a parish often diminishes in scope, leaving the parish struggling to survive.

Finally, and as important to the future of our Church in the Americas as anything else, is the need on every level of the Archdiocese of bringing our youth fully into the life of the Church. In recent years, the term "spiritual renewal" best de-

387

scribes the increasingly intense interest in the Church manifested by our youth. This interest must be met by *encouraging young people to participate in parish activities* in a wide variety of programs. Lectures themselves are a passive kind of involvement. Their inspirational value is doubtful. What our parishes need are outreach programs that offer a variety of interests which have the purpose of bringing to our youth *into* the environment of the Church, making it possible for the priest to really get to know them and for them to know and better understand their Church.

In summary, the growth of the Church and its institutions in the New World, under the spiritual leadership of then Archbishop Athenagoras, and during the past 25 years under the spiritual and intellectual guidance of His Eminence Archbishop Iakovos, has been one of the remarkable events in the history of Eastern Orthodoxy. This growth can and must continue.

The Archdiocese can, and will, continue to provide the spiritual, moral and administrative leadership of our Church. It cannot ensure its own future, its unity, or its stability. Only the body of Greek Orthodox parishes which comprise the Archdiocese can do that. Our parishes can ensure Orthodoxy's future in the Americas, for all time, but only if they understand the need for unity and stability and have the will to continue to build on Christ's legacy.